ADVANCES IN CLINICAL CHEMISTRY

VOLUME 23

Advances in CLINICAL CHEMISTRY

Edited by
A. L. LATNER
Microbiological Chemistry Research Laboratory
The University of Newcastle upon Tyne
Newcastle upon Tyne, England

MORTON K. SCHWARTZ
Department of Clinical Chemistry
The Memorial Sloan-Kettering Cancer Center
New York, New York

VOLUME 23 • 1983

ACADEMIC PRESS
A Subsidiary of Harcourt Brace Jovanovich, Publishers
New York   London
Paris   San Diego   San Francisco   São Paulo   Sydney   Tokyo   Toronto

COPYRIGHT © 1983, BY ACADEMIC PRESS, INC.
ALL RIGHTS RESERVED.
NO PART OF THIS PUBLICATION MAY BE REPRODUCED OR
TRANSMITTED IN ANY FORM OR BY ANY MEANS, ELECTRONIC
OR MECHANICAL, INCLUDING PHOTOCOPY, RECORDING, OR ANY
INFORMATION STORAGE AND RETRIEVAL SYSTEM, WITHOUT
PERMISSION IN WRITING FROM THE PUBLISHER.

ACADEMIC PRESS, INC.
111 Fifth Avenue, New York, New York 10003

United Kingdom Edition published by
ACADEMIC PRESS, INC. (LONDON) LTD.
24/28 Oval Road, London NW1 7DX

LIBRARY OF CONGRESS CATALOG CARD NUMBER: 58-12341
ISBN 0-12-010323-0

PRINTED IN THE UNITED STATES OF AMERICA

83 84 85 86   9 8 7 6 5 4 3 2 1

CONTENTS

CONTRIBUTORS vii

PREFACE ix

Clinical Chemistry of Vitamin B_6
R. GARTH WILSON AND RICHARD E. DAVIS

1. Introduction 1
2. Chemistry and Biochemistry 3
3. Methods for the Assessment of Vitamin B_6 Status 12
4. Clinical Chemistry 16
   References 52

Aluminum
ALLEN C. ALFREY

1. Introduction 69
2. Methodology 70
3. Aluminum Metabolism 74
4. Aluminum Toxicity 79
   References 85

Clinical Chemistry of Thiamin
RICHARD E. DAVIS AND GRAHAM C. ICKE

1. Introduction 93
2. Chemistry and Biochemistry 95
3. Methods for the Assessment of Thiamin Status 107
4. Clinical Chemistry 112
   References 130

Vitamin-Responsive Inborn Errors of Metabolism
K. BARTLETT

1. Introduction 142
2. Biotin (Vitamin H) 144
3. Cobalamin (Vitamin B_{12}) 158
4. Folate 163
5. Pyridoxine (Vitamin B_6) 168

6. Riboflavin 174
7. Thiamin 177
8. Conclusions 181
   References 183

Spectrophotometry of Hemoglobin and Hemoglobin Derivatives
E. J. van Kampen and W. G. Zijlstra

1. Introduction 200
2. Determination of Total Hemoglobin 202
3. Absorption Spectra and Millimolar Absorptivities of Hemoglobin and
   Hemoglobin Derivatives 220
4. Determination of Hemoglobin Derivatives 232
5. Concluding Remarks 248
6. Appendix 251
   References 253

Clinical Chemistry of Oxalate Metabolism
M. F. Laker

1. Introduction 259
2. Quantitative Analytical Methods 260
3. Oxalate Metabolism 269
4. Disorders of Oxalate Metabolism 271
5. Conclusions 285
   References 285

Desirable Performance Standards for Clinical Chemistry Tests
C. G. Fraser

1. Introduction 300
2. Definition of Analytical Goals 301
3. The Rationale for Achievement of Goals 319
4. Uses of Analytical Goals 321
5. Achievement of Analytical Goals 325
6. Concluding Remarks 332
   References 333

INDEX 341

CONTENTS OF PREVIOUS VOLUMES 345

CONTRIBUTORS

Numbers in parentheses indicate the pages on which the authors' contributions begin.

ALLEN C. ALFREY (69), *University of Colorado Medical School, Denver Veterans Administration Medical Center, Denver, Colorado 80220*

K. BARTLETT (141), *Department of Clinical Biochemistry and Metabolic Medicine, University of Newcastle upon Tyne, Royal Victoria Infirmary, Newcastle upon Tyne NE1 4LP, England*

RICHARD E. DAVIS (1, 93), *Department of Haematology, Royal Perth Hospital, Perth 6000, Western Australia*

C. G. FRASER (299), *Department of Clinical Biochemistry, Flinders Medical Centre, Bedford Park 5042, South Australia*

GRAHAM C. ICKE (93), *Department of Haematology, Royal Perth Hospital, Perth 6000, Western Australia*

M. F. LAKER (259), *Department of Clinical Biochemistry and Metabolic Medicine, Royal Victoria Infirmary, Newcastle upon Tyne NE1 4LP, England*

E. J. VAN KAMPEN (199), *Clinical Chemical Laboratory, Diaconessenhuis, Groningen, The Netherlands*

R. GARTH WILSON (1), *Department of Clinical Chemistry, Princess Margaret Hospital for Children, Perth 6001, Western Australia*

W. G. ZIJLSTRA (199), *Department of Physiology, University of Groningen, 9712 KZ Groningen, The Netherlands*

PREFACE

As in the past, the Editors have attempted to present a volume with material related to the latest methodological and technical developments in clinical chemistry and reviews of biochemical changes as they relate to disease. The Editors hope that the review articles included in this volume will provide the reader with a firm understanding of topics presented and will continue the tradition established in the previous volumes.

In this volume, Wilson and Davis have reviewed the clinical chemistry of vitamin B_6, and Davis and Icke the clinical chemistry of thiamin. In these articles, the importance of these vitamins in good health and in disease is considered, as well as the role of the clinical chemistry laboratory in providing the physician and the nutritionist with meaningful values. Bartlett has reviewed the role of vitamins in inborn errors of metabolism and has described those diseases in which vitamin deficiencies play an important role. Alfrey has considered in detail the metabolic role of aluminum, the methods of measurement of this metal, and the clinical effects of aluminum excess and deficiency. This element is the second most plentiful element in the earth's crust, and alterations in the body burden of aluminum have been recently associated with major clinical disturbances. Van Kampen and Zijlstra have reviewed the spectrophotometry of hemoglobin and hemoglobin derivatives. This is a complete and timely update of the subject previously considered by these same authors in Volume 8 of *Advances in Clinical Chemistry*. Both theoretical and pragmatic aspects of the chemistry of hemoglobin and its derivatives are considered. In another article, Laker has considered the metabolism and the clinical chemistry of oxalate. Finally, Collum Fraser has proposed performance standards for clinical chemistry tests. This is an area of considerable interest at this time as clinical chemists become aware of and concerned about the precision and accuracy of the analytical values they report. In this article, Dr. Fraser has considered the definition and use of analytical goals as well as the rationale and potential for the achievement of these goals. This will be the last volume under the leadership of the current Editors. We thank the contributors and the publishers for their help and consideration in the past and express our wish that the quality of this series is maintained.

A. L. LATNER
M. K. SCHWARTZ

CLINICAL CHEMISTRY OF VITAMIN B_6

R. Garth Wilson* and Richard E. Davis†

*Department of Clinical Chemistry,
Princess Margaret Hospital for Children
and
†Department of Haematology,
Royal Perth Hospital,
Perth, Western Australia

1. Introduction 1
   1.1. History 1
   1.2. Nomenclature 2
2. Chemistry and Biochemistry 3
   2.1. Chemistry 3
   2.2. Biochemistry 6
3. Methods for the Assessment of Vitamin B_6 Status 12
   3.1. Direct Measurement of the Vitamin 12
   3.2. Indirect Measurement of the Vitamin 15
4. Clinical Chemistry 16
   4.1. Recommended Dietary Allowances 16
   4.2. Exogenous Deficiency 17
   4.3. Conditioned Deficiency 19
   4.4. Relative Deficiency 30
   4.5. Dependency 36
   4.6. Other Clinical States Possibly Associated with Abnormal
       Vitamin B_6 Metabolism 41
   4.7. Megavitamin Therapy 51
   References 52

1. Introduction

1.1. History

Paul György was the first to recognize vitamin B_6 as a distinct vitamin entity in 1934, when he described it as the rat acrodynia-preventing factor (G1). He devised a bioassay for it (G2) and went on to extract the vitamin from wheat germ (B1). In 1938, five different laboratories claimed to have isolated a crystalline compound possessing vitamin B_6 activity (S1). The structure was determined to be 2-methyl-3-hydroxy-4,5-bis(hydroxymethyl)pyridine which was named "pyridoxine" by György. Some of the excitement and rivalry generated in this "golden age" of nutritional biochemistry is hinted at in recent articles by two of the early workers in this field (S2, L1). By 1939, the synthesis of pyridoxine had been accomplished by four different laboratories and since then, its production has constantly increased while the cost of production has fallen from several thousand dollars per kilogram to less than $50 (B2).

1.2. Nomenclature

Although the term "pyridoxine" displaced the name vitamin B_6 between 1939 and 1942, Snell showed that compounds other than pyridoxine contributed to the vitamin B_6 activity of natural materials (S1). His group was able to show that pyridoxal and pyridoxamine had the expected vitamin activity for both lactic acid bacteria and animals.

That most of the vitamin B_6 in natural materials is present as phosphorylated derivatives was discovered by the early 1950s; in the revised nomenclature of the International Union of Pure and Applied Chemistry (I1) the term "vitamin B_6" is reserved for all 3-hydroxy-2-methylpyridine derivatives exhibiting, qualitatively in rats, the biological activity of pyridoxine. "Pyridoxine," "pyridoxal," and "pyridoxamine" are recommended names for the alcohol, aldehyde, and amine, respectively (Fig. 1).

Since its discovery and synthesis over 40 years ago, vitamin B_6 has interested chemists, biochemists, pharmacists, physicians, and the general public to the extent that several hundred references a year now appear in the scientific (and not-so-scientific) literature. Since, as we shall see, the biological function of the vitamin is widespread, it is easy to implicate it in a variety

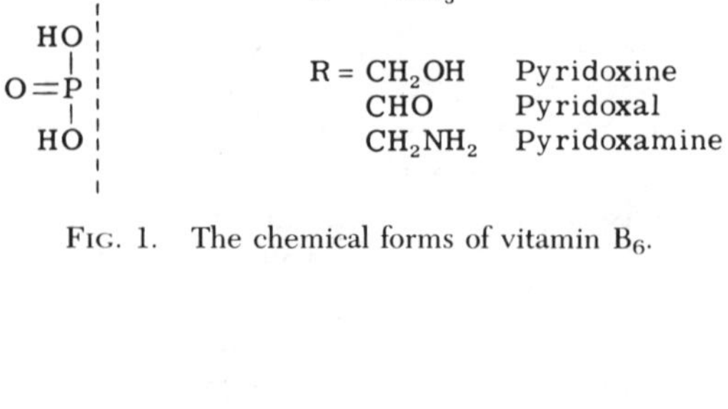

Fig. 1. The chemical forms of vitamin B_6.

of disease states both in animals and man. In addition, since pyridoxine is cheap to produce and appears to have no harmful side effects, it has been used in attempts to "cure" many diseases. The claims made for vitamin B_6 are difficult to analyze when it is realized that not only is the measurement of the vitamers in biological fluids difficult (as indicated by the numerous methods reported in the literature), but, also, the assessment of the vitamin B_6 status of the individual has many problems. This article attempts to evaluate some of these conflicts, to collate information from several disciplines into a form useful for clinical chemists, and hopefully to stimulate clinical chemists to take a greater interest in this fascinating field.

2. Chemistry and Biochemistry

2.1. CHEMISTRY

The chemistry and biochemistry of vitamin B_6 have been the subject of numerous reviews in recent years (A1, F1, G3, P1, S1, S3) and will not be dealt with in depth here. Although vitamin B_6 is not synthesized by higher animals, it is produced by all plants so far studied, and by many microorganisms.

In *Escherichia coli*, the molecule is formed from three triose units: the 2'-methyl and adjacent carbon-2 are derived from pyruvate via acetaldehyde, carbons-3, -4, and -4' from one triose, and carbons-5, -5', and -6 from another. The origin of the heterocyclic nitrogen has not yet been determined (B3, P1). A hypothetical model for the biosynthesis of vitamin B_6 in *Bacillus subtilis* has recently been suggested based on work with vitamin B_6-producing mutants (P2).

Pyridoxal, pyridoxamine, and pyridoxine are almost equally active in supporting animal growth; since pyridoxal 5-phosphate is the only generally utilizable coenzyme form of the vitamin, it follows that the various dietary forms are easily interconverted. A single enzyme, pyridoxal phosphokinase (EC 2.7.1.35), acts on all three nonphosphorylated forms and is probably activated by zinc ions in animal tissues (Fig. 2). Both pyridoxine phosphate and pyridoxal phosphate can be dephosphorylated by phosphatases; pyridoxine and its phosphate can be converted to the pyridoxal derivatives by a pyridoxine oxidase (EC 1.1.1.65), which uses riboflavin 5'-phosphate as its prosthetic group (B3, S3).

Only pyridoxal phosphate and pyridoxamine phosphate show coenzyme activity in the majority of systems. Although aminotransferases are able to use either pyridoxal phosphate or pyridoxamine phosphate, all other vitamin B_6-dependent enzymes require pyridoxal phosphate for reconstitution of an

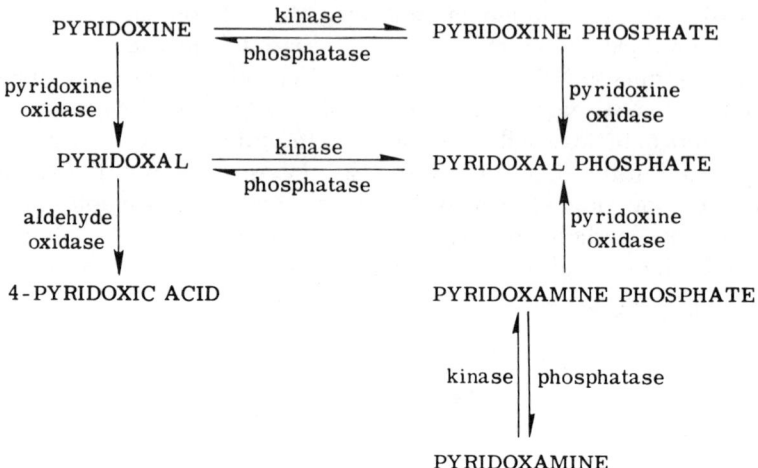

FIG. 2. The interconversions of vitamin B_6.

active holoenzyme. All enzymes requiring vitamin B_6, with the exception of glycogen phosphorylase, are involved with amino acid metabolism (B3). Reactions of these enzymes are of two main types: transaminations to form α-oxo-acids from α-amino acids and decarboxylations to form primary amines. Racemization of D-amino acids to L-amino acids is also carried out by the vitamin B_6-dependent enzymes, but this occurs mainly in bacteria.

The initial reaction of any of these enzymes is the formation of a Schiff base between the α-amino group and the 4'-aldehyde group of pyridoxal phosphate. The subsequent reaction depends on which bond of the α-carbon of the amino acid is broken prior to hydrolysis of the complex.

It has been shown that under physiological temperatures and pressures, pyridoxal phosphate will react nonenzymically to carry out the same major reactions with amino acids as occurs in biological systems (B3, G3). Using these reactions as models for the structural requirements for the coenzymic reactions of pyridoxal phosphate, Snell (S4) noted that the 2'-methyl and 5'-phosphorylated hydroxymethyl groups are necessary for coenzyme function, but if substituted at these positions, the analogs showed no reduced activity in the nonenzymic reactions. On the other hand, the heterocyclic nitrogen, 4'-formyl group, and phenolic hydroxyl group are essential for both enzymic and nonenzymic reactions and cannot be substituted without loss of activity.

The model systems require catalysis with metal ions, particularly Cu^{3+}, Fe^{3+}, and Al^{3+}, for stabilizing the intermediate in the initial reaction to form the Schiff base. In the enzymic system, it is thought that this stabilization is provided by charged groups at the catalytic site [possibly histidinyl residues (B3)], although there are some pyridoxal phosphate enzymes which require metal activators [for example, lysyl oxidase (EC 1.13.12.12) requires copper].

The formation of the initial Schiff base is much faster in the enzyme-catalyzed reactions, and, in general, these reactions proceed up to 10^6 times faster than the nonenzymic model. The increased substrate specificity of the enzyme reaction is thought to be due to the maintenance of a rigid geometry of the Schiff base at the active site of the enzyme.

The nature of the enzyme–coenzyme interaction is such that the removal of the cofactor from its enzyme is difficult; often, where this has been achieved, it has been difficult to reconstitute the holoenzyme. This suggests that protein denaturation, rather than simple loss, is the basis of the observation. Similarly, the reconstitution from an apoenzyme is slow, particularly if phosphate is present, as this is thought to interfere with the binding of the cofactor to the enzyme through the phosphate group of pyridoxal phosphate or pyridoxamine phosphate.

It has long been thought that the various forms of vitamin B_6 present in food were heat sensitive. Support was given for this idea when a group of children were referred to pediatricians following the onset of seizures with marked opisthotonos and stiff jerky convulsive movements after they had been fed a commercially processed milk food (C1, M1, R1, S5), which, in its production, had been vigorously heat sterilized and contained only 60 μg of vitamin B_6 per liter. This compared with more than 200 μg/liter for the reconstituted powder and 130 μg/liter for human breast milk. This loss of vitamin B_6 was attributed to heat denaturation during the sterilization procedure. The same liquid formula fortified with 450 μg pyridoxine per liter, fed to children for several months, produced no convulsions. Cats fed on a particular brand of commercial pet food developed an oxalate nephrocalcinosis which was thought to be due to vitamin B_6 deficiency (G4). It was subsequently shown that the heating process used in the preparation of the food destroyed much of its vitamin B_6 content.

Methods incorporating a heating stage for the measurement of vitamin B_6 (A2) appeared to give lower results than those that did not require heating (D1). Davis and Smith (D2) showed that heating for 45 minutes at 121°C resulted in a loss of approximately 30% of pyridoxal activity in serum while synthetic pyridoxal showed no such loss.

Heating of pyridoxal in the presence of amino acids may lead to condensation of the carbonyl group of pyridoxal and the amino groups of amino acids, leading to the formation of Schiff base ligands such as pyridoxylideneglutamate and pyridoxylidenealaninate. These products are unstable but the formation of a metal complex can stabilize the ligand in the Schiff base configuration (B4, C2). It could be expected, therefore, that when heating serum there would be an opportunity for these Schiff base ligands to be formed; since it has been shown that the biological activity of pyridoxal in serum is reduced following the application of heat, it appears that it is the result of the complexes formed which results in the reduction of biological activity. Pyridoxal heated in the absence of amino acids does not appear to lose any of its activity (D2).

2.2. Biochemistry

2.2.1. Absorption

Man cannot synthesize vitamin B_6 *in vivo*, hence its absorption from external sources is required. Few studies of this process have appeared in the literature. Nutritional experiments in which animals were shown to grow equally well on limited amounts of pyridoxine, pyridoxamine, or pyridoxal indicate that all three forms probably are absorbed with equal efficiency from the gastrointestinal tract (S3). Early work indicated that the absorption process in man was rapid. Scudi and co-workers showed that 8.7% of a 50-mg intravenous dose of pyridoxine appeared in the urine within 1 hour and 7.6% of a 100-mg oral dose appeared within 4 hours (S5). Booth and Brain (B5) found that the absorption of tritium-labeled pyridoxine in rats increased linearly with increasing size of the oral dose. They found no tendency for the pyridoxal to plateau, even after a dose which exceeded by 500-fold the daily requirements of the rat. The preferred site of absorption was the jejunum, although absorption would also occur if the vitamer was introduced into the ileum or cecum. There was no evidence that an active-transport process was present, since increasing the oral dose did not result in any absorption occurring lower in the gut, which might have been expected had active-transport sites been saturated. Absorption was fast; about half an oral dose of the labeled vitamin disappeared from the gastrointestinal tract within 10 minutes.

In humans, Brain and Booth (B6) observed that the 24-hour excretion of radioactivity following an oral dose of tritiated pyridoxine hydrochloride was low (12%), but that if the oral dose of labeled pyridoxine was accompanied by a relatively large parenteral dose of unlabeled pyridoxine, then a much larger amount of the dose was excreted (17%).

This work has been criticized on the grounds that the quantities of pyridoxine administered in these studies were high in relation to dietary needs in both rat and human and that absorption was tested indirectly by measuring urinary excretion of the vitamin. Recent work (M1) has shown that the absorption of tritiated pyridoxine hydrochloride, in isolated loops of rat jejunum, was rapid and linear over the range 0.2 to 1 mM. Using the same model, it has also been shown that the mechanisms involved in pyridoxal phosphate absorption are more complex than the passive diffusion of the nonphosphorylated vitamer (M2, M3). Part of the process appears to involve intraluminal hydrolysis by alkaline phosphatase, although there is some evidence that, in rats, at least, some pyridoxal phosphate may be absorbed intact by a mechanism not yet understood. Hamm *et al.* have shown that under normal physiological conditions the majority of dietary pyridoxamine phosphate is hydrolyzed to pyridoxamine in the jejunum and absorbed by passive diffusion (H1). Nonspecific alkaline phosphatases capable of hydrolyzing pyridoxal phosphate and pyridoxamine phosphate have been detected in many tissues, including human intestine (S3).

2.2.2. Transport

The absorbed pyridoxine is phosphorylated in the liver and possibly the red cells, and is subsequently oxidized to pyridoxal phosphate. In the red cell, pyridoxal is first dephosphorylated before release into the plasma; pyridoxal is also released into the plasma from the liver. Thus the plasma has to transport all three forms of the vitamin, although the majority is pyridoxal phosphate. Anderson *et al.* (A2) investigated the plasma binding of these vitamers and showed that pyridoxine added to plasma behaved exactly the same as pyridoxine in saline when fractionated by gel filtration, demonstrating an absence of protein binding. Pyridoxal phosphate, on the other hand, eluted almost entirely with a protein peak containing mainly albumin, transferrin, and 3.5 S α-glycoprotein. Although most pyridoxal eluted with the protein, some was unbound. Other workers (A3) suggested that pyridoxal phosphate was associated with a 3-lysyl residue binding site on albumin (Ser-Leu-Phe-Glu-Lys-Pro-ε-[pyridoxyl]-Lys-Lys).

Dempsey and Christensen (D3) had earlier shown that bovine serum albumin had two specific binding sites with relatively high association constants for pyridoxal phosphate. Anderson and her team (A4) suggested that if this also applies to human plasma albumin, then this strong binding would account for the inability of pyridoxal phosphate to enter the red cell from plasma. This is in contrast to its substantial uptake by red cells suspended in saline (A5). Lumeng and co-workers (L2) have confirmed that circulating pyridoxal phosphate in normal blood is bound principally to albumin in a molar ratio of 2:1 through a Schiff base formation with the ε-amino group of lysine, which not only prevents its transport into red cells, but also retards its degradation by alkaline phosphatases.

Human plasma can bind more than 800 μg pyridoxal phosphate per milliliter. In the absence of albumin, pyridoxal phosphate is incorporated into red cells by passive diffusion (M4), unlike the nonphosphorylated forms of the vitamin which are actively transported against a concentration gradient (S3), but saturation of the transport process occurs at relatively low concentrations of pyridoxine. These studies raise questions about the physiological role of circulating pyridoxal phosphate. Since pyridoxal phosphate bound to albumin cannot enter cells, the recipient tissues must possess a mechanism for its uptake. Cells capable of assimilating albumin thus may also obtain their pyridoxal phosphate at the same time. Such a process could account for up to 40% of the observed rate of the disappearance of plasma pyridoxal phosphate. Alternatively, the cells may have binding proteins with greater affinity than albumin for pyridoxal phosphate. Since pyridoxal is more freely permeable across the cell membrane (at least in blood), it is possible that dephosphorylation (by phosphatases either in the blood or at the cell membrane) occurs, and the cell rephosphorylates the pyridoxal after entry.

Many of these ideas are still speculative and further work is required to confirm them.

2.2.3. Storage

In rats and mice, most of the pyridoxal 5'-phosphate of muscle is accounted for in the phosphorylase enzyme (EC 2.4.1.1). In the absence of a clear role for the coenzyme in connection with phosphorylase function, Krebs and Fischer (K1) proposed that the muscle enzyme might serve as a physiological store for vitamin B_6. Recently Black et al. (B7) showed that in rats fed high levels of vitamin B_6, muscle phosphorylase increases independently of body size. If the rats are made vitamin B_6 deficient, then muscle phosphorylase does not fall over a period of at least 8 weeks, and losses only occurred following anorexia and weight loss (B8). This work also raises the question as to the possible cofactor role for pyridoxal phosphate in muscle phosphorylase and the possibility that the sparing effect during starvation might provide a prolonged capability for sustaining glyconeogenic enzymes during this period (B8, R1).

Recent work has suggested, however, that pyridoxal phosphate does play a catalytic role in the control of dephosphorylation of phosphorylase a (Y1).

2.2.4. Excretion

4-Pyridoxic acid (2-methyl-3-hydroxy-4-carboxy-5-hydroxymethylpyridine) has been shown to be the major metabolite and major excretion product of all forms of vitamin B_6 in man (W1). After conversion of the other vitamers to pyridoxal, this is then oxidized by the relatively nonspecific, FAD-dependent general aldehyde oxidase [aldehyde, oxygen oxidoreductase (EC 1.2.3.1)] of liver (S3). Johansson et al. found that 20% of the radiolabel of pyridoxine administered orally was excreted in the first day, followed by the remainder with a half-life of 18–28 days (J1). However, only three patients were studied, none of whom could be considered normal. The 4-pyridoxic acid excreted amounted to 20–40% of the total label excreted.

This pattern of excretion has been confirmed by many others (D4, K2, M5, R2). Conversely, patients on vitamin B_6-deficient diets excrete no 4-pyridoxic acid in their urine (B9, S6).

If dietary vitamin B_6 deficiency exists, then urinary excretion of 4-pyridoxic acid will be practically zero at the time when the increase in urinary excretion of tryptophan metabolites occurs, after a loading dose of 2 g L-tryptophan (W1). On adding vitamin B_6 to the diet of these patients, the tryptophan loading test returns to normal much earlier than the urinary excretion of 4-pyridoxic acid.

Although 4-pyridoxic acid is the main excretion product of vitamin B_6 metabolism, certain of the vitamers are excreted. Kelsay *et al.* (K2) found that when dietary intake of vitamin B_6 was adequate (about 1.5 mg/day) in a group of normal male subjects, they excreted about 120 µg total vitamin B_6, 65% as pyridoxal, 30% as pyridoxamine, and only negligible amounts of pyridoxine. On a vitamin B_6-deficient diet (0.16 mg vitamin B_6 per day) total urinary excretion of vitamin B_6 decreased to about 40 µg with about equal quantities of pyridoxal and pyridoxamine. After administration of 50 mg pyridoxine on 2 consecutive days, the subjects excreted 4% as pyridoxal, 10% as pyridoxine, 2% as pyridoxamine, and 50% as 4-pyridoxic acid on the second day. Snyderman *et al.* (S6), studying a small group of patients, also found that "vitamin B_6" (assayed by a technique which measured all the vitamers) excretion fell rapidly when the patients were on a vitamin B_6-deficient diet. In contrast to the urinary excretion of 4-pyridoxic acid, urinary vitamin B_6 values were, although small, always measurable. This has led some workers to test the usefulness of urinary vitamin B_6 excretion as an index of dietary intake. Although they showed this to be of limited value in itself as an index of the severity of a vitamin B_6 deficiency in an individual, they found it useful as a reflection of the subjects' recent dietary intake of the vitamin (S7).

Although vitamin B_6 requirement in the human is related to the level of protein intake, dietary protein appears to have little effect on the urinary excretion of vitamin B_6 (C1).

2.2.5. *Regulation*

Lumeng *et al.* (L2), working with dogs, showed that injection of pyridoxine or pyridoxal caused a rise in plasma pyridoxal phosphate. Bilateral nephrectomy or resection of the spleen, stomach, or intestine had no effect on this rise. However, following hepatectomy with or without resection of other splanchnic organs, there was no increase in plasma pyridoxal phosphate, suggesting that the liver was the primary site for the pyridoxal phosphate circulating in the plasma.

The initial reaction involves the formation of pyridoxine phosphate by the action of pyridoxal kinase (EC 2.7.1.35). The oxygen-dependent oxidation of the product of the kinase reaction by pyridoxine phosphate oxidase (EC 1.4.3.5) leads to pyridoxal phosphate. Since at unphysiological concentrations this vitamer inhibits many enzyme reactions, mechanisms must exist for regulating its cellular content *in vivo*. In rat liver, the level of pyridoxal phosphate increases with the intake of pyridoxine up to 50 µg per day, but no additional increase in liver vitamin B_6 occurs with higher intake (R3). It has been suggested that pyridoxine phosphate oxidase may be the regulator since it is sensitive to product inhibition (B10), or that various cellular phosphatases may control the breakdown of the pyridoxal phosphate (S3). Li and his colleagues (L3) showed that in rats, the pyridoxal phosphate content of liver *in vivo*, and hepatocytes *in vitro*, remained unaltered in the presence of excess pyridoxine. If the hepatocyte was fractionated to yield cytosol, then this was capable of producing increasing amounts of pyridoxal phosphate even when the concentration of pyridoxine in the reaction was much greater than in the original liver. The pyridoxal phosphate formed did not appear to be inhibiting the pyridoxine oxidase. They attributed this phenomenon to a possible binding of the pyridoxal phosphate formed to cytosol proteins, thus keeping the concentration of "free" pyridoxal phosphate low. They also showed that isolated intact hepatocytes could increase their production of pyridoxal phosphate from pyridoxine if the cells were incubated in 80 mM phosphate which inhibits membrane-bound phosphatases. Under these conditions, there was increased production of pyridoxal phosphate with pyridoxine concentrations from 0.5 to 50 µM. They suggest that binding by proteins acts to protect cellular pyridoxal phosphate, whereas phosphatases may be responsible for the degradation of any pyridoxal phosphate formed in excess of the binding capacity for the cofactor.

The liver must possess a unique transport mechanism since it is the only organ whose cells provide for the efflux of phosphorylated pyridoxal, possibly as the albumin–pyridoxal phosphate complex (L2). This may not be the only point of control of pyridoxal phosphate availability. Other evidence suggests that the rate of dissociation of pyridoxal phosphate from its apoenzymes may be a factor in determining the breakdown of enzyme protein (L4).

Katanuma and his team have shown, in rats, specific proteases to pyridoxal phosphate-dependent enzymes (K4). These appear to be specific endopeptidases which may remain inactive (possibly by binding with an unknown inhibitor) in the presence of a normal intake of vitamin B_6. Of enzymes tested so far, substrate specificity is (in decreasing order of reactivity): ornithine transaminase, homoserine dehydratase, mitochondrial AST, serine dehydratase, cytoplasmic AST, and tyrosine aminotransferase. Pyridoxal phosphate and pyridoxal protect the enzyme against proteolysis, but pyridoxine phosphate and pyridoxamine phosphate do not. These proteases are located on mitochondrial inner membranes (their activities are probably low at birth) and in areas where new mitochondria are being formed (K5).

Thus the fine regulation of the level of pyridoxal phosphate at the site of its action may well be accomplished through regulating the amount of apoenzyme available. Some pyridoxal phosphate-dependent enzymes (e.g., rat liver cysteine–sulfinate decarboxylase and mammalian liver and brain alanine and aspartate aminotransferases) appear fully saturated *in vivo*, that is, there is no increase in activity when pyridoxal phosphate is added to an assay system *in vitro*. Others, however, appear to be less than fully saturated (i.e., an increase in pyridoxal phosphate causes an increase in activity). Human brain glutamate decarboxylase, for example, is only about 60% saturated *in vivo*. Thus it would be expected that a fall in the availability of pyridoxal phosphate will, in the first instance, affect the activity of those enzymes which have such a low affinity for the cofactor. The picture is unfortunately more complicated. In studies of vitamin B_6 deficiency in rats, the fully saturated cysteine–sulfinate decarboxylase loses its activity rapidly and is absent after 2 weeks. Similarly, muscle glycogen phosphorylase loses much of its pyridoxal phosphate, but while liver cystathionase loses much of its activity, this can be restored *in vitro*, indicating that the apoenzyme has not been degraded.

In neonatal rats from vitamin B_6-deficient dams, Bayoumi *et al.* showed that the brain glutamic acid decarboxylase was very low (B11). However, *in vitro* incubation with additional pyridoxal phosphate produced an increase of apoenzyme several times above normal. This massive induction of apoenzyme appears to occur to ensure that a maximum of pyridoxal phosphate will be "trapped" by the enzyme, which is involved in the maintenance of the inhibitory substance γ-aminobutyric acid.

Lefauconnier and co-workers (L5) showed that long-term administration of cortisol to rats led to increased liver tyrosine, alanine aminotransferases, and serine deaminase, but that the total liver pool of vitamin B_6 did not alter. On the other hand, there were decreases in activity of cysteine–sulfuric decarboxylase and homoserine dehydratase, indicating a redistribution of the available pyridoxal phosphate.

This ability to "sacrifice" some less essential pyridoxal phosphate-dependent enzymes, to redistribute available coenzyme, and to increase the "scavenging" effect of more essential enzymes in response to dietary deficiency must be borne closely in mind when evaluating the mass of sometimes conflicting information on the treatment of various clinical conditions of vitamin B_6.

3. Methods for the Assessment of Vitamin B_6 Status

The majority of techniques for assessing vitamin B_6 status fall into the following categories: direct measurement of blood levels of the vitamin, measurement of the excretion rate of the vitamin, measurement of the metabolites or abnormal metabolic products resulting from a deficient state, and the measurement of some other process dependent on the concentration of the vitamin in the body. All of these approaches to the assessment of vitamin B_6 status have significant problems, either technical or physiological, and interpretation of results should not be made without an understanding of the techniques used.

We intend to survey only some of the variety of methods available and to discuss only some of their individual advantages.

3.1. DIRECT MEASUREMENT OF THE VITAMIN

3.1.1. *Biological Assays*

The first biological assay was designed by György (G2) who assayed extracts for vitamin B_6 activity by observing their ability to cure rats of a particular form of dermatitis. Stokes (S8) devised a microbiological assay in 1943 using *Neurospora sitophila;* in the same year Atkin (A6) used *Saccharomyces carlsbergensis* as an assay organism. *Saccharomyces carlsbergensis* has the advantage of being sensitive to the three nonphosphorylated forms of the vitamin, although there is some variation in its sensitivity to each form. Assays using this organism have been found particularly useful for determining the total vitamin B_6 content of foodstuffs, but are not sufficiently sensitive for measuring physiological levels of the vitamin in man. A protozoological method for detecting clinical vitamin B_6 deficiency was described by Baker *et al.* (B12). The organism used was the ciliate protozoan *Tetrahymena pyriformis*. The organism responded to all nonphosphorylated forms of the vitamin. However, it was found to be between 120 and 150 times more sensitive to the aldehyde and amine than it was to the alcohol. The assay had a sensitivity threshold of 0.5 µg/liter and results appeared to be rather high compared with those currently obtained (A2, D5). A method using *Lactobacillus casei* as the test organism was described by Rabinowitz (R4). *Lactobacillus casei* was later found to be very sensitive to pyridoxal but relatively insensitive to pyridoxine and pyridoxamine. These two forms of the vitamin were found to stimulate growth of the test organism only when used at a concentration 1000 times greater than that required for pyridoxal (D5). The organism does not respond to the phosphorylated form of the vitamin and prolonged acid hydrolysis was used to liberate the pyridoxal (S9). When used for the assay of pyridoxal in whole blood, falsely high values were obtained and it was thought that the prolonged acid hydrolysis of blood converted L-alanine to D-alanine; D-alanine can replace pyridoxal as a growth factor for *L. casei* (H2, S10). Anderson (A2) examined the effect of shortening the period of acid hydrolysis and found that autoclaving at 15 psi for 1 hour gave values which were no different from values obtained when autoclaving was continued for 2 hours. Using the shorter period of acid hydrolysis appeared to overcome the problem of converting L-alanine to D-alanine. Based on this and other modifications, Anderson (A2) described a clinically useful method for assaying pyridoxal 5'-phosphate in serum and red cells. However, the test remained tedious and time consuming and still required exposure of the sample to vigorous heat treatment.

In 1973, Davis (D1) described a method for assaying pyridoxal 5'-phosphate in serum and red cells which avoided the use of heat. Dephosphorylation of the pyridoxal was effected by the use of an acid phosphatase and this greatly simplified the procedure. Another innovation used in this method was the use of a chloramphenicol-resistant mutant of *L. casei*. Chloramphenicol was added to the assay medium and this avoided the need for sterilization or aseptic technique and permitted automation of the procedure. The automated method was able to set up 80 assays an hour and, following incubation, results could be read at a rate of 160 an hour.

Results using the automated assay were generally higher than those obtained using Anderson's method and the heating required in the latter was thought to be responsible for this. Davis (D2) observed that heating serum samples at 121°C for 1 hour, as required for acid hydrolysis, resulted in a reduction in the concentration of assayable vitamin. This was later confirmed by Baker (B4), who described a method for producing pyridoxylideneglutamate by heating monosodium glutamate and pyridoxal hydrochloride together at 120°C for 30 minutes. Heating certain amino acids in the presence of pyridoxal will cause them to form complexes with the pyridoxal and such complexes have a reduced biological activity (D6).

Because the major form of vitamin B_6 in man is pyridoxal 5-phosphate, a microbiological array using an assay organism such as *L. casei*, which is sensitive only to this form of the vitamin, had the advantages of being both specific and extremely sensitive. However, the vitamin is most commonly found in food as pyridoxine and/or pyridoxamine. These can be measured using *S. carlsbergensis* if information on the total vitamin B_6 content only is required. Alternatively, the pyridoxine and pyridoxamine can be sequentially converted to pyridoxal by treatment with manganese dioxide and glyoxylic acid (K3) which can then be assayed using *L. casei*.

3.1.2. Chemical Assays

Pyridoxal phosphate, when reacted with cyanide, gives a high yield of fluorescence which is proportional to the concentration of the vitamin. A number of methods have been devised using this approach; they are claimed to be very sensitive and able to measure concentrations as low as 0.6 nmol/liter (A7, L6, T1). However, these methods when used to measure pyridoxal concentrations in biological material may suffer from nonspecific fluorescence due to interfering substances. Chauhan and Dakshinamurti (C3) have recently described a fluorometric assay which can be applied to the measurement of pyridoxal and pyridoxal 5'-phosphate in serum. The authors claim that the method is specific and that the specificity is based on the reductive amination of pyridoxal and its 5'-phosphate with methyl anthranilate and sodium cyanoborohydride at pH 4.5 to 5.0. Separation of the highly fluorescent methyl-N-pyridoxyl anthranilate was achieved by a combination of column and thin-layer chromatography. The specificity of the method permits its use for the assay for pyridoxal and its 5'-phosphate in serum.

Although methods using thin-layer electrophoresis (C4), thin-layer chromatography (C5, M6), gas–liquid chromatography (P3, W2), high-performance liquid chromatography (G5, W3, W4, Y2), and gas chromatography (P3) have been described, few have been found suitable for use as routine assay systems and none has found a place in the routine measurement of blood levels. An advantage of some of these techniques is their ability to separate and measure all forms of the vitamin in a single run; some will, in addition, separate 5-hydroxyindoles (C5) and pyridoxic acid (V1). Most of these methods are relatively insensitive when compared with enzymic and microbiological techniques. They have a sensitivity threshold of approximately 0.5 µg.

3.1.3. Radioimmunoassay

Thanassi and Cidlowski (T2) described a radioimmunoassay for phosphorylated forms of vitamin B_6. They raised hapten-specific antibodies in rabbits by immunizing them with phosphopyridoxal bovine albumin. The purified hapten-specific antibodies were then covalently coupled to immunobeads. The antibody–immunobead preparation binds [^{3}H]pyridoxine 5'-phosphate and this provided the basis for a radioimmunoassay for some of the B_6 vitamins. [^{3}H]Pyridoxine phosphate is displaced from the matrix-attached antibodies by the three phosphorylated forms of the vitamin, but is not displaced to any significant extent by free forms. In this system, pyridoxal phosphate was found to be a less effective competing ligand than pyridoxine phosphate. The assay is extremely sensitive and is able to detect concentrations in the picomole range.

3.1.4. *Enzymic Assays*

Enzymic assays are based on the ability of pyridoxal phosphate to restore activity to an apoenzyme; tests have been devised using glutamic-oxaloacetic transaminase (S11), tyrosine decarboxylase (C6, H3, M7, R5, S12), and tryptophanase (G6, H4, M8, O1). Test procedures depend on the presence of an excess of apoenzyme and substrate; the amount of product formed is proportional to the concentrations of pyridoxal 5′-phosphate present. Haskell and Snell (H4) described a simple technique using *E. coli* apotryptophanase. The sample to be measured was added to the apoenzyme and incubated for 20 minutes at 37°C; the reaction was started by the addition of L-tryptophan. The indole produced was reacted with Erlich's reagent and the resulting color was measured in a spectrophotometer. In spite of the apparent simplicity of many of these methods, they have not been widely used in clinical laboratories.

3.2. Indirect Measurement of the Vitamin

The kynurenine pathway of tryptophan metabolism requires pyridoxal 5-phosphate at four stages. In man, tryptophan is converted to N'-formylkynurenine by tryptophan pyrrolase, which determines the quantity of tryptophan entering this pathway. The N'-formylkynurenine is converted to kynurenine by the enzyme formamidase (M9). Kynurenine and the product of the next step, 3-hydroxykynurenine, are transaminated by kynurenine transaminase to kynurenic acid and xanthurenic acid, respectively, and this requires the participation of pyridoxal 5-phosphate as coenzyme. Kynureninase splits alanine from the side chain of both kynurenine and 3-hydroxykynurenine to form anthranilic acid and 3-hydroxyanthranilic acid, respectively. Both steps require the participation of pyridoxal 5-phosphate.

A deficiency of vitamin B_6 may interrupt this pathway at one or more points, resulting in the excretion of one or more of the vitamin B_6-dependent intermediates—kynurenine, anthranilic acid, kynurenic acid, xanthurenic acid, and 5-hydroxyanthranilic acid. To overcome the complexity of measuring all of these intermediates, it has been found convenient by some workers to measure the level of urinary xanthurenic acid. Lepkovsky *et al.* (L7) first recognized that vitamin B_6-deficient rats excreted large amounts of xanthurenic acid when placed on a diet rich in tryptophan. This led to the development of numerous methods for its measurement (L8, R6, S13, W5). Patients were usually given a loading dose of tryptophan and their urine was collected over the following 24 hours. The dose of tryptophan used varied from 2 g (A8, B13, C7) to 10 g (B14, B15, W6). It was subsequently recognized that if there was a marked vitamin B_6 deprivation, kynurenine transaminase might lose its pyridoxal 5-phosphate, and this would inhibit the production of xanthurenic acid. This led to an appreciation of the fact that the measurement of a wide range of tryptophan metabolites might more accurately reflect vitamin B_6 status.

Studies by Wolf (W1) indicated that kynureninase was apparently preferentially depressed over that of kynurenine transaminase, allowing conversion of 3-hydroxykynurenine and kynurenine to xanthurenic acid and kynurenic acid, respectively, until pyridoxal 5-phosphate had been depressed to a very low level.

4-Pyridoxic acid is the chief urinary metabolite of all forms of vitamin B_6 and the measurement of this metabolite can be used to assess vitamin B_6 nutrition (F2, F3, R2, W7). The methods were tedious to perform and required preliminary separation of the 4-pyridoxic acid by column chromatography. It was then converted to the corresponding lactone and the pH brought to 9.0 with ammonium hydroxide to give maximum fluorescence. More recently, Gregory and Kirk (G7) have described a method using high-performance liquid chromatography which appears to be both rapid and sensitive.

The urinary excretion of 4-pyridoxic acid decreases as the intake of vitamin B_6 decreases. However, the exact relationship has not been satisfactorily determined. The excretion measurement now appears to be more suitable as an investigation of vitamin B_6 metabolism rather than as an indirect method of trying to establish vitamin B_6 deficiency.

4. Clinical Chemistry

4.1. Recommended Dietary Allowances

The World Health Organization (W8) has not considered the human requirements for vitamin B_6 since it felt that although deficiencies of the vitamin are known, these have been produced under highly unusual conditions and were secondary to disease states or an inborn error of metabolism. The National Academy of Sciences of the United States has, however, reviewed the field regularly (N1–N3) and has recommended dietary allowances which vary for different age-groups and situations. The requirement for vitamin B_6 in man appears to be increased when high-protein diets are consumed and, having considered average protein intakes and the availability of the vitamin in the diet, the Academy recommends a daily dietary allowance of 2.2 mg of the vitamin for young adult males and 2.0 mg for young adult females (N3).

Recommendations for infants have been based on limited information, but experience with proprietary formulas has suggested that metabolic requirements are satisfied if the vitamin is present in amounts of 0.015 mg/g protein; the Academy suggests a recommended dietary allowance of 0.3 mg of vitamin B_6/day for the young infant, rising to 0.6 mg/day in infants 0.5 to 1.0 year of age consuming a mixed diet (N3). Special requirements have been determined for the vitamin B_6 content of artificial milk for infants from birth to weaning (W9).

Recommended allowances for children and adolescents cannot confidently be made since only limited data are available for evaluation (K6, L9, M10, R7).

Additional allowances have been recommended for women during pregnancy and lactation (N3), but not for users of oral contraceptive agents.

There has been a suggestion that vitamin B_6 inadequacy may be a nutritional problem in the elderly (V2), and patients receiving total parenteral nutrition may also require special vitamin supplements (A9, K7, N4).

Recommended dietary allowances are not meant to cover the needs of individuals in whom requirements are altered by extreme environmental conditions, acquired illnesses (viral, bacterial, or parasitic infections), malabsorption syndromes, and congenital abnormalities. Nor do they take into account the concept of biochemical and nutritional individuality within the "normal range," even between members of the same race or ethnic group (W10).

Vitamin deficiency and dependency states have been reviewed by Scriver (S14). He defines "deficiency" as being exogenous or conditioned, the former occurring when the vitamin intake falls below the recommended dietary allowance and the latter being the state in which the "physiological requirement" for the vitamin can be higher than the dietary allowance. In this case, intake of pharmacological levels of the vitamin may be necessary to restore normal function to the target tissue reactions dependent on the vitamin or its derivatives.

4.2. Exogenous Deficiency

Although vitamin B_6 deprivation has been investigated in laboratory animals and shown to cause altered brain function (C8), it was not until the early 1950s that a vitamin B_6-depleted diet was shown to produce clinical disorders in some infants. Snyderman and co-workers (S6, S15) studied two infants with gross neurological malformations. They were fed vitamin B_6-free diets containing 15% protein. By the seventh day, one child had convulsive seizures following earlier biochemical evidence of reduced body stores of the vitamin. The other child developed anemia without convulsions. Intravenous administration of 50 mg of pyridoxine relieved the seizures in the first child and 1 mg per day orally corrected the anemia in the second.

However, the greatest impetus to the study of vitamin B_6 deficiency in humans was given by the referral to pediatricians in the United States of a group of children who had been fed a commercially processed milk food (S6, C9, C10, M11, M12). These children presented at about 2 months of age with a typical history of "jitteryness," nervousness, of being easily startled by noise, colic, and irritableness. They had sudden onset of seizures with marked opisthotonus and stiff jerky convulsive movements. Between attacks the children remained jumpy and overall had lost weight and become anorexic. A change in their milk formula had stopped the symptoms in some. Others were given injections of pyridoxine and also recovered. It was subsequently discovered that all the children had been fed a liquid formulation which, in its production, had been vigorously heat sterilized and contained only 60 µg of vitamin B_6 per liter. This compared with more than 200 µg/liter for the reconstituted powder and 130 µg/liter for human breast milk. This loss of vitamin B_6 was attributed to heat denaturation during the sterilization procedure. The same liquid formula fortified with 450 µg of pyridoxine per liter, fed to children for several months, produced no convulsions.

Coursin (C9) showed that both the abnormal electroencephalograms and clinical signs which were present in some of the affected children returned to normal following intramuscular injections of pyridoxine.

Perhaps one of the more interesting facts to arise from all these observations is that only about three children out of each thousand, thought to have been on the vitamin B_6-deficient diet, developed any seizures. In addition, no child had been on a totally vitamin B_6-deficient diet and, when all other factors had been taken into account, it became obvious that many of these children who developed symptoms had vitamin B_6 requirements above "normal." Bessey and others (B16) observed that while an average of 0.26 mg of pyridoxine per day was sufficient to stop convulsions in children who had been fed the offending milk preparation, four to six times this amount was required to prevent biochemical evidence of vitamin B_6 deficiency (i.e., a positive tryptophan load test). Control children required less than 0.5 mg of pyridoxine per day to normalize the tryptophan load test. Scriver and Hutchison (S16) investigated a similar case in great detail, measuring not only the effect of increasing pyridoxine supplementation of the diet on the tryptophan load test, but also on cystathionine, oxalate, and taurine metabolism, vitamin B_6 excretion, and cerebrospinal fluid vitamin levels. Their patient required between 2.25 and 2.50 mg of pyridoxine per day to return the clinical and biochemical signs to normal, well above the daily requirement of 0.2 to 0.4 mg estimated by Bessey (B16).

This demonstration of biochemical individuality in the requirement for vitamin B_6 is in keeping with Williams' ideas (W10) and, at the same time, should alert one to the possibility that some infants may develop signs of vitamin B_6 deficiency at levels of intake which may otherwise be considered adequate for most infants.

4.3. Conditioned Deficiency

Conditioned deficiency, that is, adequate intake with respect to the recommended daily allowance, arises in the following situations.

4.3.1. *Defective Intestinal Absorption of the Vitamin*

Since vitamin B_6 is rapidly absorbed from the upper jejunum by passive diffusion, it is likely that conditions which reduce the area of absorptive surface might reduce the amount of the oral load of vitamin B_6 which is absorbed. Unfortunately, many of the diseases producing this effect on the intestinal mucosa often also produce anorexia and, consequently, poor diet and inadequate intake of the vitamin. Not all investigators have made allowance for poor dietary intake, and may be criticized on this ground. However, there is sufficient evidence to suggest that, at least, a biochemical deficiency of vitamin B_6 occurs in several gut diseases.

Uncomplicated malabsorption of the vitamin has been suggested as the reason for low levels of circulating pyridoxal phosphate in celiac disease (nontropical sprue) by Brain and Booth (B6), following the earlier work of Baker and Sobotka (B17). Sideroblastic anemia with pyridoxine deficiency has been reported in an adult with celiac disease, and this completely disappeared when the patient was placed on a gluten-free diet (D7). Anderson *et al.* (A2) found low plasma pyridoxal phosphate levels in 10 out of 12 patients with untreated celiac disease. Four patients who had been treated with a gluten-free diet showed plasma pyridoxal phosphate levels within the normal range. Reinken and his team (R8) investigated pyridoxal phosphate levels and the activity of pyridoxal kinase in the serum and duodenal mucosa of 14 children with acute celiac disease and of 10 children in clinical and biochemical remission; they found reduced levels of the vitamin in both plasma and duodenal mucosa of the acutely ill children, but with pyridoxal kinase levels significantly higher in mucosa of both groups of children with celiac disease than in normal controls. The authors claim this increased activity is a partially compensatory mechanism. They did not monitor the plasma pyridoxal levels during treatment, nor was there any assessment of vitamin intake.

Biochemical evidence of vitamin B_6 deficiency in the form of increased urinary xanthurenic acid excretion has been reported in cases of kwashiorkor (A10). This deficiency was overcome with oral pyridoxine, 20 mg twice daily. Theron *et al.* (T3) measured several indirect indexes of vitamin B_6 status in 13 patients with kwashiorkor and found that although all his cases showed at least one index to be normal, very few cases showed more than two abnormal findings on repeated testing. They attributed these results to the variable sensitivity in detecting pyridoxine need; as little had been done to estimate intakes in these children over the period of the tests, the results may reflect varying intake as well as varying absorption.

Children with gastroenteritis have low levels of circulating pyridoxal phosphate (as estimated by the erythrocyte aspartate aminotransferase assay), and this is not affected by an adequate vitamin intake before the occurrence of acute bowel inflammation (A11). The mechanism of the malabsorption here is probably a combination of decreased transit time through the absorptive area of the gut, slightly reduced absorptive area, and temporarily reduced intake.

The chronic inflammatory condition, Crohn's disease, is also associated with low plasma pyridoxal phosphate levels (A2), as well as low plasma folate levels. Whether this deficiency is a result of malabsorption or due to an increased requirement induced by the inflammatory process requires further investigation.

Sanderson and Davis (S17) showed that active gastric ulceration was associated with low serum pyridoxal phosphate levels, whereas patients with active duodenal ulceration had normal values. They studied diet, alcohol intake, and drug therapy and showed that the difference was not related to these variables. However, later work (S18) showed that various other forms of gastric pathology (gastric carcinoma, gastritis, benign polyps) can be associated with low plasma pyridoxal phosphate levels. The explanation for the effect in widely differing conditions, particularly as no absorption of the vitamin occurs in the stomach, awaits further study.

As might be expected, the jejunal–ileal bypass procedure has been associated with low plasma pyridoxal phosphate levels (H5).

4.3.2. *Defective Cellular and Intercellular Transport*

Vitamin B_6 moves easily across membranes only when it occurs in the free form as the alcohol, aldehyde, or amine. Phosphorylation to the corresponding coenzyme forms impedes transport. Penetration of the phosphorylated vitamers into tissues is probably dependent on a membrane carrier function that is genetically controlled and whose efficiency can be impaired by mutation (S14). However, since the B_6 vitamins are required as cofactors in many enzyme systems, a major disturbance in biosynthesis or transport of the vitamin is unlikely to be compatible with life.

4.3.3. *Impaired Oxidation or Phosphorylation Mechanisms in Vitamin B_6 Metabolism*

Primary acquired sideroblastic anemia is a well-defined form of anemia characterized by the presence of iron-containing granules that form a cuff around the nucleus of the developing red cells. These ring sideroblasts are thought to arise from some defect in heme synthesis. Pyridoxine is necessary for heme synthesis (C11, K8, L10); it is required for the activation of glycine to form a Schiff base, which then condenses with succinyl CoA in the initial reaction, leading to formation of heme. Sideroblastic anemia has been reported in pyridoxine-deficient animals and in patients receiving pyridoxine antagonists. Treatment of the condition with oral pyridoxine may lead to hematological improvement. However, response to the pyridoxine is usually suboptimal, if it occurs, and very large doses are required.

Hines and Grasso (H6) suggest that a block in the conversion of pyridoxine to pyridoxal phosphate might be the underlying cause of sideroblastic anemia. Hines and Cowan (H7) showed that conversion was defective in alcohol-induced sideroblastic anemia and thought pyridoxine kinase activity was inhibited. Hines' small group of patients responded to intramuscular pyridoxal phosphate, but not intravenous pyridoxine. Similarly, Mason and Emerson (M13) reported a case of primary sideroblastic anemia that responded dramatically to intramuscular pyridoxal phosphate after failing to respond to oral pyridoxine (malabsorption was excluded as a possible problem, but at no time were plasma pyridoxal phosphate levels measured). Anderson et al. (A2) demonstrated subnormal serum pyridoxal phosphate levels in 16 out of 26 patients with sideroblastic anemia.

In the conversion of pyridoxine to pyridoxal phosphate, the second enzyme involved, a pyridoxine oxidase, requires flavin mononucleotide as cofactor; riboflavin deficiency has been suggested as a causative factor in the reduced conversion of pyridoxine to pyridoxal phosphate and as being responsible for the oral lesion found in pregnancy in women in some countries (I2). However, Lakshmi and Bamji (L11) failed to show any correlation between riboflavin status and blood pyridoxal phosphate in either rats or man. Riboflavin has also been implicated by Anderson et al. (A12) in improving the red cell conversion of pyridoxine phosphate to pyridoxal phosphate in a case of β-thalassemia in which riboflavin was given orally for 8 weeks. Anderson and her colleagues (A12) also claimed that both homozygotes and heterozygotes for β-thalassemia show a reduced rate of red cell conversion of pyridoxine phosphate to pyridoxal phosphate *in vitro,* and that this conversion can be increased by incubation of the blood with riboflavin.

Klieger et al. (K9) suggested that toxemia of pregnancy was a result of the inability of the toxemic placenta to convert pyridoxal to pyridoxal phosphate. The pyridoxal kinase activity was lower than normal. In addition, pyridoxine phosphate was found, rather than pyridoxal phosphate, as the major vitamer, suggesting a reduced pyridoxine phosphate oxidase activity. Gaynor and Dempsey (G8) showed similar trends in a study of 44 normal and 31 eclamptic placentas, but values for enzyme activities were not statistically significant.

Pyridoxal kinase activity is low in the serum of newborn preterm infants, as is the level of serum pyridoxal phosphate (R9). This is only one factor which may put the newborn at risk for vitamin B_6 depletion.

Mauhuren and Coburn (M14), investigating pyridoxal phosphate levels in the leukocytes and platelets from patients with Down's syndrome, postulated that this vitamer turns over more rapidly than in the cells from normal patients; earlier work (G9) showed that patients with Down's syndrome excreted significantly less xanthurenic acid, following a 5 g tryptophan load, than did controls. The significance of these findings is hard to establish.

4.3.4. Chemical Inactivation

Chemical inactivation leads to loss of available vitamin at its required sites of activity.

4.3.4.1. Bioavailability. Although total vitamin B_6 levels can be measured in foodstuffs and calculations of human intake of vitamin B_6 can be made, it is insufficient to recommend an intake of the vitamin without considering its stability and bioavailability. The latter has not been studied in much detail, compared with the biochemistry of the vitamin.

Yasumoto *et al.* (Y3) have suggested that a significant amount of vitamin B_6 in rice bran is in a bound form as $5'$-O-(β-D-glucopyranosyl)pyridoxine; although rats can utilize the complex, this has not been confirmed in humans. Nelson *et al.* (N5), using intestinal perfusion techniques with human volunteers, found that only 30% of pyridoxine from orange juice was absorbed, and that this was possibly due to natural binding to a molecule of 3000 d. Davis and Smith (D6) reported that serum contained a heat-labile component of vitamin B_6. They found a 30% loss of activity after heating to 121°C for 45 minutes, whereas synthetic pyridoxal was stable under these conditions.

The reduction in biological activity may be due to the formation of stable complexes with some amino acids, notably glutamate, when these temperatures are used (B4). However, temperatures employed in domestic cooking seldom destroy any of the three forms of the vitamin (B18), except, perhaps, in the case of milk, where there are considerable losses when milk is heat sterilized, possibly due to the formation of a vitamin-inactive complex between pyridoxal and cysteine (B19).

The forms of vitamin B_6 may alter during food processing and activity may be lost. For example, pyridoxal is largely changed to pyridoxamine during sterilization of milk; this vitamer may complex with sulfhydryl compounds, forming complexes with less than 25% of the activity of either vitamer (B18). Complexes of pyridoxal phosphate bound to dietary protein, such as ϵ-pyridoxyllysine, have in rats only 60% of the molar potency of pyridoxine

(G10). Pyridoxal phosphate also forms cyclic compounds with histamine and histidine (K10) which may interfere with absorption of the vitamin.

4.3.4.2. *Antagonists.* Drugs can impair vitamin absorption, increase vitamin excretion, or interfere with vitamin utilization. Such drug-induced deficiency can occur even when the diet is adequate for normal maintenance. In these cases, administration of a larger dose to compensate for losses may overcome the problem. Vitamin B_6 antagonists have one of two actions: either to combine with pyridoxal phosphate, rendering it inactive as a coenzyme, or to act as a substitute for the vitamer in the molecular configuration of the apoenzyme requiring pyridoxal phosphate. Aspects of this work have been reviewed by Holtz and Palm (H8), Wolf (W1), and Klosterman (K11).

There are several naturally occurring vitamin B_6 antagonists, few of which, however, could be implicated in producing a deficiency state in humans. In chickens, on the other hand, there is clear evidence that linantine from linseed meal has a detrimental effect (K11). Linantine is hydrolyzed to 1-amino-D-proline, which forms stable hydrazones with pyridoxal and pyridoxal phosphate. Similar chemical compounds are found in *Gyromytra esentensa* and *Agaricus bisporis* (edible mushroom). Gyromytrin forms an N-methyl-N-formyl hydrazone and agaritine forms an α-glutamyl derivative of the 4-hydroxymethyl phenylhydrazone, although there is no evidence that mushroom gourmands are prone to vitamin B_6 deficiency. The Leguminosae have several species which produce antagonists of the hydrazone type, which act principally by strongly inhibiting pyridoxal kinase. The *Canavalia* species produce canavanine, which is metabolized to canaline, a substituted hydroxylamine which can form hydrazones with pyridoxal phosphate. The *Mimosa* and *Leucaena* species produce mimosine [3-hydroxy-4-oxo-1(4H)-pyridinealanine], which complexes with pyridoxal phosphate. It is interesting to note that D-cycloserine, produced from *Actinomyces*, is the next lower homolog of canaline, although it is usually present as the lactam, and that when used as therapy against tuberculosis, it has produced convulsions which are relieved by pyridoxine supplementation (K11).

A widely used antituberculous drug, isonicotinic acid hydrazide, has been implicated as a cause of vitamin B_6 deficiency. It was observed that patients treated with large doses of the drug developed a peripheral neuritis similar to that seen in exogenous vitamin B_6 deficiency (B15). Administration of pyridoxine prevented the symptoms. Isonicotinic acid hydrazine forms a hydrazone with pyridoxal phosphate, which is inactive as a coenzyme and, in addition, is a potent inhibitor of pyridoxal kinase (M15); it may also inhibit kynurenine transaminase (W1), which, in turn, may account for the somewhat different pattern of excretion of tryptophan metabolites in patients taking isonicotinic acid hydrazide without vitamin B_6 supplements. Krishwaswamy (K12) observed that some patients with tuberculosis, undergoing therapy with isonicotinic acid hydrazide, varied in their sensitivity to the drug, even after accounting for the patients' inherited capacity to inactivate it (slow/fast acetylation). He loaded his patients with methionine and showed increased urinary excretion of cystathionine and an elevated ratio of cystathionine to cysteine sulfonic acid, suggesting a block in conversion of cystathionine to the cysteine sulfonic acid. Cystathionase is more sensitive to pyridoxal phosphate depletion than cystathionine synthetase, so it seems likely that he was simply observing biochemical individuality in response to vitamin B_6 deficiency, rather than to some specific action of the isonicotinic acid hydrazide.

Despite 25 years of literature on the vitamin B_6-depleting action of isoniazid, editors still find it necessary to publish reports recommending supplementation with vitamin B_6. McKenzie et al. (M16) reported the case of a neonate, the child of a mother being treated for tuberculosis, who was commenced on 20 mg of isonicotinic acid hydrazide twice daily at birth, without supplemental pyridoxine. Multifocal clonic convulsions started on the thirteenth day and continued until pyridoxine therapy was commenced 4 days later. Previously, Morales and Lincoln (M17) had reported that 20 children on isonicotinic acid hydrazide showed no signs of deficiency when the dose was less than 10 mg/kg body weight daily.

L-Penicillamine (α-amino-β-thioisovaleric acid) was shown to cause vitamin B_6 depletion in rats by Kuchinskas and his team in 1957 (K13) and Ueda et al. (U1), and it was found that both the D and L isomers can form thiazolidine derivatives with pyridoxal phosphate by nonenzymic reaction. D-Penicillamine is used in the long-term treatment of Wilson's disease and in the management of heavy metal poisoning, as well as in the treatment of some patients with cystinuria; hence, it is important to monitor the vitamin B_6 status of these patients regularly or to ensure that more than an adequate concentration of the vitamin is present in the diet (E1). Penicillamine has also been used in the treatment of multiple sclerosis, where it can be tolerated in high doses (2 to 2.5 g per day) if given with 150 mg per day pyridoxine (S19).

L-Dopa (3,4-dihydroxyphenylalanine) is often used in the treatment of Parkinson's disease. At neutral pH, this compound reacts with pyridoxal and pyridoxal phosphate to form tetrahydroisoquinoline derivatives (E2, E3), which not only exhibit no vitamin activity but also inhibit certain enzymes requiring pyridoxal phosphate. In addition, animal experiments have shown a significant reduction in the functional levels of cerebral pyridoxal phosphate following acute oral L-dopa administration (K14). However, Mars

(M18) has shown that patients treated for long periods with L-dopa showed an increased ability to synthesize pyridoxal phosphate from pyridoxine in red cells. This may explain why measurements of plasma pyridoxal phosphate levels in such patients have been inconclusive in showing coenzyme deficiency.

Disulfiram has been used in the treatment of alcoholism for many years. This drug is first metabolized to diethyldithiocarbamate, which is further metabolized to diethylamine and carbon disulfide. Carbon disulfide produces the pharmacological and toxic effects seen with disulfiram. Chronic exposure to carbon disulfide has been shown to cause, among other things, pyridoxine deficiency in animals and man (M19). It was found that disulfiram raised serum cholesterol in alcoholics, but if pyridoxine was added to the treatment this rise was not observed. Patients receiving pyridoxine alone showed a decrease in serum cholesterol after 3 weeks abstinence from alcohol. The inference of this work is that vitamin B_6 supplements should be given to alcoholics treated with disulfiram to avoid the potential atherosclerotic changes caused by raised serum cholesterol.

One vitamin B_6 antagonist widely used experimentally in animals and man is 4-deoxypyridoxine. The 4'-hydroxymethyl group of pyridoxal is essential for coenzyme activity and replacement with a methyl group abolishes activity of the compound. When administered to either animals or humans, 4-deoxypyridoxine must be phosphorylated at the 5-position before it can compete with the physiologically occurring pyridoxal phosphate at binding sites of the apoenzymes. It would appear that the phosphorylated analog has binding affinities for apoenzymes similar to those of pyridoxal phosphate; hence, the degree of inhibition of vitamin B_6 enzyme-dependent reactions will vary with the degree of disassociation of the apoenzyme and coenzyme. This explains the complex, altered metabolism of the tryptophan-to-niacin pathway which Wolf (W1) described. In addition, 4-deoxypyridoxine probably also inhibits the activity of pyridoxal kinase, thus reducing the production of pyridoxal phosphate (M15).

4.3.4.3. *Oral Contraceptive Agents.* Since the initial report by Rose (R10) describing altered tryptophan metabolism in women using oral contraceptives, there have been over a hundred reports confirming and expanding on his observations. Care must be taken in evaluating this mass of information, as there are considerable variations in each study: the number of patients investigated, their ages, types of oral contraceptive agents used, duration of taking these agents, diet, and socioeconomic status. A multiplicity of methods for assessing vitamin status and an equally wide range of treatment regimes have been used. Rose studied 14 women taking various contraceptives and observed increased urinary excretion of xanthurenic acid in response to an L-tryptophan load (5 g), compared to a control group. After supplementation with pyridoxine hydrochloride (40 mg daily for 5 days), urinary excretion of xanthurenic acid after an identical tryptophan load dropped significantly in one subject.

The accumulated data on xanthurenic acid excretion in users of oral contraceptives have been well reviewed by Wolf (W1). In most studies, about 70–80% of all users show abnormal tryptophan metabolism, with the effect of the oral contraceptive being relatively rapid and disappearing in many women during the menses interval when no oral contraceptives are being taken (L12). Investigation of the amount of oral pyridoxine required to return urinary xanthurenic acid excretion to normal showed that 2 mg/day was sufficient in only 10% of apparently deficient users and even 20 mg/day was not sufficient in some women. Luhby (L12) recommended that 30 mg/day should be given to correct the abnormal metabolism in oral contraceptive users—some 15 times the recommended dietary allowance.

The findings in women taking oral contraceptives have been explained as follows: the estrogen component of the oral contraceptive probably stimulates activity of tryptophan pyrrolase. Although an oral estrogenic contraceptive steroid has been shown to induce tryptophan pyrrolase activity in ovariectomized–adrenalectomized rats (B20), little work has been possible on human liver tryptophan pyrrolase. However, hydrocortisone is known to induce human liver tryptophan pyrrolase (G11), and the estrogens may exert their effect by increasing corticosteroid production. The increased activity of the initial enzyme in the main metabolic pathway of tryptophan to niacin would have the effect of diverting more tryptophan in this direction, thus increasing the requirement for pyridoxal phosphate for the dependent enzymes further down the pathway. If the vitamin B_6 intake is not sufficient to meet this demand, then an apparent reduction in activity of the apoenzymes with the lowest binding affinity for pyridoxal phosphate will occur. Since kynureninase is a cytoplasmic enzyme as opposed to kynurenine transaminase which is mitochondrial, the former is more likely to show signs of coenzyme depletion first, which might account for increase in urinary xanthurenic acid excretion in women taking oral contraceptives.

Estrogens induce many proteins (including other pyridoxal phosphate-dependent enzymes) not involved with tryptophan metabolism. This increased requirement for cofactor may bring about a redistribution from less essential, that is, less fully saturated, holoenzymes to the more essential enzyme systems (e.g., brain amino acid decarboxylases) (B3). Kynureninase is not induced by estrogens (B21), thus accentuating the relative deficiency of coenzyme for this enzyme. Certain estrogen conjugates, particularly sulfates, compete for pyridoxal binding sites on kynureninase, with kynurenine transaminase apoenzymes further reducing the effectiveness of the reduced availability of the coenzyme. The progesterone component of oral contraceptives has no effect on tryptophan metabolism.

Although this information may explain the abnormal tryptophan metabolism observed in oral contraceptive users, it does not suggest that there is an absolute vitamin B_6 deficiency. Other methods for assessing vitamin B_6 nutrition have also been applied to this problem. Salkeld et al. (S20) studied 233 oral contraceptive users using the test of activation of erythrocyte aspartate aminotransferase and found that 48% have a deficient or marginal vitamin B_6 status, compared with 18% in a control group of 76 nonusers, irrespective of whether a low (0.05 mg) or high (0.075–0.1 mg) dose of estrogen preparation was taken. They found that a daily intake of at least 20 mg of pyridoxine hydrochloride was necessary to prevent biochemical deficiency, and that self-medication with multivitamin preparations was insufficient protection. Driskell et al. (D8) studied a small number of young oral contraceptive users, using the test of stimulation of erythrocyte alanine aminotransferase and showed that, although their erythrocyte alanine aminotransferase indexes were significantly higher than those of female nonusers and males of similar ages, there were several subjects in the latter groups who showed indexes above normal, suggesting a subclinical deficiency. Careful dietary histories showed that all three groups (students at an American university) reported an intake of slightly less calories, more protein, and considerably less vitamin B_6 than the 1974 recommended dietary allowances. Feltkamp et al. (F3) could not show any significant difference in erythrocyte asparate aminotransferase index values for a small group of oral contraceptive users compared to controls, but did observe that when given 300 mg of pyridoxine hydrochloride daily for 3 months, there was an increase in basal erythrocyte asparate aminotransferase activity, but that the saturation index fell to normal due to saturation of the enzyme with cofactor. The increased basal enzyme activity [on a dose 10 times that recommended by Luhby (L12)] was explained by induction of the red cell enzyme, indicating yet another factor which needs careful attention if vitamin status is to be adequately assessed.

Wien (W11) studied the vitamin B_6 status of 238 Nigerian women by measuring the erythrocyte alanine aminotransferase activity, both with and without in vitro stimulation by pyridoxal 5'-phosphate. Forty-nine women used an intrauterine contraceptive device, 47 used injectable progestogen, and 48 used oral contraceptives; 94 new patients had not used any contraceptive during the preceding 6 months. On the basis of the erythrocyte alanine aminotransferase activity, both with and without the in vitro addition of pyridoxal 5'-phosphate, no significant difference was observed between the three treatment groups. The percentage of each treatment group judged to be vitamin B_6 deficient was similar to that of the control group, and although the B_6 level of the group taking oral contraceptives was judged to be lower, the difference was not significant.

With the advent, in recent years, of techniques for accurately measuring plasma levels of pyridoxal phosphate, the problems of vitamin B_6 status in oral contraceptive users have been examined. Plasma pyridoxal phosphate levels are less influenced by factors such as enzyme affinity (as in erythrocyte aminotransferase assays) or the multiple physiological problems of examining tryptophan metabolism. Lumeng and co-workers (L13) measured plasma pyridoxal phosphate, using the tyrosine decarboxylase apoenzyme system of Chabner and Livingstone (C6), in 55 women who had been taking oral contraceptives for more than 6 months; they found the mean levels of the two groups to be significantly different, with the controls having the higher value. Since plasma pyridoxal phosphate concentration varies with age (D1), the results were reanalyzed for 5-year groupings. Oral contraceptive users had lower values than the controls for groups 25–29 years old and 30–34 years old, but not for groups 20–24 years old. They also studied a small group of controls through several menstrual cycles and showed no significant variation in plasma pyridoxal phosphate levels during the cycle. By following several oral contraceptive users for 6 months after drug use was initiated, they were able to show that the greatest drop in plasma pyridoxal phosphate occurred in the first 3 months. Comparison of plasma pyridoxal phosphate using the standard tryptophan load test in 15 contraceptive users showed 11 users with raised urinary xanthurenic acid, but only four of these had reduced plasma coenzyme levels.

In the same year, Davis and Smith (D9) reported on a study of 107 women taking oral contraceptives for periods ranging from 1 month to 9 years (mean 36 months). Using a microbiological assay (D1), they were unable to show any difference in plasma pyridoxal levels in these women, compared with age-matched women not taking oral contraceptives, in any 10-year grouping from 20 to 50 years. However, 9.3% of the women taking oral contraceptives and 13.1% of the control group had low levels compared with the accepted reference range used by these workers.

In a very much smaller group, Shane and Contractor (S21) observed significantly lowered blood pyridoxal phosphate (measured fluorimetrically) in oral contraceptive users compared with controls, but unlike previous workers, they were not able to show any difference between the two groups for either erythrocyte aminotransferase or erythrocyte aminotransferase index. They also pointed out that there was no correlation between plasma pyridoxal concentrations and the erythrocyte enzyme levels in the individuals studied.

Prasad et al. (P4), however, found evidence of vitamin B_6 depletion in oral contraceptive users whether they measured serum pyridoxal phosphate (enzymically) or erythrocyte aminotransferase indexes.

Whether the conflicting reports can be explained as assay differences or experimental design differences is difficult to judge; however, it would appear that where biochemical differences have been shown to exist between users of oral contraceptives and controls, the least number of abnormalities are seen when the blood vitamin is measured, suggesting that it is only in a few cases that anything approaching vitamin depletion occurs. It is interesting that, in these studies, clinical signs which may have suggested vitamin B_6 depletion were not mentioned or, if looked for, were not found (D9).

Only two important abnormalities encountered in users of oral contraceptives may be related to vitamin B_6 deficiency: severe depression and impairment of glucose tolerance. Although in the majority of oral contraceptive users the symptoms of a neuropsychiatric disorder may be rather trivial and short-lived, a small proportion of these women complain of severe depression which seems to be related to the use of the steroids. There are theoretical reasons for suggesting that the symptoms may be related to a defect in 5-hydroxytryptamine production in the brain (R11, W12).

Estrogen-induced tryptophan pyrrolase may divert tryptophan away from the serotonin pathway to the niacin pathway, and in doing so produce a relative pyridoxal phosphate deficiency. The rate-limiting enzyme, 5-hydroxytryptophan decarboxylase, is a pyridoxal phosphate-requiring enzyme, thus further aggravating the reduced production of 5-hydroxytryptamine. However, it would seem unlikely that such an important pathway as that producing brain amines would be so vulnerable to mild changes in vitamin B_6 availability. In addition, there is no evidence for reduced brain 5-hydroxytryptamine in "true" vitamin B_6 deficiency (R11). It is possible that 5-hydroxytryptamine decarboxylase is inhibited by the increased kynurenine and 3-hydroxykynurenine found in oral contraceptive users, as has been reported to be the case in rats (G12).

Whatever the underlying mechanism, significant improvement in 19 of 39 women taking oral contraceptives and having severe depression occurred following treatment with pyridoxine hydrochloride (40 mg daily) (A13, A14). These 19 women had evidence of "absolute vitamin B_6 deficiency," as shown by the erythrocyte alanine aminotransferase index. No improvement was observed in the other 20 women who did not show a vitamin B_6 deficiency, or with placebos in either group.

A therapeutic response of some depressed patients on oral contraceptives has been achieved by giving large doses of L-tryptophan (C12), suggesting that the available pool of free tryptophan may be inadequate in these patients, rather than any abnormality associated with vitamin B_6.

Although pyridoxal phosphate plays an important role in iron metabolism, and deficiency of the coenzyme has been associated with hypochromic anemia in man, no evidence of hypochromic anemia has been noted in oral contraceptive users when it has been looked for (D9). However, there has been a report of an isolated case of megaloblastic anemia in a 46-year-old woman who had been taking oral contraceptives continuously for 7 years (T4). A clinical trial with pyridoxine hydrochloride, 50 mg twice daily for 7 days, restored the megaloblastic hematopoiesis to normal, and resulted in great clinical improvement. Unfortunately, no assessment of vitamin B_6 status was made prior to treatment.

Impaired glucose tolerance may occur in pregnancy, with glucocorticoid administration, and in some oral contraceptive users (A15). Abnormal tryptophan metabolism is noted in all three conditions, and the connection between tryptophan metabolites and glucose metabolism has variously been attributed to xanthurenic acid–insulin complexes (K15, M20) and inhibition of phosphoenolpyruvate carboxylase by quinolinic acid. In a study of 46 women taking combined estrogen–progestogen oral contraceptives (A15), all 46 showed abnormal tryptophan metabolism, although only 18 had tissue depletion as assessed by their erythrocyte transaminase indexes. In the women with the vitamin deficiency, administration of 20 mg of pyridoxine hydrochloride twice daily for 4 weeks caused elevation of fasting blood pyruvate levels; reduction in plasma glucose, insulin, and blood pyruvate response was seen after an oral glucose load. These results suggest it is unlikely that the carbohydrate intolerance is due to formation of an xanthurenic acid–insulin complex in these women, but, rather, that enhanced quinolinic acid synthesis by pyridoxal phosphate might reduce the gluconeogenic effect of phosphoenolpyruvate carboxykinase in the liver.

4.4. Relative Deficiency

Relative deficiency occurs when primary intake is inadequate in relation to demands, as in increased metabolic activity, infection, pregnancy, and infancy.

4.4.1. *Infection*

The impact of infection on vitamin metabolism has been discussed by Vitale (V3). He came to the conclusion that most infections have an indirect effect on vitamin metabolism and that most of the effects may be mediated through changes in other systems. When measuring the effect on vitamin B_6 of increased metabolism due to infection or injury, care has to be that the changes observed do not simply reflect movement of the vitamers from one compartment to another (i.e., an adaptive process rather than a deficiency). Infections produce slight to severe inflammation, with necrosis followed by repair. One can expect an increased demand for vitamin B_6 in these processes which, if not present in sufficient amounts, might lead to deficiency states, although it is doubtful whether vitamin B_6 should be given to patients with an infectious disease unless the patient was undernourished to begin with.

4.4.2. Increased Metabolic Activity

Plasma pyridoxal levels have been observed to be low in children with burns and scalds, who otherwise have normal pyridoxine intakes (B22). Levels returned to normal after 250 mg of oral pyridoxine per day. Abnormalities of tryptophan catabolism were also noted, but these did not return to normal with supplemental pyridoxine, suggesting that pyridoxine deficiency was not the only factor altering tryptophan metabolism in these children.

Vitamin B_6 deficiency has been reported in uremia, whether or not the patient was being dialyzed (S22). Until recently, there was no adequate explanation for the pathogenesis of these findings, but it seemed reasonable to account for them as response to increased stress-altered metabolism, in general, and as a redistribution of the vitamers within body pools. Recently, however, Stone's team has shown that there is increased plasma clearance of pyridoxal phosphate, and that the presence of a plasma factor capable of inhibiting or degrading the vitamer may partially explain vitamin B_6 deficiency in uremia (S23).

4.4.3. Pregnancy

The investigation of vitamin B_6 status in the pregnant woman is bedeviled not only by the problems already discussed and problems of assessment of vitamin status in any population, but also with the problem of the continually altering physiological state of pregnancy itself.

In 1951, Sprince and others observed that pregnant women excreted increased amounts of urinary xanthurenic acid in response to a tryptophan load when compared with nonpregnant controls (S24). This was interpreted as indicating deficiency of vitamin B_6 during pregnancy, since similar results were seen in vitamin B_6-deficient nonpregnant subjects; supplementation of the diet with vitamin B_6 restored urinary excretion of xanthurenic acid to normal levels in both states. Many other workers have confirmed the finding of increased urinary xanthurenic acid after a loading dose of tryptophan and the corrective action of vitamin B_6 (W1).

Hamfelt and Hahn (H9) studied plasma pyridoxal 5-phosphate concentrations and xanthurenic acid excretion following a tryptophan load during pregnancy and found that there was a negative correlation between the two at any time during pregnancy; although the excretion of xanthurenic acid was positively correlated with the duration of pregnancy, the decrease in plasma pyridoxal phosphate during the same time was more striking.

Shane and Contractor (S21) have shown that plasma pyridoxal phosphate concentrations serve as a good index of vitamin B_6 status in pregnancy. In their careful study of 10 third-trimester pregnancies, there was a highly significant ($p < 0.005$) lowering of plasma pyridoxal phosphate levels, compared with an age-matched control group; this confirmed the observations of others (A2, H9, H10, W13).

This work led to studies on the supplementation of the diet of pregnant women with pyridoxine, and to some controversy as to whether this is either necessary or, indeed, ethical. The National Research Council of the National Academy of Sciences (N3) recommends that the daily allowance of vitamin B_6 during pregnancy should be 2.6 mg, an increase of 0.6 mg over that for nonpregnant women. Hamfelt and Turemo (H10) studied two groups of pregnant women, one taking 2 mg of pyridoxine orally per day and the other taking 10 mg; and found that the 2 mg dose was insufficient to maintain their plasma pyridoxal levels at the same levels as the controls. Similarly, Cleary *et al.* (C13) found that 10 of 13 women given the recommended dietary allowance of vitamin B_6 had plasma pyridoxal values below the lower limit of normal for nonpregnant subjects, while most of those taking 10 mg of pyridoxine per day had values above this. Baker *et al.* (B23) found that of 133 women at parturition (who had been assessed as taking between 9 and 10 mg of pyridoxine per day during pregnancy), 8% had abnormally low plasma pyridoxal levels; 87% of those who had taken no additional vitamin B_6 during this time had low plasma levels. Lumeng *et al.* (L14) conducted a prospective study of vitamin B_6 supplementation during pregnancy and found that of the 26 women studied, all appeared to have an adequate diet (1800–2400 kcal with 100 g protein), yet only one had a vitamin B_6 intake greater than 2.5 mg per day and 12 had an intake less than 1.4 mg per day.

Some clinical signs during pregnancy have been attributed to a relative lack of vitamin B_6, so far as the signs can be altered by the administration of pyridoxine. Vitamin B_6 deficiency in pregnancy had been suspected as early as 1942, when Willis and others used pyridoxine to alleviate symptoms of morning sickness (W14). The General Practitioners Research Group have shown that pyridoxine is useful in the control of pregnancy sickness (G13). Unfortunately, indexes of vitamin B_6 status were not measured. However, Reinken and Gant (R12) showed that women with significant vomiting in the first trimester of pregnancy had plasma pyridoxal levels significantly lower than those of age-matched nonpregnant women, or those of pregnant women in their first trimester who were not suffering significant morning sickness. The plasma pyridoxal levels in the affected group returned to normal after treatment with 100 mg of oral pyridoxine for 7 days, and there was significant clinical improvement.

Depressive illnesses in oral contraceptive users have been related to relative pyridoxal phosphate deficiency; therefore it is not surprising that postpartum depression should be considered from this point of view, assuming that in some women the state of vitamin B_6 deficiency produced during pregnancy, coupled with a possible derangement of brain 5-hydroxytryptophan decarboxylase (a pyridoxal phosphate-dependent enzyme), leads to decreased brain 5-hydroxytryptamine and, hence, depression. The field has been reviewed by Livingston et al. (L15), but they could find no evidence of vitamin B_6 deficiency in this condition.

Following the observation that abnormal carbohydrate tolerance is often associated with a deficiency of pyridoxal phosphate, Bennink and Schreurs (B24) showed that treatment of patients with evidence of gestational diabetes with pyridoxine produced improved glucose tolerance in most cases. This observation has been confirmed by Spellacy et al. (S25), but Perkins (P5) failed to repeat the previous success, attributing the improvements noted by others to dietary manipulation during the treatment period.

The fetus itself must be considered in any discussion of the effects of pregnancy on vitamin B_6 status. Contractor and Shane (C14) demonstrated active transport of pyridoxal phosphate across the placenta by showing higher levels in cord plasma, compared with maternal plasma, and this has been confirmed by others (B23, C13, L14). Contractor and Shane (C15) had previously shown that the activity of pyridoxal kinase increases in fetal liver and kidney throughout gestation, while activity of the enzyme in the placenta remained constant, inferring an inability to be totally independent of the maternal supply. They also showed that the phosphorylated form of the vitamin peaks in fetal blood several hours after peaking in maternal blood, indicating that it is the phosphorylated forms that cross the placenta (C14). This extent of dependence of fetal circulating pyridoxal phosphate on maternal plasma pyridoxal phosphate was further investigated by Cleary et al. (C13), who found a significant venous–arterial difference in umbilical cord blood which indicated that the fetus utilizes and/or degrades the pyridoxal phosphate which is transported across the placenta.

In addition to reflecting the long-term vitamin B_6 status of the mother, the fetus also can concentrate a significant amount of an intravenous pyridoxine load given to the mother within hours of parturition (F4).

Diseases of the placenta are likely to affect vitamin B_6 metabolism. Brophy and Siiteri (B25) have shown that in preeclampsia, the maternal–fetal gradient of pyridoxal phosphate is reduced from 6.6 to 3.7; therefore, at term there is only half as much pyridoxal phosphate in the fetal circulation as in the maternal. Gaynor and Dempsey (G8) studied 31 preeclamptic placentas and showed a lower mean pyridoxal oxidase activity and pyridoxal kinase activity than in normal placentas, and the presence of pyridoxine phosphate as the major vitamer, rather than pyridoxal phosphate. Low cord plasma pyridoxal phosphate levels may thus be due to inadequate conversion of pyridoxal to its phosphate, but since maternal supplementation with pyridoxine leads to higher cord plasma pyridoxal phosphate levels even in babies of preeclamptic patients, a defective transport system does not seem to be a factor (B25).

It is reasonable to assume that maternal vitamin B_6 status will significantly affect the fetal pyridoxal phosphate level and, in the absence of significant vitamin B_6 stores, affect the vitamin status of the newborn baby. Lumeng *et al.* (L14) showed that children of mothers supplemented with 10 mg of pyridoxine daily had significantly higher cord blood pyridoxal phosphate levels than those whose mothers had been on lower pyridoxine supplements. This has been confirmed by Ejderhamn and Hamfelt (E4), who also showed that this increased level in capillary blood fell strikingly in the first week of life and was similar to adult levels by 6 weeks of age. In addition, newborn infants of mothers supplemented with pyridoxine had a significantly higher birthweight than those of the mothers taking the unsupplemented diet. However, a study of low-birthweight infants has not revealed any correlation between birthweight and plasma pyridoxal levels in mothers giving birth to babies with a normal or low birthweight (B26).

The effect of reduced maternal vitamin B_6 status on the fetus has been studied in some detail in the rat. Since the first studies showing the importance of vitamin B_6 deprivation in rat acrodynia, central nervous system symptoms have been recognized as a consequence of vitamin B_6 deficiency. These symptoms range from hyperirritability, hyperactivity, abnormal behavior pattern, performance deficits, and convulsive seizures (C8). Recent work (A16) has confirmed earlier studies that progeny of rats fed diets low in vitamin B_6 during gestation and lactation exhibit deficiency signs between 10 and 14 days after birth. However, very careful control of these experiments must be made, since it is very easy for the results due to pyridoxine deficiency to be masked by protein/calorie deficiency in either or both dam and offspring. Alton-Mackey and Walker (A17) have shown that pups of vitamin B_6-deficient dams show pronounced growth depression in the first 3 weeks of life, and even offspring of rats fed 100% of the recommended pyridoxine intakes during gestation grew less than controls fed 400%. Other effects noted in the offspring of rats made vitamin B_6 deficient during pregnancy or lactation include delay in onset of reflexes and advanced neuromotor coordination (A18), major malformations such as omphalocele, exencephaly, cleft palate, micrognathia, splenic hypoplasia (D10), alterations in DNA content of the fetal brain (D11, D12), and impaired elongation of fatty acids in brain, leading to impaired myelination (C16). Desaturation of linoleic acid (T5), reduced $2',3'$-cyclic nucleotide $3'$-phosphohydrolase activity (M21), and reduced phosphoserine aminotransferase (B3) are other causes of demyelination brought about by restricting vitamin B_6 in the neonatal period of rat development.

It is unlikely that the degree of vitamin B_6 deficiency produced in these rat experiments would occur in the pregnant woman, at least in the developed countries of the world, but it indicates that vitamin B_6 is involved with major systems, particularly brain, and since, unlike in rats, human neonatal brain studies are unlikely to be carried out, even a mild vitamin B_6 deficiency in the pregnant woman should be treated as potentially dangerous for her fetus. Roepke and Kirksey (R13) have shown that mothers of infants with unsatisfactory Apgar scores at 1 minute had significantly low intakes of vitamin B_6 and lower levels of the vitamin in the serum than mothers whose infants had satisfactory scores.

Some workers suggest that vitamin B_6 deficiency *in utero* may lead to an increased risk of developing atherosclerosis in later life, possibly due to partial lysyl oxidase deficiency and, hence, inadequate cross-linking of arterial elastin and collagen (L16).

4.4.4. Lactation

As early as 1959, Karlin noted that human milk was relatively poor in vitamin B_6 (K16). Using a biological assay (*Saccharomyces carlsbergensis*) which responds to all forms of the vitamin, and a method of hydrolysis which used a temperature below 100°C at atmospheric pressure, she found that during the first few days postpartum, human milk may contain less than 20 µg/liter of vitamin B_6. Milk from mothers 1 to 7 months postpartum had a mean vitamin B_6 content of 105 µg/liter. Karlin then showed that a single intramuscular injection of 50 mg of pyridoxine produced a sevenfold increase in the vitamin level in the milk within 3 hours; levels fell to preinjection levels over 5 to 7 days. Oral administration of 500 to 1000 mg of pyridoxine daily produced a high milk vitamin content (1 to 4 mg/liter) which fell to pretreatment levels within 6 days of discontinuing the treatment (K17).

West and Kirksey (W15), using similar assay techniques, have also looked at the influence vitamin B_6 intake has on levels in the milk of human subjects. Their subjects were 3 to 30 months postpartum. Those consuming less than 2.5 mg of vitamin B_6 per day in their diet had significantly less vitamin B_6 per liter of milk (129 µg) than groups consuming 2.5 and 5.0 mg (239 and 314 µg, respectively). Vitamin B_6 intake two to five times the recommended daily allowance did not significantly elevate the level of the vitamin in the milk, compared to values of subjects whose intakes approximated the allowance.

Thomas *et al.* (T6) also showed that lactating women who were taking vitamin B_6 supplements produced milk with a higher vitamin level than the unsupplemented group, but that in her group of well-nourished women, no subject produced milk with less than the norm for vitamin B_6 content.

Since pregnancy may be complicated by a degree of vitamin B_6 deficiency, and even normal breast feeding cannot be guaranteed to contain adequate vitamin B_6 levels, there is the potential for marginal deficiency in the first few days of life, particularly if alternative sources of the vitamin are not provided. In conditions where muscle mass and, hence, the phosphorylase store of pyridoxal phosphate may be reduced, the chances of deficiency in the first few days of life would be increased. Whether or not these babies develop convulsions will depend on the relative deficiency of brain glutamic acid decarboxylase or its coenzyme. Up to 30% of neonates who have convulsions have no cause found (K18). It is possible that some of these may be due, in part, to reduced vitamin B_6 status. However, few studies have been carried out to determine normal vitamin B_6 levels in this age-group (E4, R9, R14) and further work is required.

4.5. Dependency

Scriver (S14) described the dependency state as occurring in individuals "having a constant, lifelong specific requirement for a particular vitamin that greatly exceeds the recommended dietary allowance." The prevention of the disease or, at least, its biochemical abnormalities usually involves a constant intake of the vitamin, which may be as much as 500 times the recommended daily allowance. These vitamin-dependent conditions are very similar to classical "inborn errors of metabolism" in their biochemical manifestations and also in their mode of inheritance.

The interaction between pyridoxal phosphate and its apoenzyme is highly specific and the binding is complex, more so than that normally seen in enzyme–substrate systems. It is not difficult to imagine that changes in the molecular structure of the apoenzyme could alter the normal steric relationship between coenzyme and apoenzyme. In general, the vitamin-responsive inborn errors of metabolism are likely to reflect three fundamental modes of interruption to the normal coenzyme–apoenzyme interaction (S14).

1. The mutant allele may affect one of the steps in the pathway of coenzyme biosynthesis. Vitamin responsiveness would imply that the block is partial and that it can be overcome by raising the effective concentration of the vitamin at the step modified by mutation.

2. The mutant allele may affect the apoenzyme directly, altering its normal interaction with the coenzyme. Here, vitamin responsiveness implies that the unfavorable binding between mutant apoenzyme and coenzyme can be overcome by a higher concentration of coenzyme.

3. The mutant allele may modify the primary catalytic (reactive) site or the amount of functioning enzyme in the cell. Vitamin responsiveness may indicate either a rise in the basal concentration of the mutant apoenzyme

when its turnover is reduced by enzyme saturation, or the activity of another enzyme requiring the coenzyme may be augmented to allow redistribution of the metabolites.

As these inborn errors of metabolism have been extensively reviewed in recent years (N6, S26, S27), only a brief comment on the role of the vitamin will be given here.

4.5.1. Cystathioninuria

Cystathionine is a key intermediate in the transsulfuration pathway of methionine metabolism, where it fills its only known biological function in the transfer of the sulfur atom from methionine to cysteine. Cystathionine-cleaving enzyme (cystathionase EC 4.4.1.1) requires pyridoxal phosphate as a cofactor. Clinically, patients with cystathioninuria show no consistent signs or symptoms, although several have been retarded. On the other hand, sibs of affected cases who have been shown to have the biochemical disease have been clinically normal.

Frimpter (F5) reported that cystathionase activity in the liver of two affected patients was less than 10% of "normal," and that the addition of pyridoxal phosphate *in vitro* restored activity toward normal. Previous work had shown that there was biochemical improvement *in vivo* after daily supplementation with pyridoxine, and it had been proposed that the mutation in this condition altered the apoenzyme in such a way that the affinity for the coenzyme was markedly reduced.

However, Tada *et al.* (T7) have described a child with much reduced hepatic cystathionase which did not increase with *in vitro* addition of pyridoxal phosphate, nor did the patient benefit from oral pyridoxine treatment.

Two pedigrees have been described in which the vitamin responsiveness was dose specific for each pedigree, and quite different pyridoxine intakes were required to satisfy the requirements of each pedigree (S14). This suggests a significant degree of genetic heterogeneity but, as the exact nature of the binding site on the apoenzyme has yet to be worked out, the significance of these observations is not clear.

Recent work on lymphoid cell lines from cystathioninuric patients who respond to pyridoxine shows that the cells produced cystathionase molecules altered in their ability to combine with the vitamer, but with their antigenic identity unaltered. Cells from a patient unresponsive to vitamin B_6 either contained no apocystathionase, or produced a protein lacking both activity or antigenic determinants necessary for cross-reactivity to normal liver cystathionase (P6).

4.5.2. Homocystinuria

Homocystine is also a metabolite of methionine, and its accumulation may be caused by a deficiency in cystathionine synthetase (EC 4.2.1.22) and also by both acquired and inherited defects in the methyltetrahydrofolate–homocysteine methyltransferase reaction. In pyridoxine-responsive cystathionine synthetase deficiency, the residual activity of the hepatic enzyme is 1–2% of normal (M22). The mechanism of vitamin responsiveness in these patients involves a small but significant increase in the activity of the mutant enzyme, from 1–2 to 2–4% in response to massive (250–500 mg) daily doses of pyridoxine. This is sufficient to increase the conversion of methionine to its metabolites from 8 to 16 mmol per day (i.e., the amount consumed in an adult diet). Much work has been done on this variant of homocystinemia, but there is little yet to correlate the multiple clinical findings (involving the eye, vascular system, mental retardation, and other neurological defects) with the mutant enzyme activity which leads to increased plasma cystathionine. However, Mudd (M22) suggested that there was an imbalance between synthesis and degradation of the mutant enzyme, and that saturation of the apoenzyme with pyridoxal phosphate retards the inactivation, leading to a higher concentration of active enzyme. This was demonstrated experimentally by Kim and Rosenberg (K19).

Although complete reversal of biochemical abnormalities has been reported in cases treated with these massive doses of pyridoxine, not all patients with proven hepatic cystathionine synthetase deficiency respond; in fact, in some cases actual harm has occurred due to the vitamin therapy (S28).

Gaull and co-workers (G14) disagreed with Mudd's interpretation of vitamin B_6 action in homocystinuria on the grounds that the assays he used to measure cystathionine synthetase, methionine-activating enzyme, and cystathionase depended for their sensitivity on rate-limiting substrate. Using redesigned assays, they measured cystathionine synthetase activity in two patients with the enzyme deficiency, biopsied after biochemically successful pyridoxine treatment, and established that the effect of the vitamer was not on the enzyme. The ability to respond biochemically to pyridoxine seemed to these workers to be related to the dietary intake of methionine and the dose of pyridoxine, and might be related to the formation of a hemithioacetate between pyridoxal phosphate and methionine.

4.5.3. Xanthurenicaciduria

This is another vitamin-responsive inherited disorder which is now thought to be due to a defective coenzyme binding mechanism (S14). The disorder was first recognized by Knapp (K20) when surveying individuals for vitamin B_6 deficiency. His cases showed no clinical abnormalities, but responded to an oral tryptophan load with a markedly exaggerated urinary excretion of xanthurenic acid, kynurenine, and 3-hydroxykynurenine. The other important finding was that none of these patients had any other evidence of vitamin B_6 deficiency, yet when treated with 10–100 mg of pyridoxine per day, they showed marked improvement of their biochemical abnormality.

Although most cases of this rare disease seem to be benign, Tada et al. (T8) reported four members of two kindreds, referred to them because of mental retardation; it was not possible to relate the abnormality of tryptophan metabolism to the mental abnormality. Tada's team measured the activity of the suspected defective enzyme (kynureninase EC 3.7.1.3) in liver biopsies from three of their four cases and found that it was markedly reduced in the absence of pyridoxal phosphate, but that it rose to nearly normal values when the system was saturated with the vitamer. These workers concluded from the evidence available that the disorder is due to a mutant kynureninase apoenzyme which has a much reduced affinity for its coenzyme.

4.5.4. Oxaluria

Oxalate is the major constituent of 65–75% of all kidney stones, though most patients who form oxalate stones have no apparent defect in oxalate metabolism. A small number of patients with calcium oxalate stones have "primary hyperoxalosis." This is a general term for two rare autosomally inherited disorders: type 1 (glycolic aciduria), which is due to a defect in glyoxalate metabolism, and type 2 (D-glyceric aciduria), which is caused by a defect in hydroxypyruvate metabolism.

Pyridoxine deficiency in experimental animals leads to hyperoxaluria and oxalosis (G15, R15), probably due to impaired transamination of glyoxylate to glycine, and in man, pyridoxine deficiency may result in an increase in oxalate excretion (F6). The administration of pyridoxine to normal volunteers receiving a diet adequate in vitamin B_6 produced a marked reduction in oxalate excretion (G16), and patients with recurring oxalate kidney stones excrete less xanthurenic acid and 4-pyridoxic acid than "normals." It is interesting that in cases of pyridoxine-responsive seizures in children, oxalate excretion appeared normal at all times, irrespective of pyridoxine treatment (S16).

Several groups have used pyridoxine in the treatment of hyperoxaluria in the hope of reducing the formation of oxalic acid (which leads to nephrolithiasis and progressive renal failure) by increasing its conversion to glycine (G17, S29, W16). However, it is not at all clear what dose level constitutes an adequate trial of treatment, since regimens ranging from 100 to 1000 mg of pyridoxine per day have been used with both biochemical and clinical improvement in some patients, while others have failed to respond. More knowledge of the basic defects in these conditions may produce a more rational approach to therapy.

4.5.5. Hypochromic Anemia

Harris et al. (H11) gave the first complete description of this condition in a man with a hypochromic, microcytic anemia which improved dramatically following treatment with oral pyridoxine hydrochloride. Associated abnormalities of iron and tryptophan metabolism also reverted to normal. Horrigan and Harris (H12) reviewed 62 cases of spontaneously occurring anemia, responsive in varying degrees to vitamin B_6. Although the hematological abnormalities varied considerably, with only half the patients having "characteristic" hypochromic and microcytic changes of the mature erythrocyte, there was sufficient information overall to suggest the existence of a genetically determined inborn error of metabolism affecting erythropoietic tissue which may respond to treatment with vitamin B_6.

The necessity for pyridoxine in heme synthesis was shown by Kikuchi et al. (K8) and Levere and Granick (L10), who demonstrated that it is required for the activation of glycine in the first step in the production of protoporphyrin IX via δ-aminolevulinic acid. However, experiments using ducklings (S30) suggested that other reactions in the heme synthetic pathway may be affected. This and the investigation of some patients with pyridoxine-responsive anemia (H12) suggest that a block at the step involving δ-aminolevulinic acid synthesis may be the defect in these conditions. Although the requirement for pyridoxal phosphate in heme synthesis has been established, it is important to note that treatment with pyridoxine hydrochloride has failed to produce a complete hematological remission in any patient. These patients show a conspicuous absence of the neurological signs associated with vitamin B_6 deficiency.

4.5.6. Vitamin B_6-Responsive Neonatal Convulsions

The concept of vitamin B_6 dependency was first introduced by Hunt and his colleagues in 1954 (H13). They studied a 13-day-old girl with intractable seizures since birth. They found, after several trials, that intramuscular pyridoxine was capable of controlling the seizures; it was subsequently shown that an oral dose of 2 mg daily kept the infant free of seizures. Since then, many other cases have been reported (D12, G18), all of which confirm Hunt's original description, although these reports usually indicate a requirement for larger amounts of pyridoxine to maintain seizure control (10–25 mg per day). The dependency may be so great that convulsions can occur in utero (B27, H14).

Scriver (S31) postulated that the vitamin was acting as the cofactor for glutamic acid decarboxylase (EC 4.1.1.15), catalyzing the formation of γ-aminobutyric acid from glutamic acid in the brain. He suggested that it was the reduction in the amount of this inhibitory transmitter in the central nervous system which was causing the seizures. Scriver and Whelan (S32) demonstrated the presence of glutamic acid decarboxylase in rat kidney, and suggested that renal tissue might be used to determine whether children suffering from pyridoxine-responsive convulsions were unable to convert glutamic acid to δ-aminobutyric acid at a normal rate.

Yoshida and co-workers (Y4) confirmed this hypothesis by demonstrating that *in vitro* kidney tissue from a patient with vitamin B_6-dependent convulsions was less able to convert L-[U-^{14}C]glutamic acid to γ-aminobutyric acid than was tissue from controls. However, the rate of conversion was restored to a normal level when pyridoxal phosphate was added in large amounts to the reaction system.

Recently, Lott *et al.* (L17) have had the opportunity to examine the brain and spinal cord in a 13-year-old boy dying with vitamin B_6-dependent seizures in progress. Biochemical findings include decreased γ-aminobutyric acid concentration and increased glutamate concentration in the frontal and occipital cortices, but not in the cord. Pyridoxal 5'-phosphate concentration was reduced in the frontal cortex. Virtually no glutamic acid decarboxylase was detected until the concentration (*in vitro*) was greater than 0.05 mM, then the enzyme behaved like that in controls. This boy also showed evidence of vitamin B_6-dependent brain cystathionase deficiency, a finding which has not been seen before in cases of vitamin B_6 dependency.

Bayon *et al.* (B28) have carefully examined the kinetic behavior of glutamic acid decarboxylase from mouse brain and suggest that there may be two different enzyme activities, one dependent and the other independent of pyridoxal phosphate. Whether this may also occur in humans has yet to be determined.

Heeley *et al.* (H14) have put forward an alternative suggestion based on a study of four patients. They showed that these vitamin B_6-responsive patients were able to synthesize pyridoxal phosphate normally, but were unable to maintain the prolonged high levels normally found in plasma after pyridoxine loading. They suggested such findings might be explained by an instability of the albumin–pyridoxal phosphate complex or by a greatly increased phosphatase activity. The latter alternative did not seem likely in view of the normal pyridoxic acid excretion of these patients.

4.6. OTHER CLINICAL STATES POSSIBLY ASSOCIATED WITH ABNORMAL VITAMIN B_6 METABOLISM

A number of clinical syndromes have been associated with an abnormality of pyridoxal 5-phosphate metabolism or an abnormality of one or more of its associated enzymes, and these have been studied in some depth. However, there are a number of other clinical conditions which have been associated with a deficiency of vitamin B_6. Such an association may be due to increased demand or a reduced intake occasioned by the particular disease state, or it may be iatrogenic; this has already been discussed. Finally, it may be due to a failure of one of the many enzymes associated with the various pyridoxal 5-phosphate metabolic pathways as a result of the disease process; some of these conditions are discussed in the following pages.

4.6.1. *Rheumatoid Arthritis*

Abnormalities of tryptophan metabolism have been noted in patients with rheumatoid arthritis (P7), and McKusick and others (M23) showed objective clinical improvement after therapy with pyridoxine hydrochloride. On the other hand, Schumacher *et al.* (S33) noted that although 32% of 55 patients with rheumatoid arthritis had plasma pyridoxal phosphate levels of less than 5 ng/ml (compared with only 7% of a similar-aged group of controls), treatment with 50–150 mg of pyridoxine hydrochloride per day for 3 months had no significant effect clinically. The plasma pyridoxal phosphate levels rose up to three times pretreatment levels, but fell to pretreatment values after stopping supplementation.

Anderson and co-workers (A2) showed that plasma pyridoxal phosphate values were lower than normal in 30 of 31 affected patients studied. Sanderson *et al.* (S34) found 35 of 42 affected patients had serum pyridoxal concentrations below the lower limit of the reference range; this could not be related to the age, sex, or drug therapy of the individuals. However, indomethacin was thought to play a role in the etiology of low serum pyridoxal levels in three of seven patients with osteoarthritis. Neither group investigated the effect of pyridoxine treatment in their subjects.

Explanations for the possible role of low vitamin B_6 in the pathology of rheumatoid arthritis include (1) decreased phagocyte effectiveness, perhaps related to vitamin B_6-dependent myeloperoxidase synthesis (V4) and (2) suppression of the immune response (D13); as the etiology of rheumatoid arthritis is unknown, these ideas must remain speculative. Many investigators fail to investigate adequately the dietary factors involved which may well be important in this debilitating condition and which may account for the low plasma pyridoxal. However, McKusick and others (M23) reported that when standard diets were fed to patients with rheumatoid arthritis and controls, the former excreted significantly less pyridoxine in their urine, suggesting that dietary insufficiency is not a major cause of the low serum pyridoxal found in these patients. Many patients with rheumatoid arthritis have a history of dyspepsia and peptic ulceration, and complications of ulceration are not infrequent. These conditions may be associated with the drug therapy of arthritis, or it may be that the agents used merely augment conditions already present (S34).

4.6.2. *Vitamin B_6 and Immunological Response*

Vitamin B_6 deficiency in rats has been shown to reduce their ability to respond to immunological challenge. Deficient rats responded poorly when challenged with influenza virus and diphtheria toxoid. In transplantation experiments, skin homografts from Wistar rats were grafted onto vitamin B_6-deficient rats of the Long–Evans strain. Of the 50 rats that received grafts, 92% had intact grafts after 2 weeks and 60% after 10 weeks, compared with 9 and 1%, respectively, of 71 controls (A19).

Rats fed on vitamin B_6-deficient diets for 2 weeks were reported to have shown evidence of failure to gain weight, a decrease in thymus weight, and a reduction in the number of cells present in both the peripheral blood and bone marrow. The number of small lymphocytes in the bone marrow was also reduced, and both B cells and null cells were equally affected. All of these defects were corrected following the return of the animals to a normal diet (C17). Lymphocytes from mice fed on a vitamin B_6-free diet for 5–6 weeks showed a reduced capacity to respond to foreign lymphoid cells *in vitro* and showed a significant reduction in cytotoxicity. Pyridoxal 5-phosphate, but not pyridoxal, added to the culture medium *in vitro* partially restored the impaired functions of T lymphocytes (S35). Robson and Schwarz (R16) found that feeding a vitamin B_6-deficient diet to pregnant rats led to reduced immunological competence in the offspring. Thoracic duct lymphocytes from deficient rats were found to have a reduced capacity to respond in mixed lymphocyte and normal lymphocyte transfer reactions.

Davis (D10) observed that congenital pyridoxine deficiency was not only teratogenic in the rat, but also had an adverse effect on immunological function. This was demonstrated as a defect in the capacity for expression of delayed hypersensitivity in 8-week-old rats.

Willis-Carr and St. Pierre (W17) found that young Lewis rats fed on a pyridoxine-deficient diet had a reduction of T lymphocyte numbers and defects of cellular immunocompetence. They found that maintenance on a vitamin B_6-deficient diet for 2 weeks produced a severe defect in thymic epithelial cell function. These cells are responsible for inducing T lymphocyte differentiation. Exposure of lymphoid precursors from vitamin B_6-deficient donors to normal thymic epithelial cell monolayers resulted in their conversion to functional T lymphocytes. Thymic epithelial cell monolayers from vitamin B_6-deficient animals were unable to effect maturation of T lymphocytes.

The precise mechanism behind these events is not known. Pyridoxal phosphate is required for DNA biosynthesis, hence this may be reduced in the deficient animal. Vitamin B_6 deficiency has the potential for causing significant damage to the immune process in the fetus and newborn.

4.6.3. Malignant Disease

A reduced level of plasma pyridoxal phosphate has been associated with Hodgkin's disease (C16), carcinoma of the urinary bladder (P8), breast cancer (B29), and enhanced activity of the tumor promoter 12-O-tetradecanoylphorbol-13-acetate when applied to mice maintained on a vitamin-free diet (M24).

Chabner et al. (C18) studied plasma pyridoxal phosphate levels and L-tryptophan metabolism in 43 patients with Hodgkin's disease. Reduced levels of pyridoxal phosphate were found in 8 of 14 untreated patients; 14 of 21 patients excreted increased quantities of one or more of the intermediates of the tryptophan pathway. They found that both tests gave abnormal results, most frequently in patients with advanced disease. Fourteen patients in complete remission following chemotherapy had results which were within the normal range. The eight patients with a reduced level of plasma pyridoxal phosphate were anergic, and this is of interest in view of the effect that a deficiency of pyridoxal phosphate has upon immunological response (C17, D13, R16, S35). In the case of Hodgkin's disease, it appears that the pyridoxal deficiency may be associated with increased demand because of increased cell proliferation or malabsorption. Tests of folic acid absorption in two patients with advanced Hodgkin's disease showed poor absorption of the vitamin; (P8) it seems likely that the results of these assays reflect the absorption pattern of a number of water-soluble vitamins.

An interesting relationship between vitamin B_6 deficiency and carcinoma of the urinary bladder has been suggested by Price and Brown (P9). They examined the excretion of tryptophan metabolites in 45 patients; 29 had spontaneous tumors and 16 had tumors associated with an industrial cause. Patients with spontaneous tumors excreted significantly more acetylkynurenine, kynurenine, and 3-hydroxykynurenine than a normal control group and significantly more of these three metabolites (plus kynurenic and xanthurenic acids) than the group with industrial bladder tumors. Administration of pyridoxine to some of the patients with spontaneous tumors considerably reduced the difference in the excretion pattern. Some tryptophan metabolites have previously been shown to have carcinogenic properties in the bladders of mice (A20). Byar and Blackard, in a prospective clinical trial, randomized 121 patients with stage 1 bladder cancer to treatment with placebo, pyridoxine (25 mg/day), or intravesical thiotepa. Excluding patients having a recurrence in the first 10 months and those followed up for less than 10 months, they found that treatment with pyridoxine was significantly better than placebo. Thiotepa was significantly better than placebo or pyridoxine (B29). This was an interesting result because although pyridoxine would reduce the excretion of tryptophan metabolites, it would only do so after the malignancy had been established. To be of benefit, a recrudescence would need to depend in some patients at least on the continuing presence of the carcinogen (in this case, metabolites of tryptophan). Alternatively, it may be that pyridoxine could exert its effect in a quite different manner.

Bell (B30) studied the urinary excretion of 4-pyridoxic acid in 79 patients with early breast cancer. Her results indicated that patients with below-median levels of urinary 4-pyridoxic acid had a significantly greater probability of recurrence than those who excreted amounts greater than the median. A reduced excretion of 4-pyridoxic acid is found in vitamin B_6 deficiency, but how this could affect the recrudescence rate of breast cancer is not clear. The results of other studies have been paradoxical in that those women who excreted a small amount of tryptophan metabolites following a 5 g load of L-tryptophan had an increased tendency to recurrence, whereas the opposite would have been expected (B31).

Studies on women with either colonic or gynecological cancers showed that there was a significant lowering of plasma pyridoxal levels in those patients with systemic metastases, but normal levels in patients with primary cancers (P10). Studies also excluded dietary deficiency in those patients with low pyridoxal phosphate levels on the basis of normal urinary 4-pyridoxal acid levels. Impaired liver function due to metastases did not seem to be a factor since plasma pyridoxal levels were low in breast cancer patients having only locally recurrent disease.

Foy et al. (F7) found that baboons maintained on vitamin B_6-deficient diets showed elevated serum α-fetoprotein levels, and suggested that marginal dietary vitamin B_6 may be related to the higher incidence of primary carcinoma of the liver in Africans; however, Ajdukiewicz and Pollitt (A21) were not able to show biochemical evidence of vitamin B_6 deficiency in a small group of Africans with primary liver carcinoma.

4.6.4. Liver Disease

Abnormal metabolism of vitamin B_6 has often been observed in patients with liver disease. Labadarios et al. (L18) showed abnormally low plasma pyridoxal levels in a number of patients with a variety of liver diseases. There was no difference in the values between groups who were considered alcoholics and those who were nondrinkers. These investigators came to the same conclusion as Mitchell et al. (M25), that the problem of maintaining plasma pyridoxal levels in patients with liver disease was due to an enhanced degradation of the vitamer, since urinary 4-pyridoxic acid excretion, following either oral pyridoxine or intravenous pyridoxal phosphate, was highest in patients with the lowest plasma pyridoxal levels. Neither group of investigators found evidence for abnormal synthesis of pyridoxal phosphate. Since pyridoxal phosphate is involved in the synthesis of nucleic acids, hepatic regeneration may be effected; in patients with decompensated cirrhosis, there seems to be good reason to supplement with vitamin B_6; pyridoxal phosphate may have some advantages over pyridoxine.

Since serum aspartate and alanine aminotransferase activities are commonly used to detect hepatocellular injury and to monitor the progress of liver diseases, the lower plasma pyridoxal phosphate level in patients with liver disease may cause falsely low values and treatment with either supplemental vitamin B_6 or improved diet may in turn produce artifactual elevations. Rossouw et al. (R17), however, showed that the vitamin B_6 deficiency is unlikely to be an important determinant of serum aspartate aminotransferase activity in patients with chronic liver disease. Rosalki and Bayoumi (R18), on the other hand, suggest that the serum enzyme activity is enhanced by in vitro addition of pyridoxal phosphate, but that this is greater in patients with postmyocardial infarct than in those with chronic liver disease.

4.6.5. Vitamin B_6 and Alcohol

Davis and Smith (D14) studied the serum pyridoxal and folate levels in 50 alcoholics. Twenty-five had serum pyridoxal levels below the lower limit of the reference range. The percentage of males and females affected was similar. Fourteen patients had a reduced serum folate concentration; this included nine of the patients with a reduced level of serum pyridoxal. There was a significant correlation between a low serum pyridoxal and a low serum folate, and, also, between a low serum pyridoxal and a reduced level of hemoglobin. Six patients were disoriented on admission and three had a low serum pyridoxal. In five patients, alcoholism was associated with epilepsy, and three had a low serum pyridoxal. Of five patients with Wernicke–Korsakoff syndrome, only one had a low serum pyridoxal.

Vitamin B_6 depletion has previously been observed in alcoholics (L19), based on measuring the excretion of xanthurenic acid following a 10-g loading dose of tryptophan.

Although the direct measurement of serum pyridoxal provides information on the available vitamin, the indirect methods dependent on tryptophan loading provide information on the availability of various enzymes required for normal pyridoxal function. In a study of folate metabolism in alcoholics, Davis and Leake (D15) found that four patients excreted an excess of formiminoglutamic acid in the presence of a normal concentration of serum folate. Failure to synthesize the enzymes required for the metabolism of folate was probably due to alcohol-associated liver dysfunction, and indirect assays such as the histidine loading test for folate and the tryptophan loading test for vitamin B_6 provide a useful means of examining these pathways. On the other hand, Solomon and Hillman (S36), in a study of 26 alcoholics, found increased erythrocyte pyridoxine kinase activity. The increase could not be explained on the basis of iron deficiency; only two alcoholics had a low serum iron with absent iron stores. Moreover, a number had evidence of active liver disease, which would tend to reduce the pyridoxine kinase level.

4.6.6. Pyridoxal and Diabetes Mellitus

The recent introduction of an automated technique for the measurement of pyridoxal (D1) has made it possible to examine the pyridoxal status of a number of population groups. Davis et al. (D16, D17) studied the levels in 518 patients with diabetes mellitus. There were 185 males and 333 females ranging in age from 16 to 87 years. Two hundred and forty-one were being treated with insulin, 154 with oral hypoglycemic agents, and 123 by diet alone. Twenty-five percent of these diabetics were found to have a serum pyridoxal level below the lower limit of the reference range. The mean levels of both sexes and all age-groups were, with two exceptions, significantly lower than those found in a large group of control subjects. In a study of diabetic children, 24% of 63 children were found to have a serum pyridoxal level below the lower limit of the reference range (W18).

Pyridoxine has been found to improve oral glucose tolerance in gestational diabetes. Bennink and Schreurs (B24) treated 14 patients with evidence of gestational diabetes and vitamin B_6 deficiency with 100 mg of pyridoxine daily for 14 days. The glucose tolerance test was then repeated and showed that only 2 of the 14 patients still had evidence of abnormal carbohydrate metabolism. It has also been found that pyridoxine can favorably influence carbohydrate metabolism in some women taking oral contraceptive agents (A15, S37). A deficiency of vitamin B_6 interferes with tryptophan catabolism and results in the accumulation of xanthurenic acid and other intermediates in the formylkynurenine pathway. Xanthurenic acid has been shown to bind to insulin, and the resulting insulin–xanthurenic acid complex was found to have a greatly reduced ability to lower blood glucose levels, compared with native insulin (K15, M20). Doubts have been expressed about the validity of this hypothesis (A15, C19), but there is no doubt that abnormal carbohydrate tolerance is often associated with a deficiency of pyridoxal phosphate (D16, W18).

Patients with diabetic neuropathy appear to have significantly lower serum concentration of pyridoxal phosphate than diabetics without this complication (M26), and have also been reported to show abnormal tryptophan metabolism following a 2-g L-tryptophan metabolic load (J2). McCann and Davis (M26) observed a significantly reduced serum pyridoxal concentration for 50 patients with diabetic neuropathy, compared with 50 diabetics without this complication. The duration of diabetes was not significantly different in patients with neuropathy and in those without neuropathy. The severity of the diabetes varied considerably in patients with neuropathy, some having mild, easily controlled disease, while others were labile and at times poorly controlled. The lack of correlation of diabetic neuropathy with duration and severity of diabetes suggests the possibility of some other factor in the production of diabetic neuropathy. The association of low serum pyridoxal concentrations represents a possible etiological factor. The frequency of low serum pyridoxal concentrations in diabetic children was similar to that observed in diabetic adults, but neuropathy is an unusual finding in diabetic children. It appears likely from the results of these observations that there is an increased demand for pyridoxal phosphate in patients with diabetes mellitus above that which can be supplied by their diet. This results in a chronic, mild deficiency state which may be a factor in the development of diabetic neuropathy.

Serum pyridoxal levels have been found to be significantly reduced in diabetic patients with bacteriuria (D18). In a study of 1017 diabetic patients, 142 were found to have significant bacteriuria. These patients had significantly reduced levels of serum pyridoxal, when compared with 142 diabetic patients matched for age and sex, but without infection of the urinary tract. Measurements were repeated up to 6 months after antibacterial treatment and serum pyridoxal concentrations were still low. This supports the suggestion that these patients had low serum pyridoxal levels before they developed their urinary tract infection. An increased frequency of infection of the urinary tract in patients with diabetes may have an immunological basis associated with a deficiency of pyridoxal.

4.6.7. *Atherosclerosis*

This disease has a multifactional etiology, but morphological investigations reveal splitting and fraying of the internal lamina of the arteries at the site where the atherosclerosis develops most severely. Artery walls are composed largely of elastin and collagen, which are cross-linked to provide the necessary strength and elasticity required of these vessels. The mechanism of cross-linking has been thoroughly investigated; it is initiated by an amino oxidase, lysyl oxidase (EC 1.13.12.12), which is extracellular and has a specific requirement for copper and molecular oxygen (L16). There is also strong evidence to suggest that the enzyme, when derived from vitamin B_6-deficient chicks, at least, requires pyridoxal phosphate as a cofactor (M27). Vitamin B_6 deficiency in chickens produces abnormalities of the arterial wall elastin and is associated with markedly reduced lysyl oxidase activity (M28). Levene and Murray (L16) suggest that pregnancy induces a measurable and functional vitamin B_6 deficiency, and that the focal splitting of the internal elastic lamina, observed commonly in the coronary arteries of infants at or before birth, may result from the inadequate cross-linking of elastin due to a deficiency of vitamin B_6. Whether this explanation is adequate or whether an absolute or relative deficiency of copper (one of the other requirements of lysyl oxidase) has to be considered remains to be investigated (K21).

Vitamin B_6 has also been implicated in the etiology of atherosclerosis in the rat through the metabolism of lipids and glycosaminoglycans (V5), and possibly through the control of cholesterol metabolism (O2).

4.6.8. Vitamin B_6 and the Pituitary

In 1973, Foukas (F8) reported that pyridoxine (600 mg/daily) effectively suppressed puerperal lactation and breast engorgement when compared with diethylstilbestrol and placebos. McIntosh (M29) studied three women with primary galactorrhea–amenorrhea syndrome with hyperprolactinemia. They experienced a return to regular ovulatory menses after taking oral pyridoxine (200–600 mg/daily). This improvement was reversed when pyridoxine was discontinued. On the other hand, in two women with prolonged secondary amenorrhea, but without hyperprolactinemia or galactorrhea, pyridoxine hydrochloride at doses up to 600 mg/daily did not restore ovulatory menses.

McIntosh (M29) hypothesized that pyridoxine might decrease prolactin by increasing conversion of DOPA to dopamine, an inhibitor of prolactin in the hypothalamus.

Delitala et al. (D19) claimed that a single 300-mg intravenous dose of pyridoxine produced a significant decrease in prolactin in normal healthy subjects and Harris and co-workers (H15) have demonstrated a similar result in rats and attributed it to a direct effect on the pituitary gland. However, these observations have not been substantiated by other workers (L20, S38, T9).

Inhibition of lactation by pharmacological doses of vitamin B_6 has not been seen by Canales et al. (C20); the vitamin did not appear to have any effect on basal prolactin levels in postpartum women (D20) or in normal volunteers (P11).

Attempts have been made to extrapolate the reputed positive effects of pharmacological doses of vitamin B_6 on the inhibition of lactation to the suspected positive effect of supplementation with near-physiological amounts of the vitamin to nursing mothers (G19). The positive advantages to the fetus and the newborn of mothers with good vitamin B_6 status have already been discussed.

Delitala et al. (D21) have claimed that pyridoxine inhibits thyrotropin release in patients with primary hypothyroidism by a direct action on either the hypothalamus or pituitary. In normal women, the effect of a high dose of vitamin B_6 on thyrotropin is highly variable and unpredictable (P11).

Both pituitary and serum growth hormone levels have been found to be low in rats raised on vitamin B_6-deficient diets (R19). Delitala et al. (D21) have observed an increase in growth hormone response to injection of pyridoxine in adults, but a decrease in neonates (D22). Reiter and Root (R20) also showed the variability of the effect of pyridoxine (300 mg iv) on serum growth hormones in older children with short stature.

4.6.9. Vitamin B_6 and Hyperkinetic Syndromes

In-depth studies of children with autistic syndromes (C21) showed that supplemental pyridoxine therapy was able to modify behavior in some children. Rimland and others (R21) carried out a double-blind crossover study of high-dose vitamin B_6 therapy in 20 children with autistic syndromes who had previously benefited from megavitamin therapy. In the study, each child's vitamin B_6 supplement was replaced during two separate experimental trial periods with either pyridoxine or a matched placebo. Behavior was rated as deteriorating significantly during the pyridoxine withdrawal. The authors were cautious in interpreting their results because the number of uncontrolled variables, not least of which was the pyridoxine dose, ranged from 75 to 3000 mg per day (2 to 95 mg/kg body weight).

Hyperactive children have also been treated with oral pyridoxine (ranging from 5 to 40 mg/kg) with a resultant significant fall in blood 5-hydroxytryptamine levels (B32), suggesting that a defect in the action of 5-hydroxytryptophan decarboxylase may account for the abnormally high blood levels usually seen in this condition. However, Coleman et al. (C22) have cautioned that the effect of pyridoxine on low-serotonin hyperkinetic children should not be attributed to any single pathway.

This small subpopulation of autistic children did appear to be one of the rare examples in which megavitamin treatment has been proved, after careful study, to show some effectiveness (L21).

Higher dose pyridoxine treatment has been shown to be effective in the treatment of some cases of tardive dyskinesia (D23), possibly by reducing dopaminergic influences in the CNS.

4.6.10. Other Conditions

Vitamin B_6 has also been claimed to be of value in the treatment of carpal-tunnel syndrome (E5), schizophrenia (B33), asthma (C23), and Huntington's disease (B34). Vitamin B_6 deficiency has been suggested as a possible contributing factor in multiple sclerosis (M30), carbon disulfide poisoning (G20), carbon monoxide poisoning (N7), cadmium toxicity (S39), and phenylketonuria (L22, S40).

4.7. MEGAVITAMIN THERAPY

In many of the treatment regimes mentioned in this article, vitamin B_6 has been given to the patient in amounts ranging from 5 to 2000 times the recommended dietary allowance. Little work has been carried out with respect to the adverse features of such megavitamin therapy. The problem of vitamin abuse has become apparent in the United States, since vitamins may be sold as dietary supplements and not drugs; hence, the Food and Drug Administration cannot control the upper dose limit of these vitamins unless their safe use requires medical supervision (E6). Water-soluble vitamins have come to be considered safe, since they are stored in a limited manner, but recent work with vitamin B_6 (B7) on muscle phosphorylase as a reservoir of vitamin B_6 in the rat suggests that this is no longer so clear cut.

Cohen and others (C24) investigated rats fed 500 mg of pyridoxine per kg body weight and compared them with rats fed 50 mg/kg. They found that the mean body weight of the animals on the high pyridoxine dose was 20% higher than that of the other groups, and that there was increased peritoneal and subcutaneous fat. This carefully controlled study could not attribute this increase in fat to any other factor. The LD_{50} for vitamin B_6 for mice and rats is 5.5 g/kg, yet these effects were seen at 10% of this level.

Alder and Zbinden (A22) reported that rats treated with 400–650 mg of pyridoxine per kg body weight per day developed clinical signs of neurotoxicity; demyelination of dorsal nerve roots has been observed in dogs on as little as 50 mg/kg/day (P12). Dogs receiving oral pyridoxine at the rate of 200 mg/kg/day showed clinical signs of ataxia and muscle weakness, as well as histopathological evidence of bilateral loss of myelin and axons in dorsal nerve roots (P12).

The doses used in these studies are large and it may be premature to extrapolate from other species to man. However, it would appear desirable to acquire neurological information from patients who are prescribed large amounts of pyridoxine for extended periods of time. Cohen et al. (C24) reviewed previous work and commented on increases in the specific activities of some enzymes during megavitamin B_6 therapy (for example, cystathioninase during treatment for cystathionine synthetase deficiency). This increased activity might cause an imbalance of metabolic systems and produce effects unrelated to the condition for which the vitamin was prescribed.

Schimke et al. (S28) reported such a situation in two patients with homocystinuria receiving 250 and 500 mg of pyridoxine daily. Both developed convulsions when neither had a previous history of seizures. Wilcken and Turner (W19) have reported a decrease in levels of blood folate in homocystinuric children receiving large doses of vitamin B_6 over a long time, but there were no clinical or hematological changes.

Less well-documented and well-identified effects of megadoses of pyridoxine have been reported. Rimland et al. (R21) mention increased irritability, sound sensitivity, and enuresis in some autistic children on high doses of pyridoxine; the symptoms disappeared on the addition of extra magnesium to the diet. Coleman (C21) claimed the only side effect of vitamin B_6 in large amounts in her study of autistic children was insomnia.

Williams (W10) has put forward the case for biochemical individuality, including the daily requirements for vitamins, and it is clear that some people have higher requirements than most. However, the promotion of sensational anecdotes regarding the efficiency of megavitamin therapy has led some health professionals to forget the axiom that "in matters of health, no substance is safe until proved safe, or effective until proved effective" (H16).

References

A1. Axelrod, A. E., and Martin, C. J., Water-soluble vitamins. Part 1. *Annu. Rev. Biochem.* **30**, 383–408 (1966).
A2. Anderson, B. B., Peart, M. B., and Fulford-Jones, C. E., The measurement of serum pyridoxal by a microbiological assay using *Lactobacillus casei*. *J. Clin. Pathol.* **23**, 232–242 (1970).
A3. Anderson, J. A., Chang, H. W., and Grandjean, C. J., Nature of the binding site of pyridoxal-5'-phosphate to bovine serum albumin. *Biochemistry* **10**, 2408–2415 (1971).
A4. Anderson, B. B., Newmark, P. A., Rawlins, M., and Green, R., Plasma binding of vitamin B6 compounds. *Nature (London)* **250**, 502–504 (1974).
A5. Anderson, B. B., Fulford-Jones, C. E., Child, J. A., Beard, M. E. J., and Bateman, C. J. T., Conversion of vitamin B6 compounds to active forms in the red blood cell. *J. Clin. Invest.* **50**, 1901–1909 (1971).
A6. Atkin, L., Shultz, A. S., and Williams, W. L., Yeast biological methods for determination of vitamins. Pyridoxine. *Ind. Eng. Chem., Anal. Ed.* **15**, 141–144 (1943).
A7. Adams, E., Fluorometric determination of pyridoxal phosphate in enzymes. *Anal. Biochem.* **31**, 118–122 (1969).
A8. Altman, K., and Greengard, O., Correlation of kynurenine excretion with liver tryptophan pyrrolase levels in disease after hydrocortisone induction. *J. Clin. Invest.* **45**, 1525–1534 (1966).
A9. American Medical Association, "Guidelines for Multivitamin Preparations for Parenteral Use." Department of Foods and Nutrition, Chicago, Illinois, 1975.
A10. Abbasy, A. S., Zeitoun, M. M., and Abouinfa, M. H., The state of vitamin B6 deficiency as measured by urinary xanthurenic acid. *J. Trop. Pediatr.* **5**, 45–50 (1959).
A11. Araya, M., Silink, S. J., Nobile, S., and Walker-Smith, J. A., Blood vitamin levels in children with gastroenteritis. *Aust. N. Z. J. Med.* **51**, 239–250 (1975).
A12. Anderson, B. B., Saary, M., Stephens, A. D., Perry, G. M., Lersundi, I. C., and Horn, J., Effect of riboflavin on red cell metabolism of vitamin B6. *Nature (London)* **264**, 574–575 (1976).
A13. Adams, P. W., Rose, D. P., Folkard, J., Wynn, V., Seed, M., and Strong, R., Effect of pyridoxine hydrochloride (vitamin B6) upon depression associated with oral contraception. *Lancet* **1**, 897–904 (1973).

A14. Adams, P. W., Wynn, V., Seed, M., and Folkard, J., Vitamin B6 depression and oral contraception. *Lancet* **2**, 516–517 (1974).
A15. Adams, P. W., Wynn, V., Folkard, J., and Seed, M., Influence of oral contraceptives, pyridoxine (vitamin B6) and tryptophan on carbohydrate metabolism. *Lancet* **1**, 759–764 (1976).
A16. Aycock, J. E., and Kirksey, A., Influence of different levels of dietary pyridoxine on certain parameters of developing and mature brains in rats. *J. Nutr.* **106**, 680–688 (1976).
A17. Alton-Mackey, M. G., and Walker, B. L., The physical and neuromotor development of progeny of female rats fed graded levels of pyridoxine during lactation. *Am. J. Clin. Nutr.* **31**, 76–81 (1978).
A18. Alton-Mackey, M. G., and Walker, B. L., Physical and neuromotor development of progeny of pyridoxine restricted rats cross-fostered with control or isonutritional dams. *Am. J. Clin. Nutr.* **31**, 241–246 (1978).
A19. Axelrod, A. E., and Trakatellis, C., Relationship of pyridoxine to immunological phenomena. *Vitam. Horm. (N.Y.)* **22**, 591–607 (1964).
A20. Allen, M. J., Boyland, E., Dukes, C. E., Horning, E. S., and Watson, J. G., Cancer of the urinary bladder induced in mice with metabolites of aromatic amines and tryptophane. *Br. J. Cancer* **11**, 212–228 (1957).
A21. Ajdukiewicz, A., and Pollitt, N., Liver carcinoma and pyridoxine deficiency. *Am. J. Clin. Nutr.* **29**, 813 (1976).
A22. Alder, S., and Zbinden, G., Use of pharmacological screening tests in subacute neurotoxicity studies of isoniazid, pyridoxine HCl and hexachlorophene. *Agents Actions* **3**, 233–243 (1973).
B1. Birch, T. W., and György, P., A study of the chemical nature of vitamin B_6 and methods for its preparation in a concentrated state. *Biochem. J.* **30**, 304–315 (1936).
B2. Bavernfiend, J. C., and Miller, O. N., Vitamin B6: Nutritional and pharmaceutical usage, stability, bioavailability, antagonists, and safety. *In* "Human Vitamin B6 Requirements," pp. 78–108. Natl. Acad. Sci., Washington, D. C., 1978.
B3. Bender, D. A., "Amino Acid Metabolism." Wiley, New York. 1975.
B4. Baker, R. J., Bellen, J. C., and Ronai, P. M., Technetium ^{99m}-pyridoxylideneglutamate: A new hepatobiliary radiopharmaceutical. I. Experimental aspects. *J. Nucl. Med.* **16**, 720–727 (1975).
B5. Booth, C. C., and Brain, M. C., The absorption of tritium-labelled pyridoxine hydrochloride in the rat. *J. Physiol. (London)* **164**, 282–294 (1962).
B6. Brain, M. C., and Booth, C. C., The absorption of tritium-labelled pyridoxine HCl in control subjects and in patients with intestinal malabsorption. *Gut* **51**, 241–247 (1964).
B7. Black, A. L., Guirard, B. M., and Snell, E. E., Increased muscle phosphorylase in rats fed high levels of vitamin B6. *J. Nutr.* **107**, 1962–1968 (1977).
B8. Black, A. L., Guirard, B. M., and Snell, E. E., The behaviour of muscle phosphorylase as a reservoir for vitamin B6 in the rat. *J. Nutr.* **108**, 670–677 (1978).
B9. Baysal, A., Johnson, B. A., and Linkswiler, H., Vitamin B6 depletion in man: Blood vitamin B6, plasma pyridoxal phosphate, serum cholesterol, serum transaminases and urinary vitamin B6 and 4-pyridoxic acid. *J. Nutr.* **89**, 19–23 (1966).
B10. Brown, G. M., and Reynolds, J. J., Biogenesis of the water-soluble vitamins. *Annu. Rev. Biochem.* **32**, 419–462 (1963).
B11. Bayoumi, R. A., Kirwan, J. R., and Smith, W. R. D., Some effects of dietary vitamin B6 deficiency and 4-deoxypyridoxine on γ-aminobutyric acid metabolism in rat brain. *J. Neurochem.* **19**, 569–576 (1972).
B12. Baker, H., Frank, O., Ning, M., Gellene, R. A., Hutner, S. H., and Leevy, C. M., A protozoological method for detecting clinical vitamin B6 deficiency. *Am. J. Clin. Med.* **18**, 123–133 (1966).

B13. Brown, R. R., and Price, J. M., Quantitative studies on metabolites of tryptophan in the urine of dog, cat, rat and man. *J. Biol. Chem.* **219**, 985–997 (1956).
B14. Baker, E. M., Canham, J. E., Nunes, W. T., Sauberlich, H. E., and McDowell, M. E., Vitamin B6 requirement for adult men. *Am. J. Clin. Nutr.* **15**, 59–66 (1964).
B15. Biehl, J. P., and Vilter, R. W., Effect of isoniazid on vitamin B6 metabolism; its possible significance in producing isoniazid neuritis. *Proc. Soc. Exp. Biol. Med.* **85**, 389–392 (1954).
B16. Bessey, O. A., Adam, D. J. D., and Hansen, A. E., Intake of vitamin B6 and infantile convulsions: A first approximation of requirements of pyridoxine in infants. *Pediatrics* **20**, 33–44 (1957).
B17. Baker, H., and Sobotka, H., Microbiological assay methods for vitamins. *Adv. Clin. Chem.* **5**, 173–235 (1962).
B18. Bender, A. E., "Food Processing and Nutrition," p. 42. Academic Press, New York, 1978.
B19. Bernhart, F. W., D'Amato, E., and Tomarelli, R. M., The vitamin B6 activity of heat-sterilised milk. *Arch. Biochem. Biophys.* **88**, 267–269 (1960).
B20. Brin, M., Abnormal tryptophan metabolism in pregnancy and with the oral contraceptive pill. *Am. J. Clin. Nutr.* **24**, 699–703 (1971).
B21. Brown, R. R., Normal and pathological conditions which may alter the human requirement for vitamin B6. *J. Agric. Food Chem.* **20**, 498–505 (1972).
B22. Barlow, J. B., and Wilkinson, A. W., Plasma pyridoxal phosphate levels and tryptophan metabolism in children with burns and scalds. *Clin. Chim. Acta* **64**, 79–82 (1975).
B23. Baker, H., Frank, O., Thomson, A. D., Langer, A., Munves, E., De Angelis, B., and Kaminetzky, H. A., Vitamin profile of 174 mothers and newborns at parturition. *Am. J. Clin. Nutr.* **28**, 59–65 (1975).
B24. Bennink, H. J.T. C., and Schreurs, W. H. P., Improvement of oral glucose tolerance in gestational diabetes by pyridoxine. *Br. Med. J.* **3**, 13–15 (1975).
B25. Brophy, M. H., and Siiteri, P. K., Pyridoxal phosphate and hypertensive disorders of pregnancy. *Am. J. Obstet. Gynecol.* **121**, 1075–1079 (1975).
B26. Baker, H., Thind, I. S., Frank, O., De Angelis, B., Caterine, H., and Louria, D. B., Vitamin levels in low-birth-weight newborn infants and their mothers. *Am. J. Obstet. Gynecol.* **129**, 521–524 (1977).
B27. Bejsovec, M., Kulenda, Z., and Ponca, E., Familial intrauterine convulsions in pyridoxine dependency. *Arch. Dis. Child.* **42**, 201–207 (1967).
B28. Bayon, E., Possani, L. D., Tapia, M., and Tapia, R., Kinetics of brain glutamate decarboxylase. Interaction with glutamate, pyridoxal 5'-phosphate, and glutamate-pyridoxal 5'-phosphate Schiff base. *J. Neurochem.* **25**, 519–525 (1977).
B29. Byar, D., and Blackard, C., Comparisons of placebo, pyridoxine and topical thiotepa in preventing recurrence of stage 1 bladder cancer. *Urology* **10**, 556–561 (1977).
B30. Bell, E., The excretion of a vitamin B6 metabolite and the probability of recurrence of early breast cancer. *Eur. J. Cancer* **16**, 297–298 (1980).
B31. Bell, E. D., Bulbrook, R. D., Hayward, J. L., and Tong, D., Tryptophan metabolism and recurrence rates of patients with breast cancer after mastectomy. *Acta Vitaminol. Enzymol.* **29**, 104–107 (1975).
B32. Bhagavan, H. M., Coleman, M., and Coursin, D. B., The effect of pyridoxine hydrochloride on blood serotonin and pyridoxal phosphate contents in hyperactive children. *Pediatrics* **55**, 437–441 (1975).
B33. Bucci, L., Pyridoxine and schizophrenia. *Br. J. Psychiatry* **122**, 240 (1973).
B34. Barr, A. N., Heinze, W., Mendoza, J. E., and Perlik, S., Longterm treatment of Huntington disease with L-glutamate and pyridoxine. *Neurology* **28**, 1280–1280 (1978).

C1. Canham, J. E., Baker, E. M., Raica, N., and Sauberlich, H. E., Vitamin B6 requirements of adult men. *Proc. Int. Congr. Nutr.*, 7th, 1966 (1967).
C2. Chiotellis, E., Subramanian, G., and McAfee, J. G., Preparation of Tc-99m labeled pyridoxal-amino acid complexes and their evaluation. *Int. J. Nucl. Med. Biol.* **4**, 29–41 (1977).
C3. Chauhan, M. S., and Dakshinamurti, K., Fluorometric assay of pyridoxal and pyridoxal 5′-phosphate. *Anal. Biochem.* **96**, 426–432 (1979).
C4. Colombini, C. E., and McCoy, E. E., Rapid thin-layer electrophoretic separation and estimation of all vitamin B6 compounds and of some 5-hydroxyindoles. *Anal. Biochem.* **34**, 451–458 (1970).
C5. Contractor, S. E., and Shane, B., Estimation of vitamin B6 compounds in human blood and urine. *Clin. Chim. Acta* **21**, 71–77 (1968).
C6. Chabner, B., and Livingstone, D., A simple enzymatic assay for pyridoxal phosphate. *Anal. Biochem.* **34**, 413–423 (1970).
C7. Chabner, B. A., De Vita, V. T., Livingstone, D. M., and Oliverio, V. T., Abnormalities of tryptophan metabolism and plasma pyridoxal phosphate in Hodgkin's Disease. *N. Engl. J. Med.* **282**, 838–843 (1970).
C8. Coursin, D. B., Vitamin B6 and brain function in animals and man. *Ann. N. Y. Acad. Sci.* **166**, 7–15 (1969).
C9. Coursin, D. B., Convulsive seizures in infants with pyridoxine-deficient diet. *JAMA, J. Am. Med. Assoc.* **154**, 406–408 (1954).
C10. Coursin, D. B., Vitamin B6 deficiency in infants. *Am. J. Dis. Child.* **90**, 344–348 (1955).
C11. Chillar, R. K., Johnson, C. S., Beutler, E., and Sonne, K., Erythrocyte pyridoxine kinase levels in patients with sideroblastic anaemia. *N. Engl. J. Med.* **295**, 881–883 (1976).
C12. Coppen, A., Shaw, D. M., Herzberg, B., and Maggs, R., Tryptophan in the treatment of depression. *Lancet* **2**, 1178–1180 (1967).
C13. Cleary, R. E., Lumeng, L., and Li, T. K., Maternal and foetal plasma levels of pyridoxal phosphate at term. Adequacy of vitamin B6 supplementation during pregnancy. *Am. J. Obstet. Gynecol.* **121**, 25–31 (1975).
C14. Contractor, S. F., and Shane, B., Blood and urine levels of vitamin B6 in the mother and foetus before and after loading of the mother with vitamin B6. *Am. J. Obstet. Gynecol.* **107**, 635–640 (1970).
C15. Contractor, S. F., and Shane, B., Pyridoxal kinase in the human placenta and foetus through gestation. *Clin. Chim. Acta* **25**, 465–474 (1969).
C16. Chauhan, M. S., and Dakshinamurti, K., The elongation of fatty acids by microsomes and mitochondria from normal and pyridoxine-deficient rat brains. *Exp. Brain Res.* **36**, 265–273 (1979).
C17. Cassell, S., Robson, L., and Rose, C., The effects of vitamin B6 deficiency on the bone marrow of the rat. *Anat. Rec.* **191**, 47–54 (1978).
C18. Chabner, B. A., De Vita, V. T., Livingston, D. M., and Oliverio, V. T., Abnormalities of tryptophan metabolism and plasma pyridoxal phosphate in Hodgkin's Disease. *N. Engl. J. Med.* **282**, 838–834 (1970).
C19. Cornish, E. J., and Tesoriero, W., Pyridoxine and oestrogen-induced glucose tolerance. *Br. Med. J.* **3**, 649–650 (1975).
C20. Canales, E. S., Soria, J., and Zarate, A., The influence of pyridoxine on prolactin secretion and milk production in women. *Br. J. Obstet. Gynaecol.* **83**, 387–388 (1976).
C21. Coleman, M., "The Autistic Syndromes." North-Holland Publ., Amsterdam. 1976.
C22. Coleman, M., Steinberg, G., Tippett, J., Bhagavam, H. N., Coursin, D. B., Gross, M., Lewis, C., and De Veau, L., A preliminary study on the effect of pyridoxine administra-

tion in a subgroup of hyperkinetic children: A double-blind cross-over comparison with methylphenidate. *Biol. Psychiatry* **14**, 741–751 (1979).
C23. Collip, P. J., Goldzier, S., Weiss, N., Soleymani, Y., and Snyder, R., Pyridoxine treatment of childhood bronchial asthma. *Ann. Allergy* **35**, 93–97 (1975).
C24. Cohen, P. A., Schneidman, K., Ginsberg-Fellner, R., Sturman, J. A., Knittle, J., and Gaull, G. E., High pyridoxine diet in the rat: Possible implications for megavitamin therapy. *J. Nutr.* **103**, 143–151 (1973).
D1. Davis, R. E., Smith, B. K., and Curnow, D. H., An automated method for the microbiological assay of serum pyridoxal. *J. Clin. Pathol.* **26**, 871–874 (1973).
D2. Davis, R. E., and Smith, B. K., The heat stability of synthetic and natural pyridoxal. *Lab. Prac.* **24**, 515–516 (1975).
D3. Dempsey, W. B., and Christensen, H. N., The specific binding of pyridoxal 5′-phosphate to bovine plasma albumin. *J. Biol. Chem.* **237**, 1113–1120 (1962).
D4. Donald, E. A., McBean, L. D., Simpson, M. H. W., Sun, M. F., and Aly, H. E. Vitamin B6 requirement in young adult women. *Am. J. Clin. Nutr.* **22**, 10–13 (1971).
D5. Davis, R. E., The measurement of pyridoxal and its clinical significance. Ph.D. Thesis, University of Western Australia, Perth (1974).
D6. Davis, R. E., and Smith, B. K., The heat stability of synthetic and natural pyridoxal. *Am. J. Clin. Nutr.* **27**, 323 (1974).
D7. Dawson, A. M., Holdsworth, C. D., and Pitcher, C. S., Sideroblastic anaemia in adult coeliac disease. *Gut* **5**, 304–308 (1964).
D8. Driskell, J. A., Gedders, J. M., and Urban, M. C., Vitamin B6 status of young men, women, and women using oral contraceptives. *J. Lab. Clin. Med.* **87**, 813–821 (1976).
D9. Davis, R. E., and Smith, B. K., Pyridoxal, vitamin B12 and folate metabolism in women taking oral contraceptive agents. *S. Afr. Med. J.* **48**, 1937–1940 (1974).
D10. Davis, S. D., Nelson, T., and Sheppard, T. H., Teratogenicity of vitamin B6 deficiency: Omphalocele, skeletal and neural defects and splenic hypoplasia. *Science* **169**, 1329–1330 (1970).
D11. Driskell, J. A., and Kirksey, A., The cellular approach to the determination of pyridoxine requirements in pregnant and non-pregnant rats. *J. Nutr.* **101**, 661–667 (1971).
D12. Dodge, P. R., Prensky, A. L., and Feigin, *In* "Nutrition and the Developing Nervous System" (S. J. Holmes, ed.), p. 450. Mosby, St. Louis, Missouri, 1975.
D13. Davis, S. D., Immune deficiency and runting syndrome in rats with pyridoxine deficiency. *Nature (London)* **251**, 548 (1974).
D14. Davis, R. E., and Smith, B. K., Pyridoxal and folate deficiency in alcoholics. *Med. J. Aust.* **2**, 357–360 (1974).
D15. Davis, R. E., and Leake, E., Histidine Metabolite excretion and serum folate levels. *Aust. J. Exp. Biol. Med. Sci.* **43**, 185–190 (1965).
D16. Davis, R. E., Calder, J. S., and Curnow, D. H., Serum pyridoxal and folate concentrations in diabetics. *Pathology* **8**, 151–156 (1976).
D17. Davis, R. E., McCann, V. J., and Calder, J. S., Serum pyridoxal concentrations in diabetic patients. *Aust. N. Z. J. Med.* **7**, 685 (1977).
D18. Davis, R. E., McCann, V. J., Ormonde, N. W., and Goodwin, C. S., The association of bacteriuria and reduced serum pyridoxal concentrations in patients with diabetes mellitus. *Pathology* **13**, 587–591 (1981).
D19. Delitala, G., Massala, A., Alagna, S., and De Villa, L., Effect of pyridoxine on human hypophyseal trophic hormone release: A possible stimulation of hypothalamic dopaminergic pathway. *J. Clin. Endocrinol. Metab.* **42**, 603–605 (1976).
D20. Del Pozo, E., and Del Re, R. B., Vitamin B6 in nursing mothers. *N. Engl. J. Med.* **301**, 107 (1979).

D21. Delitala, G., Rovasio, P., and Lotti, G., Suppression of thyrotropin and prolactin release by pyridoxine in primary hypothyroidism. *J. Clin. Endocrinol. Metab.* **45**, 1019–1022 (1977).
D22. Delitalia, G., Meloni, T., Massala, A., Alagna, S., Devilla, L., and Costa, R., Action of somatostatin, levodopa, and pyridoxine on growth hormone secretion in newborn infants. *Biomedicine* **27**, 13–15 (1978).
D23. De Veaugh-Geiss, J., and Manion, L., High-dose pyridoxine in tardive dyskinesis. *J. Clin. Psychiatry* **39**, 573–575 (1978).
E1. Evered, D. F., Penicillamine and cystinuria. *Br. Med. J.* **2**, 120 (1963).
E2. Evered, D. F., L-Dopa as a vitamin B6 antagonist. *Lancet* **1**, 914 (1971).
E3. Evered, D. F., L-Dopa and its combination with pyridoxal 5'-phosphate. *Lancet* **2**, 46 (1971).
E4. Ejderhamn, J., and Hamfelt, A., Pyridoxal phosphate concentrations in blood in infants and their mothers compared with the amount of extra pyridoxal taken during pregnancy and breast feeding. *Acta Paediatr. Scand.* **69**, 327–339 (1980).
E5. Ellis, J. M., Kishi, T., Azuma, J., and Folkers, K., Vitamin B6 deficiency in patients with a clinical syndrome including carpal tunnel defect. Biochemical and clinical response to therapy with pyridoxal. *Res. Commun. Chem. Pathol. Pharmacol.* **13**, 743–757 (1976).
E6. Editorial, The lid is off. Vitamin abuse. *JAMA, J. Am. Med. Assoc.* **238**, 1761–1762 (1977).
F1. Fasella, P., Pyridoxal phosphate. *Annu. Rev. Biochem.* **36**, 185–203 (1967).
F2. Fujita, A., Fujita, D., and Fujino, K. J., Fluorometric determination of vitamin B_6. III. Fractional determination of pyridoxal and 4-pyridoxic acid. *J. Vitaminol.* **1**, 279–289 (1955).
F3. Feltkamp, H., Patt, V., and Zilliken, F., Aktivitatsmesung der erythrozytaren glutamatoxalacetat transaminase unter der einnahme von hormonalen kontrazeptiva und pyridoxin beim menschen. *Clin. Chim. Acta* **53**, 305–310 (1974).
F4. Frank, O., Walbroehl, G., Thomson, A., Kaminetzky, H., Kubes, Z., and Baker, H., Placental transfer: Feotal retention of some vitamins. *Am. J. Clin. Nutr.* **23**, 662–663 (1970).
F5. Frimpter, G. W., Cystathioninuria: Nature of the defect. *Science* **149**, 1095–1096 (1965).
F6. Faber, S. R., Feiltler, W. W., Bleiler, R. E., Ohlson, M. A., and Hodges, R. E., The effects of an induced pyridoxine and pantothenic acid deficiency on excretions of oxalic and xanthurenic acids in the urine. *Am. J. Clin. Nutr.* **12**, 406–412 (1963).
F7. Foy, H., Kondi, A., and Linsell, C. A., Positive alpha$_1$-foetoprotein tests in pyridoxine deprived baboons: Relevance to liver carcinoma in Africans. *Nature (London)* **225**, 952–953 (1970).
F8. Foukas, M. D., An antilactogenic effect of pyridoxine. *Br. J. Obstet. Gynaecol.* **80**, 718–720 (1973).
G1. György, P., Vitamin B2 and the pellagra-like dermatitis in rats. *Nature (London)* **133**, 498–499 (1934).
G2. György, P., Investigations on the vitamin B2 complex. 1. The differentiation of lactoflavin and the "rat antipellagra" factor. *Biochem. J.* **29**, 741–759 (1935).
G3. Guirard, B. M., and Snell, E. E., Vitamin B6 function in transamination and decarboxylation reactions. *Compr. Biochem.* **15**, 138–199 (1964).
G4. Gershoff, S. N., Faragalla, F. F., Nelson, D. A., and Andrus, S. B., Vitamin B6 deficiency and oxalate nephrocalcinosis in the cat. *Am. J. Med.* **27**, 72–80 (1959).
G5. Gregory, J. F., and Kirk, J. R., Improved chromatographic separation and fluorometric determination of vitamin B6 compounds in foods. *J. Food Sci.* **42**, 1073–1076 (1977).

G6. Gailani, S., Pyridoxal phosphate determination in isolated leucocytes and tissue by *E. coli* apotryptophanase. *Anal. Biochem.* **13**, 19–27 (1965).
G7. Gregory, J. F., and Kirk, J. R., Determination of urinary 4-pyridoxal acid using high performance liquid chromatography. *Am. J. Clin. Nutr.* **32**, 879–883 (1979).
G8. Gaynor, R., and Dempsey, W. B., Vitamin B6 enzymes in normal-preeclamptic human placentae. *Clin. Chim. Acta* **37**, 411–416 (1972).
G9. Gershoff, S. N., Hegsted, D. M., and Trulson, M. F., Metabolic studies in mongoloids. *Am. J. Clin. Nutr.* **6**, 526–530 (1958).
G10. Gregory, J. F., and Kirk, J. R., Biological activity in rats of ϵ-pyridoxyllysine bound to dietary protein. *Fed. Proc., Fed. Am. Soc. Exp. Biol.* **37**, 588 (1978).
G11. Greengard, O., Relationship between urinary excretion of kynurenine and liver tryptophan oxygenase activity. *Am. J. Clin. Nutr.* **24**, 709–711 (1971).
G12. Green, A. R., and Curzon, G., The effect of tryptophan metabolites on brain 5-hydroxytryptamine metabolism. *Biochem. Pharmacol.* **19**, 2061–2068 (1970).
G13. General Practitioners Research Group, Report 35. Meclozine and pyridoxine in pregnancy sickness. *Practitioner* **190**, 251–253 (1963).
G14. Gaull, G. E., Sturman, J. A., and Rassim, D. K., Enzymatic and metabolic studies of homocystinuria: Effects of pyridoxine. *Neuropaediatrie* **1**, 199–226 (1969).
G15. Gershoff, S. N., The formation of urinary stones. *Metab., Clin. Exp.* **13**, 875–881 (1964).
G16. Gershoff, S. N., Mayer, A. L., and Kulczycki, L. L., Effect of pyridoxine administration on the urinary excretion of oxalic acid, pyridoxine, and related compounds in mongoloids and non-mongoloids. *Am. J. Clin. Nutr.* **7**, 76–79 (1959).
G17. Gibbs, D. A., and Watts, R. W. E., The action of pyridoxine in primary hyperoxaluria. *Clin. Sci.* **38**, 277–286 (1970).
G18. Gentz, J., Hamfelt, A., Johansson, S., Lindstedt, S., Persson, B., and Zetterström, R., Vitamin B6 metabolism in pyridoxine dependency with seizures. *Acta Paediatr. Scand.* **56**, 17–26 (1967).
G19. Greentree, L. B., Dangers of vitamin B6 in nursing mothers. *N. Engl. J. Med.* **300**, 141–142 (1979).
G20. Gorny, R., The level of pyridoxal phosphate in the blood plasma of rats exposed to carbon disulphide. *Biochem. Pharmacol.* **20**, 2114–2115 (1971).
H1. Hamm, M. W., Mehansho, H., and Henderson, L. M., Transport and metabolism of pyridoxamine and pyridoxamine phosphate in the small intestine of the rat. *J. Nutr.* **109**, 1552–1559 (1979).
H2. Haskell, B. E., and Wallnofer, U., D-Alanine interference in microbiological assays of vitamin B6 in human blood. *Anal. Biochem.* **19**, 569–577 (1967).
H3. Hamfelt, A., A method of determining pyricosal phosphate in blood by decarboxylation of L-tyrosine-^{14}C (U). *Clin. Chim. Acta* **7**, 746–748 (1962).
H4. Haskell, B. E., and Snell, E. E., An improved apotryptophanase assay for pyridoxal phosphate. *Anal. Biochem.* **45**, 567–576 (1972).
H5. Howard, L., Oldendorf, M., and Chu, R., Pyridoxine deficiency: Another potential sequel to jejunal-ileal bypass procedure. *N. Engl. J. Med.* **295**, 733 (1976).
H6. Hines, J. D., and Grasso, J. A., The sideroblastic anaemias. *Semin. Hematol.* **7**, 86–106 (1970).
H7. Hines, J. D., and Cowan, D. H., Studies on the pathogenesis of alcohol-induced sideroblastic bone marrow abnormalities. *N. Engl. J. Med.* **283**, 441–446 (1970).
H8. Holtz, P., and Palm, D., Pharmacological aspects of vitamin B6. *Pharmacol. Rev.* **16**, 113–178 (1964).
H9. Hamfelt, A., and Hahn, L., Pyridoxal phosphate concentration in plasma and tryptophan load test during pregnancy. *Clin. Chim. Acta* **25**, 91–96 (1969).

H10. Hamfelt, A., and Turemo, T., Pyridoxal phosphate and folic acid concentrations in blood and erythrocyte aspartate aminotransferase activity during pregnancy. *Clin. Chim. Acta* **41**, 287–298 (1972).
H11. Harris, J. W., Whittington, R. M., Weisman, R., and Horrigan, D. L., Pyridoxine responsive anaemia in the human adult. *Proc. Soc. Exp. Biol. Med.* **91**, 427–432 (1956).
H12. Horrigan, D. L., and Harris, J. W., Pyridoxine-responsive anaemia: Analysis of 62 cases. *Adv. Intern. Med.* **12**, 103–174 (1964).
H13. Hunt, A. D., Stokes, J., McCrory, W. W., and Stroud, H. H., Pyridoxine dependency: Report of a case of intractable convulsions in an infant controlled by pyridoxine. *Pediatrics* **13**, 140–145 (1954).
H14. Heeley, A., Pugh, R. J. P., Clayton, B. E., Shepherd, J., and Wilson, J., Pyridoxal metabolism in vitamin B6-responsive convulsions of early infancy. *Arch. Dis. Child.* **53**, 794–802 (1978).
H15. Harris, A. R. C., Smith, M. S., Salhanick, S. A. H. A., Vagenakis, A. G., and Braverman, L. E., Pyridoxine-induced inhibition of prolactin release in the female rat. *Endocrinology* **102**, 362–366 (1978).
H16. Herbert, V., The vitamin craze. *Arch. Intern. Med.* **140**, 173–176 (1980).
I1. IUPAC-IUB, Commission on Biochemical Nomenclature, Nomenclature for vitamin B6 and related compounds. *Eur. J. Biochem.* **40**, 325–327 (1973).
I2. Iyengar, L., Oral lesions in pregnancy. *Lancet* **1**, 680 (1973).
J1. Johansson, S., Lindstedt, S., Register, U., and Wadström, L., Studies on the metabolism of labelled pyridoxine in man. *Am. J. Clin. Nutr.* **18**, 185–196 (1966).
J2. Jones, C. L., and Gonzalez, V., Pyridoxine deficiency: A new factor in diabetic neuropathy. *J. Am. Podiatry Assoc.* **68**, 646–653 (1978).
K1. Krebs, E. G., and Fischer, E. H., Phosphorylase and related enzymes of glycogen metabolism. *Vitamin Horm. (N.Y.)* **22**, 399–410 (1964).
K2. Kelsay, J., Baysal, A., and Linkswiler, H., Effect of vitamin B6 depletion on the pyridoxal, pyridoxamine, and pyridoxine content of the blood and urine of men. *J. Nut.* **94**, 490–494 (1968).
K3. Kelly, A., A differential microbiological assay for pyridoxine, pyridoxamine, and pyridoxal, and its application to the determination of these vitamers in foodstuffs. M.Sc. Thesis, University of Western Australia, Perth (1981).
K4. Katanuma, N., Kominami, E., Kominami, S., and Kito, K., Mode of action of specific inactivating enzymes for pyridoxal enzymes and NAD-dependent enzymes and their biological significance. *Adv. Enzyme Regul.* **10**, 289–306 (1972).
K5. Katunuma, N., New intracellular proteases and their role in intracellular enzyme degradation. *Trends Biochem. Sci.* **2**, 122–236 (1977).
K6. Kirksey, A., Keaton, K., Abernathy, R. P., and Gregor, J. L., Vitamin B6 nutritional status of a group of female adolescents. *Am. J. Clin. Nutr.* **31**, 946–954 (1978).
K7. Kishi, H., Nishii, S., Ono, T., Yamaji, A., Kasahara, N., Hiraoko, E., Okada, A., Itakura, T., and Takagi, Y., Thiamin and pyridoxine requirements during intravenous hyperalimentation. *Am. J. Clin. Nutr.* **32**, 332–338 (1979).
K8. Kikuchi, G., Kumar, A., Talmage, P., and Shemin, D., The enzymatic synthesis of δ-aminolevulinic acid. *J. Biol. Chem.* **233**, 1214–1219 (1958).
K9. Klieger, J. A., Altshuler, C. H., and Krakow, G., Abnormal pyridoxine metabolism in toxaemia of pregnancy. *Ann. N. Y. Acad. Sci.* **166**, 288–296 (1969).
K10. Kierska, D., Sasiak, K., and Maslinski, G., Phosphopyridoxal cyclic compounds with histamine and histidine. 6. The formation of phosphopyridoxal cyclic compounds with histamine and histidine in the presence of biological material. *Agents Actions* **8**, 470–473 (1978).

K11. Klosterman, H. J., Vitamin B6 antagonists of natural origin. *J. Agric. Food Chem.* **22**, 13–16 (1974).
K12. Krishnaswamy, K., Isonicotinic acid hydrazide and pyridoxine deficiency. *Int. J. Vitam. Nutr. Res.* **44**, 457–465 (1974).
K13. Kuchinskas, E. J., Horvath, A., and Du Vigneaud, V., An anti-vitamin B6 action of L-pencillamine. *Arch. Biochem. Biophys.* **68**, 69–75 (1957).
K14. Kurtz, D. J., and Kanfer, J. N., L-Dopa: Effect on cerebral pyridoxal phosphate content and coenzyme activity. *J. Neurochem.* **18**, 2235–2236 (1971).
K15. Kotake, Y., and Murakami, E., A possible diabetogenic role for tryptophan metabolites and effects of xanthurenic acid on insulin. *Am. J. Clin. Nutr.* **24**, 826–829 (1971).
K16. Karlin, R., Study of the vitamins in the milk of women: Changes in vitamin B6 content during lactation and comparison with bovine milk. *C. R. Seances Soc. Biol. Ses Fil.* **153**, 127–131 (1959).
K17. Karlin, R., Effet d'un enrichissement en pyridoxine sur la teneur vitamine B6 du lait de femme. *Bull. Soc. Chim. Biol.* **41**, 1085–1091 (1959).
K18. Keen, J. H., and Lee, D., Sequelae of neonatal convulsions. *Arch. Dis. Child.* **48**, 542–546 (1973).
K19. Kim, Y. J., and Rosenberg, L. E., On the mechanism of pyridoxine responsive homocystinuria. 2. Properties of normal and mutant cystathlonine beta-synthase from cultured fibroblasts. *Proc. Natl. Acad. Sci. U.S.A.* **71**, 4821–4825 (1974).
K20. Knapp, A., On a new, hereditary disorder in tryptophan metabolism dependent on vitamin B6. *Clin. Chim. Acta* **5**, 6–13 (1960).
K21. Klevay, L. M., and Allen, K. G. D., Vitamin B6, copper and atherosclerosis. *Lancet* **1**, 1209 (1977).
L1. Lepkovsky, S., The isolation of pyridoxine. *Fed. Proc., Fed. Am. Soc. Exp. Biol.* **38**, 2699–2670 (1979).
L2. Lumeng, L., Brashear, R. E., and Li, T. K., Pyridoxal 5'-phosphate in plasma: Source, protein binding, and cellular transport. *J. Lab. Clin. Med.* **84**, 334–343 (1974).
L3. Li, T. K., Lumeng, L., and Veitch, R. L., Regulation of pyridoxal 5'-phosphate metabolism in liver. *Biochem. Biophys. Res. Commun.* **61**, 677–684 (1974).
L4. Litwack, G., and Rosenfield, S., Coenzyme dissociation, a possible determinent of short half-life of inducible enzymes in mammalian liver. *Biochem. Biophys. Res. Commun.* **52**, 181–188 (1973).
L5. Lefauconnier, J.-M., Portemer, C., De Billy, G., Ipaktchi, M., and Chatagner, F., Redistribution of pyridoxal phosphate in the male rat liver as a result of repeated injections of hydrocortisone. *Biochim. Biophys. Acta* **297**, 135–141 (1973).
L6. Loo, Y. H., and Badger, L., Spectrofluorimetric assay of vitamin B6 analogues in brain tissue. *J. Neurochem.* **16**, 801–804 (1969).
L7. Lepkovsky, S., Roboz, E., and Haagen-Smit, A. J., Xanthurenic acid and its role in the tryptophan metabolism of pyridoxine-deficient rats. *J. Biol. Chem.* **149**, 195–201 (1943).
L8. Looyé, A., Kwarts, E. W., and Groen, A., A new automated determination of xanthurenic acid in human urine. *Clin. Chem. (Winston-Salem, N.C.)* **14**, 890–897 (1968).
L9. Lewis, J. S., and Nunn, K. P., Vitamin B6 intakes and 24-hour 4-pyridoxic acid excretions of children. *Am. J. Clin. Nutr.* **36**, 2023–2027 (1977).
L10. Levere, R. D., and Granick, S., Control of haemoglobin synthesis in the cultivated chick blastoderm by delta-aminolevulinic acid synthetase: Increase in the rate of haemoglobin formation with delta-aminolevulinic acid. *Proc. Soc. Exp. Biol. Med.* **54**, 134–137 (1966).
L11. Lakshmi, A. V., and Bamji, M. S., Tissue pyridoxal phosphate concentration and pyridoxamine phosphate oxidase activity in riboflavin deficiency in rats and man. *Br. J. Nutr.* **32**, 249–255 (1974).

L12. Luhby, A. L., Brin, M., Gordon, M., Davis, P., Murphy, M., and Spiegel, H., Vitamin B6 metabolism in users of oral contraceptive agents. 1. Abnormal urinary xanthurenic acid excretion and its correction by pyridoxine. *Am. J. Clin. Nutr.* **24**, 684–693 (1971).
L13. Lumeng, L., Cleary, R. E., and Li, T. K., Effect of oral contraceptives on the plasma concentration of pyridoxal phosphate. *Am. J. Clin. Nutr.* **27**, 326–333 (1974).
L14. Lumeng, L., Cleary, R. E., Wagner, R., Yu, P.-L., and Li, T. K., Adequacy of vitamin B6 during pregnancy: A prospective study. *Am. J. Clin. Nutr.* **29**, 1376–1383 (1976).
L15. Livingston, J. E., MacLeod, P. M., and Applegarth, D. A., Vitamin B6 status in women with postpartum depression. *Am. J. Clin. Nutr.* **31**, 886–891 (1978).
L16. Levene, C. I., and Murray, J. C., The aetiological role of maternal vitamin B6 deficiency in the development of atherosclerosis. *Lancet* **1**, 628–629 (1977).
L17. Lott, I. T., Coulombe, T., Di Paulo, R. V., Richardson, E. P., and Levy, H. L., Vitamin B6 dependent seizures: Pathology and chemical findings in brain. *Neurology* **28**, 47–54 (1978).
L18. Labadarios, D., Rossouw, J. E., McConnell, J. B., Davis, M., and Williams, R., Vitamin B6 deficiency in chronic liver disease. *Gut* **18**, 23–27 (1977).
L19. Lerner, A. M., De Carli, L. M., and Davidson, C. S., Association of pyridoxine deficiency and convulsions in alcoholics. *Proc. Soc. Exp. Biol. Med.* **98**, 841–843 (1958).
L20. Lehtovirta, P., Ranta, T., and Seppära, M., Pyridoxine treatment of galactorrhoea-amenorrhoea syndromes. *Acta Endocrinol. (Copenhagen)* **87**, 682–686 (1978).
L21. Lipton, M. A., Mailman, R. B., and Nemeroff, C. B., *Nutr. Brain* **3**, 251 (1979).
L22. Loo, Y. H., Scotto, L., and Horning, M. G., Aromatic acid metabolites of phenylalanine in the brain of the hyperphenylalaninaemic rat: Effect of pyridoxamine. *J. Neurochem.* **29**, 411–415 (1977).
M1. Middleton, H. M., In vivo absorption and phosphorylation of pyridoxine · HCl in rat jejunum. *Gastroenterology* **70**, 43–49 (1979).
M2. Middleton, H. M., Intestinal absorption of pyridoxal 5'-phosphate: Disappearance from perfused segments of rat jejunum in vivo. *J. Nutr.* **109**, 975–981 (1979).
M3. Mehansho, H., Mamm, W., and Henderson, L. M., Transport and metabolism of pyridoxal and pyridoxal phosphate in the small intestine of the rat. *J. Nutr.* **109**, 1542–1551 (1979).
M4. Maeda, M., Takanashi, K., Aono, K., and Shiga, T., Effect of pyridoxal 5'-phosphate on the oxygen affinity of human erythrocytes. *Br. J. Haematol.* **34**, 501–509 (1976).
M5. McCoy, E. E., and England, J., Excretion of 4-pyridoxic acid during deoxypyridoxine and pyridoxine administration to mongoloid and non-mongoloid subjects. *J. Nutr.* **96**, 525–528 (1968).
M6. Mahuren, J. D., and Coburn, S. P., Separation of seven B6 vitamers by two-dimensional thin-layer chromatography. *Anal. Biochem.* **82**, 246–249 (1977).
M7. Maruyama, H., and Coursin, B. D., Enzymatic assay of pyridoxal phosphate using tyrosine apodecarboxylase and tyrosine-1-^{14}C. *Anal. Biochem.* **26**, 420–429 (1968).
M8. McCormick, D. B., Gregory, M. E., and Snell, E. E., Pyridoxal phosphokinases. I. Assay, distribution, purification and properties. *J. Biol. Chem.* **236**, 2076–2084 (1961).
M9. Mehler, A. H., and Knox, W. E., Conversion of tryptophan to kynurenine in liver. II. The enzymatic hydrolysis of formylkynurenine. *J. Biol. Chem.* **187**, 431–438 (1950).
M10. McCoy, E. E., *In* "Human Vitamin B6 Requirements," pp. 257–271. Natl. Acad. Sci., Washington, D. C., 1978.
M11. May, C. D., Vitamin B6 in human nutrition: A critique and an object lesson. *Pediatrics* **14**, 269–275 (1954).
M12. Molony, C. J., and Parmalee, A. H., Convulsions in young infants as a result of pyridoxine (vitamin B6) deficiency. *JAMA, J. Am. Med. Assoc.* **154**, 405–406 (1954).

M13. Mason, D. Y., and Emerson, P. M., Primary acquired sideroblastic anaemia: Response to treatment with pyridoxal 5'-phosphate. *Br. Med. J.* **1**, 389–390 (1973).
M14. Mauhuren, J. D., and Coburn, S. P., Pyridoxal phosphate in lymphocytes, polymorphonuclear leucocytes and platelets in Down's syndrome. *Am. J. Clin. Nutr.* **27**, 521–527 (1974).
M15. McCormick, D. B., and Snell, E. E., Pyridoxal phosphokinases. *J. Biol. Chem.* **236**, 2085–2088 (1961).
M16. McKenzie, S. A., MacNab, A. J., and Katz, G., Neonatal pyridoxine responsive convulsions due to isoniazid therapy. *Arch. Dis. Child.* **51**, 567–568 (1976).
M17. Morales, S. M., and Lincoln, E. M., The effect of isoniazed therapy on pyridoxine metabolism in children. *Annu. Rev. Tuberc.* **75**, 594–600 (1957).
M18. Mars, H., Effect of chronic levodopa treatment on pyridoxine metabolism. *Neurology* **25**, 263–265 (1975).
M19. Major, L. F., and Goyer, P. F., Effects of disulfiram and pyridoxine on serum cholesterol. *Ann. Intern. Med.* **88**, 53–56 (1978).
M20. Murakami, E., and Kotake, Y., Studies on the xanthurenic acid-insulin complex. *J. Biochem. (Tokyo)* **72**, 251–259 (1972).
M21. Morré, D. M., and Kirksey, A., The effect of a deficiency of vitamin B6 on the specific activity of 2'3'-cyclic nucleotide 3'-phosphohydrolase of neonatal rat brain. *Brain Res.* **146**, 200–204 (1972).
M22. Mudd, S. H., Edwards, W. A., Loeb, P. M., Brown, M. S., and Laster, L., Homocystinuria due to cystathionine synthase deficiency: The effect of pyridoxine. *J. Clin. Invest.* **49**, 1762–1773 (1970).
M23. McKusick, A. E., Sherwin, R. W., Jones, L. G., and Hsu, J. M., Urinary excretion of pyridoxine and 4-pyridoxic acid in rheumatoid arthritis. *Arthritis Rheum.* **7**, 636–653 (1964).
M24. Murray, A. W., Induction of epidermal cell proliferation by a tumour promotor in vitamin B6-deficient mice. *Experientia* **34**, 691 (1978).
M25. Mitchell, D., Wagner, C., Stone, W. J., Wilkinson, G. R., and Schenker, S., Abnormal regulation of plasma pyridoxal 5'-phosphate in patients with liver disease. *Gastroenterology* **71**, 1043–1049 (1976).
M26. McCann, V. J., and Davis, R. E., Serum pyridoxal concentrations in patients with diabetic neuropathy. *Aust. N. Z. J. Med.* **8**, 259–261 (1978).
M27. Murray, J. C., Frazer, D. R., and Levene, C. I., The effect of pyridoxine deficiency on lysyl oxidase activity in the chick. *Exp. Mol. Pathol.* **28**, 301–308 (1978).
M28. Murray, J. C., and Levene, C. I., Evidence for the role of vitamin B6 as a cofactor of lysyl oxidase. *Biochem. J.* **167**, 463–467 (1977).
M29. McIntosh, E. N., Treatment of women with the galactorrheaamenorrhea syndrome with pyridoxine (vitamin B6). *J. Clin. Endocrinol. Metab.* **42**, 1192–1193 (1976).
M30. Mitchell, D. A., and Schandl, E. K., Carbon monixide, vitamin B6 and multiple sclerosis: A theory of interrelationship. *Am. J. Clin. Nutr.* **26**, 890–896 (1973).
N1. National Academy of Sciences, National Research Council, "Recommended Dietary Allowances," 8th ed. Natl. Acad. Sci., Washington, D. C., 1974.
N2. National Academy of Sciences, Committee on Dietary Allowances, Food and Nutrition Board National Research Council, "Human Vitamin B_6 Requirements." Natl. Acad. Sci., Washington, D. C., 1978.
N3. National Academy of Sciences, National Research Council, "Recommended Dietary Allowances," 9th ed., pp. 96–106. Natl. Acad. Sci., Washington, D. C., 1980.
N4. Nichoalds, G. E., Meng, H. C., and Caldwell, M. D., Vitamin requirements in patients receiving total parenteral nutrition. *Arch. Surg. (Chicago)* **112**, 1061–1064 (1977).

N5. Nelson, E. W., Lane, H., and Cerda, J. J., Comparative human intestinal bioavailability of vitamin B6 from a synthetic and natural source. *J. Nutr.* **106**, 1433–1437 (1976).
N6. Nyhan, W. L., "Heritable Disorders of Amino Acid Metabolism: Patterns of Clinical Expression and Genetic Variation." Wiley, New York, 1974.
N7. Nizhegorodov, V. M., and Markhotskii, Y. L., Effect of prolonged combined exposure to carbon monoxide, nitrogen peroxide and ammonia on the vitamin B6 requirements of albino rats. *Hyg. Sanit. (USSR)* **36**, 137–139 (1971).
O1. Okuda, K., Fujii, S., and Wada, M., The microassay of pyridoxal phosphate using ^{14}C-tryptophan with tryptophanase. *In* "Methods in Enzymology" (D. B. McCormick and L. D. Wright, eds.), Vol. 18A, p. 505. Academic Press, New York, 1970.
O2. Okada, M., and Iwami, T., Effect of pyridoxine deficiency on cholesterogenesis in rats fed different levels of protein. *J. Nutr. Sci. Vitaminol.* **23**, 505–512 (1977).
P1. Plaut, G. W. E., Smith, C. M., and Alworth, W. L., Biosynthesis of water-soluble vitamins. *Annu. Rev. Biochem.* **43**, 915–922 (1974).
P2. Pflug, W., and Lingens, F., Vitamin-B6-biosynthesis in *Bacillus subtilis*. *Hoppe-Seyler's Z. Physiol. Chem.* **359**, 559–570 (1978).
P3. Patzer, E. M., and Hilker, D. M., New reagent for vitamin B6 derivative formation in gas chromatography. *J. Chromatogr.* **135**, 489–492 (1977).
P4. Prasad, A. S., Lei, K. Y., Moghissi, K. S., Stryker, J. C., and Oberleas, D., Effect of oral contraceptives on nutrients: 3 vitamins B6 B12 and folic acid. *Am. J. Obstet. Gynecol.* **125**, 1065–1069 (1976).
P5. Perkins, R. P., Failure of pyridoxine to improve glucose tolerance in gestational diabetes mellitus. *Obstet. Gynecol.* **50**, 370–372 (1977).
P6. Pascal, T. A., Gaull, G. E., Beratis, N. G., Gillam, B. M., and Tallan, H. H., Cystathionase deficiency: Evidence for genetic heterogeneity in primary cystathioninuria. *Pediatr. Res.* **12**, 125–133 (1978).
P7. Pinas, R. S., Tryptophan metabolism in rheumatic disease. *Arthritis Rheum.* **7**, 662–669 (1964).
P8. Pitney, W. R., Joske, R. A., and MacKinnon, N. L., Folic acid and other absorption tests in lymphosarcoma, chronic lymphocytic leukaemia, and some related conditions. *J. Clin. Pathol.* **13**, 440–447 (1960).
P9. Price, J. M., and Brown, R. R., Studies on the etiology of carcinoma of the urinary bladder. *Int. J. Cancer* **18**, 684–688 (1962).
P10. Potera, C., Rose, D. P., and Brown, R. R., Vitamin B6 deficiency in cancer patients. *Am. J. Clin. Nutr.* **30**, 1677–1679 (1977).
P11. Peters, F., Zimmermann, G., and Breckwoldt, M., Effects of pyridoxine on plasma levels of HCG, PRL and TSH in normal women. *Acta Endocrinol. (Copenhagen)* **89**, 217–220 (1978).
P12. Phillips, W. E. J., Mills, J. H. L., Charbonneau, S. M., Tryphonas, L., Hatina, G. V., Zawidzka, Z., Bryce, F. R., and Munro, I. C., Subacute toxicity of pyridoxine hydrochloride in the beagle dog. *Toxicol. Appl. Pharmacol.* **44**, 323–333 (1978).
R1. Review, Phosphorylase response to vitamin B6 feeding. *Nutr. Rev.* **36**, 55–57 (1978).
R2. Reddy, S. K., Reynolds, M. S., and Price, J. M., The determination of 4-pyridoxic acid in human urine. *J. Biol. Chem.* **233**, 691–696 (1958).
R3. Review, Regulation of liver metabolism on pyridoxal phosphate. *Nutr. Rev.* **33**, 214–217 (1975).
R4. Rabinowitz, J. C., Mondy, N. I., and Snell, E. E., The vitamin B6 group. XIII. An improved procedure for determination of pyridoxal with *Lactobacillus casei*. *J. Biol. Chem.* **175**, 147–153 (1948).
R5. Reinken, L., Eine mikromethode zur bestimmung von pyridoxal-5-phosphat im serum

mittels decarboxylierung von tyrosin-1-^{14}C. *Int. Z. Vitam.-Ernaehrungsforsch.* **42**, 476–481 (1972).

R6. Rosen, F., Lowy, R. S., and Sprince, H., A rapid assay for Xanthurenic acid in urine. *Proc. Soc. Exp. Biol. Med.* **77**, 399–401 (1951).

R7. Ritchey, S. J., Johnson, F. S., and Korsland, M. K., *In* "Human Vitamin B6 Requirements," pp. 272–278. Natl. Acad. Sci., Washington, D. C., 1978.

R8. Reinken, L., Zeiglauer, H., and Berger, H., Vitamin B6 nutriture of children with acute coeliac disease, coeliac disease in remission and of children with normal duodenal mucosa. *Am. J. Clin. Nutr.* **29**, 750–753 (1976).

R9. Reinken, L., and Czerny, N., Pyridoxalphosphat und pyridoxalkinaseaktivität im serum von früh- und neugenborennen. *Paediatr. Paedol.* **10**, 324–329 (1975).

R10. Rose, D. P., Excretion of xanthurenic acid in the urine of women taking progestogen-oestrogen preparations. *Nature (London)* **210**, 196–197 (1966).

R11. Rose, D. P., Aspects of tryptophan metabolism in health and disease: A review. *J. Clin. Pathol.* **25**, 17–25 (1972).

R12. Reinken, L., and Gant, H., Vitamin B6 nutrition in women with hyperemesis gravidarum during the first trimester of pregnancy. *Clin. Chim. Acta* **55**, 101–102 (1974).

R13. Roepke, J. L. B., and Kirksey, A., Vitamin B6 nutriture during pregnancy and lactation. *Am. J. Clin. Nutr.* **32**, 2249–2256 (1979).

R14. Reinken, L., and Mangold, B., Pyridoxal phosphate values in premature infants. *Int. J. Vitam. Nutr. Res.* **43**, 472–478 (1973).

R15. Ribaya, J. D., and Gershoff, S. N., Interrelationships in rats among dietary vitamin B6, glycine, and hydroxyproline. Effects of oxalate, glyoxalate, glycolate, and glycine on liver enzymes. *J. Nutr.* **109**, 171–183 (1979).

R16. Robson, L. C., and Schwarz, M. R., Vitamin B6 deficiency and the lymphoid system. II. Effects of vitamin B6 deficiency in utero on the immunological competence of the offspring. *Cell. Immunol.* **16**, 145–152 (1975).

R17. Rossouw, J. E., Labadarios, D., Davis, M., and Williams, R., Vitamin B6 and aspartate aminotransferase activity in chornic liver disease. *S. Afr. Med. J.* **53**, 436–438 (1978).

R18. Rosalki, S. B., and Bayoumi, R. A., Activations by pyridoxal phosphate of aspartate aminotransferase in serum of patients with heart and liver disease. *Clin. Chim. Acta* **59**, 357–360 (1975).

R19. Review. The role of growth hormone in the action of vitamin B6 on cellular transfer of amino acids. *Nutr. Rev.* **37**, 300–301 (1979).

R20. Reiter, E. O., and Root, A. W., Effect of pyridoxine on pituitary release of growth hormone and prolactin in childhood and adolescence. *J. Clin. Endocrinol. Metab.* **47**, 689–691 (1978).

R21. Rimland, B., Calloway, E., and Dreyfus, P., The effect of high dose of vitamin B6 on autistic children. A double-blind cross-over study. *Am. J. Psychiatry* **135**, 472–475 (1978).

S1. Snell, E. E., Vitamin B6. *Compr. Biochem.* **11**, 48–58 (1963).

S2. Snell, E. E., Lactic acid bacteria and identification of B-vitamins: Some historical notes, 1937–1940. *Fed. Proc., Fed. Am. Soc. Exp. Biol.* **38**, 2690–2693 (1979).

S3. Snell, E. E., and Haskell, B. E., The metabolism of vitamin B6. *Compr. Biochem.* **21**, 47–71 (1971).

S4. Snell, E. E., Relation of chemical structure to metabolic activity of vitamin B6. *Adv. Biochem. Psychopharmacol.* **4**, 1–22 (1972).

S5. Scudi, J. V., Unna, K., and Antapol, W., A study of the urinary excretion of vitamin B6 by a colorimetric method. *J. Biol. Chem.* **135**, 371–376 (1940).

S6. Snyderman, S. E., Holt, L. E., Carretevo, R., and Jacobs, K., Pyridoxine deficiency in the human infant. *Am. J. Clin. Nutr.* **1**, 200–207 (1953).

S7. Sauberlich, H. E., Canham, J. E., Baker, E. M., Raica, N., and Herman, Y. F., Biochemical assessment of the nutritional status of vitamin B6 in the human. *Am. J. Clin. Nutr.* **25**, 629–642 (1972).
S8. Stokes, J. L., Larsen, A., Woodward, C. R., and Foster, J. W., A Neurospora assay for pyridoxine. *J. Biol. Chem.* **150**, 17–24 (1943).
S9. Storvick, C. A., and Peters, J. M., Methods for the determination of vitamin B6 in biological materials. *Vitam. Horm. (N.Y.)* **22**, 833–854 (1964).
S10. Snell, E. E., The vitamin B6 group. VII. Replacement of vitamin B6 for some microorganisms by d(−)alanine and an unidentified factor from casein. *J. Biol. Chem.* **158**, 497–503 (1945).
S11. Schreiber, G., and Holzer, H., *In* "Methods of Enzymatic Analysis" (H. U. Bergmeyer, ed.), p. 606. Academic Press, New York, 1965.
S12. Sundaresan, P. R., and Coursin, D. B., Microassay of pyridoxal phosphate using L-tyrosine-1-^{14}C and tyrosine apodecarboxylase. *In* "Methods in Enzymology" (D. B. McCormick and L. D. Wright, eds.), Vol. 18A, p. 509. Academic Press, New York, 1970.
S13. Satoh, K., and Price, J. M., Fluorometric determination of kynurenic acid and xanthurenic acid in human urine. *J. Biol. Chem.* **230**, 781–789 (1958).
S14. Scriver, C. R., Vitamin-responsive inborn errors of metabolism. *Metab., Clin. Exp.* **22**, 1319–1344 (1973).
S15. Snyderman, S. E., Carretero, R., and Holt, L. E., Pyridoxine deficiency in the human being. *Fed. Proc., Fed. Am. Soc. Exp. Biol.* **9**, 371–372 (1950).
S16. Scriver, C. R., and Hutchison, J. H., The vitamin B6 deficiency syndrome in human infancy: Biochemical and clinical observations. *Pediatrics* **31**, 240–250 (1963).
S17. Sanderson, C. R., and Davis, R. E., Serum pyridoxal in active peptic ulceration. *Gut* **16**, 177–180 (1975).
S18. Sanderson, C. R., and Davis, R. E., Serum pyridoxal in patients with gastric pathology. *Gut* **17**, 371–374 (1976).
S19. Seelig, M. S., Alba, A., Berger, A. R., Rudez, A., and Tarlau, M., Pilot study of D-penicillamine, vitamins and minerals in multiple sclerosis. *J. Clin. Pyschiatry* **39**, 170–174 (1978).
S20. Salkeld, R. M., Knörr, K., and Körner, W. F., The effect of oral contraceptives on vitamin B6 status. *Clin. Chim. Acta* **49**, 195–199 (1973).
S21. Shane, B., and Contractor, S. F., Assessment of vitamin B6 status. Studies on pregnant women and oral contraceptive users. *Am. J. Clin. Nutr.* **28**, 739–747 (1975).
S22. Stone, W. J., Warnock, L. G., and Wagner, C., Vitamin B6 deficiency in uremia. *Am. J. Clin. Nutr.* **28**, 950–957 (1975).
S23. Spannuth, C. L., Warnock, L. G., Wagner, C., and Stone, W. J., Increased plasma clearance of pyridoxal 5'-phosphate in B6-deficient uraemic man. *J. Lab. Clin. Med.* **90**, 632–637 (1977).
S24. Sprince, H., Lowy, R. S., Folsome, C. E., Benram, J. S., and Raritan, N. J., Studies on the urinary excretion of xanthurenic acid during normal and abnormal pregnancy. *Am. J. Obstet. Gynecol.* **62**, 84–92 (1951).
S25. Spellacy, W. N., Buhi, W. C., and Birk, S. A., Vitamin B6 treatment of gestational diabetes mellitus. *Am. J. Obstet. Gynecol.* **127**, 599–602 (1977).
S26. Scriver, C. R., and Rosenberg, L. E., "Amino Acid Metabolism and its Disorders." Saunders, Philadelphia, Pennsylvania, 1973.
S27. Stanbury, J. B., Wyngaarden, J. B., and Fredrickson, D. S., eds., "The Metabolic Basis of Inherited Disease," 4th ed. McGraw-Hill, New York, 1978.
S28. Schimke, R. N., McKusick, V. A., and Weilbaecher, R. G., Homocystinuria. *In* "Amino Acid Metabolism and Genetic Variation" (W. L. Nyhan, ed.), pp. 297–313. McGraw-Hill, New York, 1967.

S29. Smith, L. H., and Williams, H. E., Treatment of primary hyperoxaluria. *Mod. Treat.* **4**, 522–530 (1967).
S30. Schulman, M. P., and Richert, D. A., Utilisation of glycine succinate δ-amino-levulinic acid for heme synthesis. *Fed. Proc., Fed. Am. Soc. Exp. Biol.* **15**, 349–351 (1956).
S31. Scriver, C. R., Vitamin B6-dependency and infantile convulsions. *Pediatrics* **26**, 62–74 (1960).
S32. Scriver, C. R., and Whelan, D. T., Glutamic acid decarboxylase in mammalian tissues outside the central nervous system and its possible relevance to hereditary vitamin B6 dependency with seizures. *Ann. N. Y. Acad. Sci.* **166**, 83–96 (1969).
S33. Schumacher, H. R., Bernhart, F. W., and György, P., Vitamin B6 levels in rheumatoid arthritis: Effect of treatment. *Am. J. Clin. Nutr.* **22**, 1200–1203 (1975).
S34. Sanderson, C. R., Davis, R. E., and Bayliss, C. E., Serum pyridoxal in patients with rheumatoid arthritis. *Ann. Rheum.* **35**, 177–180 (1976).
S35. Sergeev, A. V., Bykovskaja, S. N., Luchanskaja, L. M., and Rauschenbach, M. O., Pyridoxine deficiency and cytotoxicity of T lymphocytes in vitro. *Cell. Immunol.* **38**, 189–192 (1978).
S36. Solomon, L. R., and Hillman, R. S., Vitamin B6 metabolism in anaemia and alcoholic man. *Br. J. Haematol.* **41**, 343–356 (1979).
S37. Spellacy, W. N., Buhi, W. C., and Birk, S. A., The effects of vitamin B6 on carbohydrate metabolism in women taking oral contraceptives. Preliminary report. *Contraception* **6**, 265–273 (1972).
S38. Spiegel, A. M., Rosen, S. W., Weintraub, B. D., and Marynick, S. P., Effect of intravenous pyridoxine on plasma prolactin in hyperprolactinenic subjects. *J. Clin. Endocrinol. Metab.* **46**, 686–688 (1978).
S39. Stowe, H. D., Goyer, R. A., and Medley, P., Influence of dietary pyridoxine on cadmium toxicity in rats. *Arch. Environ. Health* **28**, 209–216 (1974).
S40. Stephens, M. C., and Dakshinamurti, K., Galactolipid fatty acids in brain of pyridoxine-deficient young rats. *Exp. Brain Res.* **25**, 465–468 (1976).
T1. Takanashi, S., and Tamura, Z., Preliminary studies for fluorometric determination of pyridoxal and of its 5′-phosphate. *J. Vitaminol.* **16**, 129–131 (1970).
T2. Thanassi, J. W., and Cidlowski, J. A., A radioimmunoassay for phosphorylated forms of vitamin B6. *J. Immunol. Methods* **33**, 261–266 (1980).
T3. Theron, J. J., Pretorius, P. J., and Joubert, C. P., The state of pyridoxine nutrition in patients with Kwashiorkor, *J. Pediatr.* **59**, 440–441 (1961).
T4. Tant, D., Megaloblastic anaemia due to pyridoxine deficiency associated with prolonged ingestion of an oestrogen-containing oral contraceptive. *Br. Med. J.* **2**, 979–980 (1976).
T5. Thomas, M. R., and Kirksey, A., Postnatal patterns of brain lipids in progeny of vitamin B6 deficient rats before and after pyridoxine supplementation. *J. Nutr.* **106**, 1404–1414 (1976).
T6. Thomas, R. T., Kawamoto, J., Sneed, S. M., and Eakin, R., The effects of vitamin C, vitamin B6 and vitamin B12 supplementation on the breast milk and maternal status of well-nourished women. *Am. J. Clin. Nutr.* **32**, 1679–1685 (1979).
T7. Tada, K., Yoshida, T., Yokoyama, Y., Sato, T., Nakagawa, H., and Arakawa, T., Cystathioninuria not associated with vitamin B6 dependency: A probable new type of cystathioninuria. *Tohoku J. Exp. Med.* **95**, 235–242 (1968).
T8. Tada, K., Yokoyama, Y., Nakagawa, H., and Arakawa, T., Vitamin B6 dependent xanthurenic aciduria (the second report). *Tohoku J. Exp. Med.* **95**, 107–114 (1968).
T9. Tolis, G., Laliberté, R., Guyda, H., and Naftolin, F., Ineffectiveness of pyridoxine (B6) to alter secretion of growth hormone and prolactin and absence of therapeutic effects on galactorrhea-amenorrhea syndrome. *J. Clin. Endocrinol. Metab.* **44**, 1197 (1977).

U1. Ueda, K., Akeda, H., and Suda, M., Intestinal absorption of amino acids. *J. Biochem. (Tokyo)* **48**, 584–592 (1960).

V1. Vanderslice, J. T., Stewart, K. K., and Yarmas, M. M., Liquid chromatographic separation and quantification of B6 vitamers and their metabolite pyridoxic acid. *J. Chromatogr.* **176**, 280–285 (1979).

V2. Vir, S. C., and Love, A. H. G., Vitamin B6 status of the hospitalised aged. *Am. J. Clin. Nutr.* **31**, 1383–1391 (1978).

V3. Vitale, J. J., The impact of infection on vitamin metabolism: an unexplored area. *Am. J. Clin. Nutr.* **30**, 1473–1477 (1977).

V4. Van Bijsterveld, O. P., The digestive capacity of pyridoxine deficient phagocytes in vitro. *J. Med. Microbiol.* **4**, 337–347 (1971).

V5. Vijayammal, P. L., and Kurup, P. A., Pyridoxine and atherosclerosis: Role of pyridoxine in the metabolism of lipids and glycosaminoglycans in rats fed normal and high fat, high cholesterol diets containing 16% casein. *Aust. J. Biol. Sci.* **31**, 7–20 (1978).

W1. Wolf, H., Studies on tryptophan metabolism in man: The effects of hormones and vitamin B6 on urinary excretion of metabolites of the kynurenine pathway. *Scand. J. Clin. Lab. Invest.* **33**, Suppl. 136 (1974).

W2. Williams, A. K., Vitamin B6: Gas-liquid chromatography of pyridoxol, pyridoxal and pyridoxamine. *J. Agric. Food Chem.* **22**, 107–109 (1974).

W3. Williams, R. C., Baker, D. R., and Schmit, J. A., Analysis of water-soluble vitamins by high-speed ion-exchange chromatography. *J. Chromatogr. Sci.* **11**, 618–624 (1973).

W4. Williams, A. K., and Cole, P. D., Vitamin B6. Ion exchange chromatography of pyridoxal and pyridoxamine. *J. Agric. Food Chem.* **23**, 915–916 (1975).

W5. Wachstein, M., and Gudaitis, A., Detection of vitamin B6 deficiency: Utilization of an improved method for rapid determination of xanthurenic acid in urine. *Am. J. Clin. Pathol.* **22**, 652–655 (1952).

W6. Wachstein, M., and Gudatis, A., Disturbance of vitamin B6 metabolism in pregnancy. II. The influence of various amounts of pyridoxine hydrochloride upon the abnormal tryptophane load test in pregnant women. *J. Lab. Clin. Med.* **42**, 98–107 (1953).

W7. Woodring, M. I., Fisher, D. H., and Storvick, C. A., A microprocedure for the determination of 4-pyridoxic acid in urine. *Clin. Chem. (Winston-Salem, N.C.)* **10**, 479–489 (1964).

W8. World Health Organization, "Handbook on Human Nutritional Requirements," W. H. O. Monogr. Ser. 61. W.H.O., Geneva, 1974.

W9. World Health Organization, In "Codex Alimentarius Commission," Report of the 9th session of the Codex Committee on Foods for special dietary uses. W.H.O., Geneva, 1975.

W10. Williams, R. J., "Biochemical Variation: Its Significance in Biology and Medicine." Wiley, New York, 1956.

W11. Wien, E. M., Vitamin B6 status of Nigerian women using various methods of contraception. *Am. J. Clin. Nutr.* **31**, 1392–1396 (1978).

W12. Wynn, V., Vitamins and oral contraceptive use. *Lancet* **1**, 561–564 (1975).

W13. Wachstein, M., Kellner, J. D., and Ortiz, J. M., Pyridoxal phosphate in plasma and leukocytes of normal and pregnant subjects following vitamin B6 load tests. *Proc. Soc. Exp. Biol. Med.* **103**, 350–353 (1960).

W14. Willis, R. S., Winn, W. W., Morris, A. T., Newsom, A. A., and Massey, A. E., Clinical observations in treatment of nausea and vomiting in pregnancy with vitamins B1 and B6. *Am. J. Obstet. Gynecol.* **44**, 265–271 (1942).

W15. West, K. D., and Kirksey, A., Influence of vitamin B6 intake on the content of the vitamin in human milk. *Am. J. Clin. Nutr.* **29**, 961–969.

W16. Will, E. J., and Bijvoet, O. L. M., Primary oxalosis: Clinical and biochemical responses to high-dose pyridoxine therapy. *Metab., Clin. Exp.* **28,** 542–545 (1979).

W17. Willis-Carr, J. I., and St. Pierre, R. L., Effects of vitamin B6 deficiency on thymic epithelial cells and T lymphocyte differentiation. *J. Immunol.* **120,** 1153–1159 (1979).

W18. Wilson, R. G., and Davis, R. E., Serum pyridoxal concentrations in children with diabetes mellitus. *Pathology* **9,** 95–99 (1977).

W19. Wilcken, B., and Turner, B., Homocystinuria: Reduced folate levels during pyridoxine treatment. *Arch. Dis. Child.* **48,** 58–62 (1973).

Y1. Yan, S. C. B., Uhing, R. J., Parrish, R. F., Metzler, D. E., and Graves, D. J., A role for pyridoxal phosphate in the control of dephosphorylation of phosphorylase a. *J. Biol. Chem.* **254,** 8263–8269 (1979).

Y2. Yasumoto, K., Tadera, K., Tsuji, H., and Mitsuda, H., Semi-automated system for analysis of vitamin B6 complex by ion-exchange column chromatography. *J. Nutr. Sci. Vitaminol.* **21,** 117–127 (1975).

Y3. Yasumoto, K., Tsuji, H., Iwami, K., and Mitsuda, H., Isolation from rice bran of a bound form of vitamin B6 and its identification as 5'-*O*'-(β-D-glucopyranosyl) pyridoxine. *Agric. Biol. Chem.* **41,** 1061–1067 (1977).

Y4. Yoshida, T., Tada, K., and Arakawa, T., Vitamin B6-dependency of glutamic acid decarboxylase in the kidney from a patient with vitamin B6 dependent convulsions. *Tohoku J. Exp. Med.* **104,** 195–198 (1971).

ALUMINUM

Allen C. Alfrey

University of Colorado Medical School,
Denver Veterans Administration Medical Center,
Denver, Colorado

1. Introduction 69
2. Methodology... 70
3. Aluminum Metabolism 74
4. Aluminum Toxicity 79
   References 85

1. Introduction

Aluminum occurs in great abundance, being the second most plentiful element in the earth's crust (E3). Because of its reactive nature it does not normally exist in the metallic state but rather in combination with oxygen, silicon, fluorine, and other elements. Besides the aluminum naturally present in the environment, man is also exposed to this element through aluminum cans and cooking utensils, various aluminum-containing medications, drinking water, baking powder, and even deodorants. In spite of this large and continuing exposure to aluminum, the body burden of aluminum has been found to be quite small and it has been assumed that the skin, lung, and gastrointestinal tract largely if not completely prevent any aluminum absorption. However, recent studies suggest that this may not always be the case (A2, A6). In addition, increasing evidence has been presented suggesting that alterations in the body burden of aluminum can be associated with major clinical disturbances (A4, E2, P4).

Even though aluminum was the subject of two large review articles, one in 1957 (C2) and the second in 1974 (S8), it is apparent that little knowledge is actually available on the metabolism and toxicity of aluminum. Furthermore, these earlier studies were predicated on the determination of aluminum in biological samples using analytical methods that suffered from lack of sensitivity and major interference problems which resulted in erroneous and inconsistent findings. With the recent development of neutron activation analysis and flameless atomic absorption methodologies for the analysis of aluminum, these problems have been largely overcome. In view of the data recently generated with these newer analytical methods and the increased interest in aluminum in uremic patients and its possible role in the pathogenesis of a variety of neurological syndromes, a review article on this element at this time seems appropriate.

2. Methodology

A variety of different methods, including atomic absorption spectrophotometry (B3), flameless atomic absorption spectrophotometry (B4–B7, E1, F3, G2, K2, M12, S1, T4, V2, W5, Z1, Z2), neutron activation analysis (B3, C6, W1), emission spectroscopy (B9, L7), and inductively coupled plasma emission spectrometry (L4), have been employed to measure aluminum levels in biological samples. Recently, the majority of investigators have used flameless atomic absorption spectrophotometry because of its sensitivity, the simplicity of sample preparation, and its availability at a reasonable cost. Analyses in the author's laboratory are performed on a model 305B atomic absorption spectrophotometer equipped with a deuterium background corrector, an HGA-2100 flameless system, and an AS-1 automatic sampling system (all from Perkin-Elmer, Norwalk, Connecticut 06856). All absorption measurements are monitored on a strip-chart recorder. Settings for the 305B and the HGA-2100 are as follows:

305B	HGA-2100
Al hollow-cathode lamp: 309.3 nm at 25 mA	Dry cycle: 30 seconds at 100°C
Slit width: 1 mm (0.7 nm bandpass)	Char cycle: 55 seconds to 1550°C
Absorption mode	Atomize cycle: 8 seconds at 2650°C
	Argon gas purge: 60 ml/minute
	Sample vol: 20 µl
	Gas flow: continuous
	Nonpyrolyzed graphite rod

Aluminum standards (25, 50, 100, and 200 µg/liter) are made from a 1 mg/liter aluminum stock and are prepared in either saturated Na_2EDTA for tissues or a phosphate solution of 4.8% saturated Na_2HPO_4 for biological fluids. Na_2EDTA enhances the signal by 45% and Na_2HPO_4 by 58% above DI standards (Fig. 1). Preliminary studies using pyrolyzed tubes suggested that phosphate suppressed the signal in these tubes, in contrast to the en-

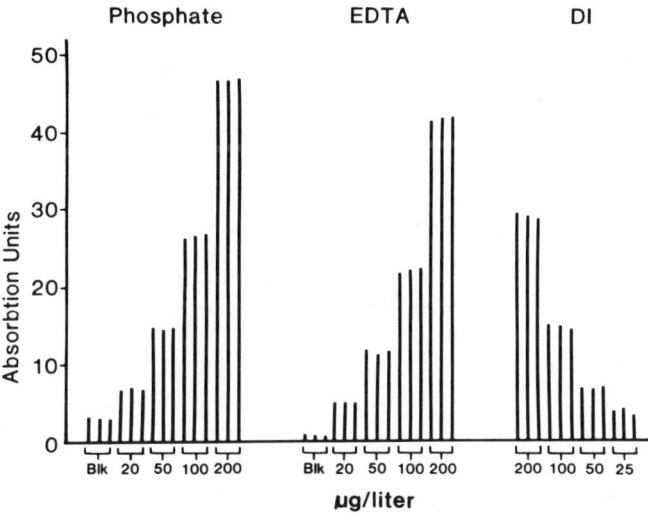

FIG. 1. Reproduction of original chart recording of aluminum standards, demonstrating the reproducibility and enhancement by phosphate and EDTA.

hancement observed in standard tubes. Therefore, the described method does not pertain to pyrolyzed tubes. The sensitivity of this method is 86 pg and precision, expressed as coefficient of variation, is 0.5%. For measurement of the aluminum in plasma, 5 µl of saturated solution of Na_2HPO_4 and 20 µl of Triton X-100 (0.5%, Harleco) are added to 100 µl of plasma. For other fluids (urine, CSF, bile, water, and dialysate), only 5 µl of saturated solution of Na_2HPO_4 is added to 100 µl of the sample. Without the addition of Triton X-100 to plasma samples there was frequently an organic residue following atomization which would commonly give a nonspecific signal, reduce tube life, and necessitate repeat tube cleaning for reuse. The addition of Triton X-100 completely eliminated these problems, and with this method approximately 60 plasma determinations can consistently be made with the same tube.

The phosphate addition to the various biological samples and standards markedly improved recoveries of standard additions, as can be appreciated in Table 1. Without phosphate addition recoveries varied between 57 and 164% in the various fluids. In contrast, with phosphate addition recoveries were found to be between 95 and 105% in the same fluids. Normal values obtained with this technique are shown in Table 2 and are compared with other reported values. It can be appreciated that most reported flameless atomic absorption values are quite similar, although ours are somewhat lower than those usually found by others. A possible reason for this could be

TABLE 1
RECOVERY OF STANDARD ADDITION

Sample	PO_4 standards (%)	DI standards (%)
Plasma	95 ± 8	57 ± 12
CSF	105 ± 9	116 ± 19
Bile	106 ± 8	156 ± 8
Dialysate	108 ± 5	123 ± 33
Urine	105 ± 8	164 ± 28

the use of Triton X-100 in the plasma sample. In general, plasma values found by neutron activation analysis are somewhat higher. This may result from failure of and/or lack of pretreatment in some studies, which would allow phosphate and silicon to contribute to the aluminum values. The highest plasma values have been found with emission spectroscopy and flame atomic absorption. Because of the lack of sensitivity and the interferences with these methods and the availability of newer methodology, these techniques are probably no longer applicable for plasma aluminum determinations.

Tissues, with the exception of bone and brain, are dried for 16 hours at 130°C. The dried tissue is ground to a fine powder and extracted four times with 2 ml of equal volumes of ether and petroleum ether (solvents are Al free). The tissue is then redried and weighed and 10 to 20 mg is placed in 5 ml of a saturated solution of EDTA in a small plastic tube. After agitating the tubes for 4 hours, the supernatant is then analyzed for aluminum. Bone is sampled from the iliac crest and adhering tissue is removed with a scalpel. The bone marrow is removed by washing in a jet of deionized water. After air drying at room temperature, the bones are gound in a Wiley mill, passed through a 2-mesh sieve, and partially defatted as described above. The bone is then reground and passed through a 60-mesh sieve. It is then extracted and analyzed as described above. For brain analysis, 25-mg aliquots of gray and white matter are separately ashed at 400°C for 16 hours in 10-ml Pyrex beakers. The ashed brain is then resuspended in 2–5 ml of Na_2EDTA and analyzed for aluminum. As can be appreciated, this method for analysis minimizes the handling of tissue samples and most potential sources of external aluminum contamination are avoided. The major criticism of this method has been in regard to the completeness of the extraction. EDTA extraction was selected for several reasons. First, we found that tissues prepared in a nitric acid digest gave erratic results for aluminum with the graphite furnace. This has recently been suggested to result from the nitrogen residue in the tube, which affects the aluminum signal. Second, EDTA was selected because the ethylenediaminetetraacetate anion forms nondissociated water-soluble complexes with a number of metals and has a high formation constant with aluminim ($10^{-16.13}$), as compared to Ca ($10^{-10.70}$) and Mg ($10^{-8.69}$) (D4). In addition, several studies have been conducted in order to verify the completeness of the extraction. First, similar aluminum values were obtained from muscle (L3), bone (L3), spleen, and liver (Table 3) when extraction was performed on each tissue in both dry and ashed states. Second, values for aluminum on duplicate samples of extracted tissues are very similar (Table 3). Third, extraction of different weights of each tissue yielded similar aluminum concentrations (L3). Fourth, tissues from control subjects were found to have aluminum values similar to that reported by other investigators. Thus, it would seem unlikely that such reproducible values would have been obtained under the various conditions if extraction was less than 100% complete.

TABLE 2
NORMAL ALUMINUM VALUES

Ours	Reported
Plasma[a]	Flameless atomic absorption
6 ± 3 µg/liter	3.7 ± 1.1–49 ± 11 µg/liter (A7, B4–B7, E1, F3, G2, K2, M12, S1, T4, V2, W5, Z1, Z2)
	Atomic absorption
	240 µg/liter (B3)
	Neutron activation analysis
	25–1460 µg/liter (B3, C6, W1)
	Emission spectroscopy
	400 ± 277 µg/liter (B9, L7)
CSF[a]	Flameless atomic absorption
6.8 ± 4.5 µg/liter	31 ± 11 µg/liter (D3, S4)
	Atomic absorption
	7000 µg/liter
Urine[a]	Flameless atomic absorption
13 ± 6 µg/day	61 ± 22 µg/day (G2)
	Neutron activation analysis
	86 ± 65 µg/day (R1)
	Emission spectroscopy
	720–1000 µg/day (T1, T2)
Bile[b]	Not available
3 ± 1 µg/liter	

[a] Normal human.
[b] Normal dog.

TABLE 3
Extraction of Duplicate Samples[a]

Sample number[b]	Ashed		Dry	
1	14.5	15.0	12.0	12.0
2	3.2	2.8	5.4	5.4
3	96.0	100.8	93.0	93.0
4	24.0	22.0	18.0	18.0
5	94.2	94.2	87.0	88.8
6	4.0	3.5	1.8	1.8
7	102.0	99.0	97.2	97.2
8	48.0	52.5	50.4	50.4
9	37.2	37.2	45.0	44.4
10	49.8	49.2	57.0	58.5
11	18.0	17.0	24.6	24.6
12	46.2	48.0	51.0	52.5

[a] Al mg/kg sample weight.
[b] Bone, muscle, spleen, or liver tissue.

3. Aluminum Metabolism

It has been suggested that as a result of aluminum-contaminated food, 10 to 90 mg of this element is ingested daily (C2). However, recent studies would suggest that these earlier estimates of aluminum contamination may be excessive and that only 2–3 mg/day of aluminum may be ingested from this source (A6, G3). If this latter value is correct, it would mean that other sources of aluminum could markedly affect the quantity of aluminum ingested. It is common practice to use aluminum compounds as a coagulum in municipal water supplies. Since there are no water standards for aluminum and some of the added aluminum goes into solution, drinking water can have aluminum levels of up to 1000 µg/liter (K6). Another potential source of aluminum is the leaching of aluminum from aluminum cans, containers, and cooking utensils. How much this could add to the daily intake is unknown. Irrespective of the amount of aluminum normally ingested, it would appear that very little is actually absorbed. However, a number of factors have been suggested to modulate aluminum absorption. Since vitamin D has been shown to increase the absorption of a number of elements besides calcium, including magnesium, strontium, lead, beryllium, barium, zinc, cadmium, cesium, and cobalt (K4, S5, W2, W6), it has been suggested that it might also affect aluminum absorption (L8). Furthermore, in experimental animals it has been shown that tissue stores of aluminum are increased by the administration of vitamin D (L8) and parathyroid hormone (M3, M4). However, it has not been directly tested as to whether this is a result of enhanced absorption. It has been suggested that another factor which affects aluminum absorption is the fluoride content of the diet, since aluminum forms strong complexes with fluoride (S10, S11). Spencer and associates found that the administration of aluminum compounds markedly decreased urinary fluoride by 10- to 30-fold (S10, S11). In addition, aluminum compounds have been administered to counteract dental fluorosis (N2). Furthermore, an additional study in rats showed that the administration of fluoride in association with aluminum prevented the increased tissue aluminum found in animals receiving a comparable amount of aluminum without fluoride, suggesting that fluoride either increases aluminum excretion or prevents its absorption (O1).

Irrespective of the various factors that could affect aluminum absorption, the data on urinary aluminum indicate that little aluminum is normally absorbed from the gastrointestinal tract. Tipton and associates (O2, T1), using emission spectroscopy, found urine aluminum excretion to be 700 to 1000 µg/day. However, more recent studies using neutron activation analysis (R1) and flameless atomic absorption spectrophotometry (G2, K2) would indicate considerably lower values (i.e., 86 ± 65, 61 ± 22, and 13 ± 6 µg/day).

There appears to be little doubt that when large oral loads of aluminum in the form of antacids are ingested, some of this excess aluminum is absorbed. Recker et al. (R1) found that urinary aluminum increased from 86 to 495 µg/day in normal subjects fed 3.8 g/day of aluminum. Similarly, Kaehny et al. (K2) found an increase in urinary aluminum excretion from 16 to 275 µg/day in subjects ingesting 2.2 g/day of aluminum and Gorsky et al. (G3) reported an increase in urinary aluminum from 65 to 280 µg/day in patients receiving 1.08 to 2.9 g/day of aluminum. The major question with these studies is not whether some of the excess aluminum is absorbed but whether all of the absorbed aluminum is eliminated by the kidney. Balance studies carried out in normal individuals ingesting 1 to 3 g of aluminum would suggest that these individuals absorb and retain 23 to 313 mg/day of aluminum (C6, G3, C1). These studies would further suggest that the kidney does not play a major role in the elimination of the absorbed aluminum. However, there are several facts that cast some doubt on the results of these balance studies. First, although the renal handling of aluminum has not been well defined, it would appear that the kidney has considerable capabilities for excreting aluminum since aluminum clearance can approach 50% of the glomerular filtration rate (K7). In addition, during the intravenous loading of aluminum, renal excretion accounts for the elimination of 50% of the administered aluminum (K7). Second, urinary aluminum rapidly decreases following the discontinuation of ingestion of aluminum compounds, suggesting that little aluminum is retained (K7). Finally, tissue aluminum levels are consistently low in subjects with normal renal function (A2, A6) and are low even in individuals with renal impairment who have been on large amounts of antacids for years and who do not have tissue aluminum levels increased to the extent that would be expected to result from the very large positive aluminum balances that have been reported (A2, A6). Thus, the reasons for these discrepancies are not readily apparent. This could probably be resolved by determining tissue aluminum levels in patients or animals with normal renal function who have been on oral aluminum compounds for extended periods of time.

Other avenues besides renal excretion for the elimination of aluminum from the body have not been identified. Although endogenous secretion of aluminum into the gastrointestinal tract has not been excluded, there is no evidence in favor of such a mechanism. Similarly, biliary aluminum levels have been found to be low, and even during acute aluminum loading they increase only slightly and are not a major excretory pathway for aluminum (K7).

Other exposures to aluminum besides the gastrointestinal tract, such as through the lung and skin, probably have little effect on the total body burden of aluminum. Although not well studied, especially in relation to the exposure to aluminum compounds present in deodorants, it would appear that the skin is largely impervious to aluminum absorption. Some inhaled aluminum is retained in pulmonary tissue and in the peribronchial lymph nodes but it is largely excluded from other tissues (S12). This is further suggested by the fact that lung aluminum concentration is greater than that of other tissues, it increases with age, unlike other tissues' aluminum levels, and pulmonary aluminum levels do not correlate with the aluminum levels of other tissues (A2, A6).

Normal tissue aluminum levels are shown in Table 4. These values were largely obtained from individuals who had no known medical disorders and experienced sudden, usually violent deaths. It is apparent that normal tissue aluminum levels are quite low. With the exception of lung, most tissue aluminum levels were found not to be affected by aging (A2). However, McDermott has recently found that brain aluminum is higher in patients older than age 70, as compared to those younger than this age (M6). Thus, with extensive aging, aluminum concentration may increase in certain tissues.

It has been suggested that specific tissue aluminum levels may increase in a number of pathological states. There has recently been a great deal of interest in the neurotoxicity of aluminum and aluminum levels have been found to be increased in the brains of patients dying of Alzheimer's dementia (C7, M8) and Guam and Kii peninsula amyotrophic lateral sclerosis (Y1). In

TABLE 4
Control Tissue Aluminum Levels[a]

	Liver	Spleen	Bone	Heart	Muscle	Lung	Brain gray matter
Mean	4.1	2.6	3.3	1.0	1.2	43	2.2
SD	1.7	2.1	2.9	0.8	1.2	43	1.3

[a] Al mg/kg dry defatted tissue.

addition, one patient with idiopathic osteoporosis who had ingested large quantities of aluminum was found to have elevated bone aluminum levels (R1). Aluminum levels have been measured only in isolated tissues from these patients, thus it is impossible to determine whether total body aluminum was increased in any of these conditions. However, it has been firmly established that in at least one condition, chronic renal failure, total body aluminum can be markedly increased (A2, A6). Although aluminum concentration can be increased in virtually any tissue in nondialyzed uremic patients, the tissues most frequently affected and having the highest concentration are liver and bone (A6). In this group of patients, bone aluminum concentration was increased in 82% and liver aluminum concentration in 56% of the patients studied (A6). The reason for this alteration in total body aluminum in patients with renal failure has not been fully defined. The most apparent explanation for this finding is the inability to excrete aluminum normally absorbed from the gastrointestinal tract as a result of the loss of renal function. However, in a number of patients this alone cannot account for the amount of aluminum retained. Normally, approximately 15 μg/day of aluminum is excreted in the urine (K2). If because of renal failure an individual was unable to excrete any of this aluminum over a 10-year period, this would result in retaining only 54 mg of aluminum. Since a number of patients had aluminum concentrations in the skeleton of over 50 mg/kg, this would mean (assuming a skeletal weight of 5 kg) that in the skeleton alone there would be over 250 mg of aluminum (A6).

On the basis of animal studies it has been suggested that the parathyroid hormone may augment aluminum absorption from the gut (M3, M4). Similarly, Mayor et al., using an emission spectroscopic method, showed a correlation between plasma aluminum and PTH levels in uremic patients (M3). This would suggest that the secondary hyperparathyroidism found in uremic patients could result in increased aluminum absorption. However, opposed to this is the finding that others have found no correlation between PTH levels and bone or plasma aluminum levels (A5). Furthermore, Ellis et al. (E2) found no correlation between either parathyroid size or degree of hyperparathyroid bone disease and bone aluminum levels. Similarly, we found aluminum levels to be less in hyperparathyroid bone disease than in the osteomalacic bone disease present in uremic patients, casting further doubt on this relationship (H5). Therefore, from the data available it seems unlikely that PTH has any effect on aluminum metabolism.

A more likely explanation for the increased body aluminum burden found in uremic patients is the orally administered aluminum compounds which are commonly given to bind phosphate in the gut and to normalize serum phosphate levels. It has been shown that the administration of these agents to normal individuals can increase urine aluminum levels to as much as 500 μg/day (G2, K2, R1). In turn, if someone with renal failure was unable to excrete this aluminum it could appreciably alter body aluminum stores.

Another mechanism of aluminum loading is performing dialysis on uremic patients with aluminum-contaminated dialysate. This contamination comes from water used to prepare the dialysate. Most of the aluminum found in municipal water supplies results from the addition of aluminum in the process of water treatment, since aluminum is a commonly used coagulum. Therefore, depending on the water source and type of treatment required, aluminum content of different water supplies can vary markedly (K6). Furthermore, since there are no public health or EPA standards for aluminum in drinking water, many water treatment facilities may not be able to supply information in regard to the aluminum concentration present in the treated water. A single determination may have little meaning since there can be marked seasonal variations in aluminum levels.

It would appear that virtually any aluminum present in the water used to prepare the dialysate represents a potential hazard to the patients. Aluminum has been shown to be bound in plasma (K1), thus any free aluminum in the dialysate would maintain a gradient between dialysate and blood and would be transported to the patients. In addition, because of the plasma binding, even with aluminum-free dialysate, dialysis is ineffective in removing aluminum from the patients (K1).

Since plasma is a transport system for aluminum, plasma aluminum measurements are useful in evaluating uremic patients undergoing active aluminum loading. Whereas normal plasma aluminum levels are 6 ± 3 μg/liter, uremic patients receiving oral aluminum-containing phosphate-binding gels (but undergoing dialysis with aluminum-free dialysate) have plasma aluminum levels of 55 ± 28 μg/liter. Uremic patients receiving dialysis with dialysate containing 80 μg/liter aluminum have plasma aluminum levels of 162 ± 66 μg/liter. This also documents the plasma binding of aluminum, since immediately following dialysis with contaminated dialysate, aluminum levels are even higher (303 ± 76 μg/liter) and are markedly above levels found in the dialysate. In the same patient population, plasma aluminum levels were measured 3 months later, the patients having been maintained on aluminum-free dialysate. At this time, plasma aluminum levels had decreased to 84 ± 28 µg/liter, a value similar to that found in patients who had always been maintained on aluminum-free dialysate (K1).

In further support that plasma aluminum levels reflect active aluminum loading and not total body aluminum is the finding that plasma aluminum does not correlate with bone aluminum content, although bone is the major storage tissue for aluminum (A2). Two other readily available tissues whose aluminum concentrations do not correlate with other tissue stores are hair and skin (A2). This probably, at least in part, is a consequence of environmental aluminum contamination of these exposed tissues.

During aluminum excess the tissues most affected are bone and liver concentrations, which can increase up to 100-fold over normal levels. Another tissue in uremic patients where aluminum levels are markedly increased is the heart (Fig. 2) (A2, A6).

4. Aluminum Toxicity

It has recently been suggested, based on its position in the periodic table and its abundance in the solar system, that aluminum is an essential element (V1). However, direct evidence that aluminum has any essential function in animals is lacking. Aluminum is present in virtually all plants. It is especially high in a number of species of subtropical plants (Theaceae, Euphorbiaceae,

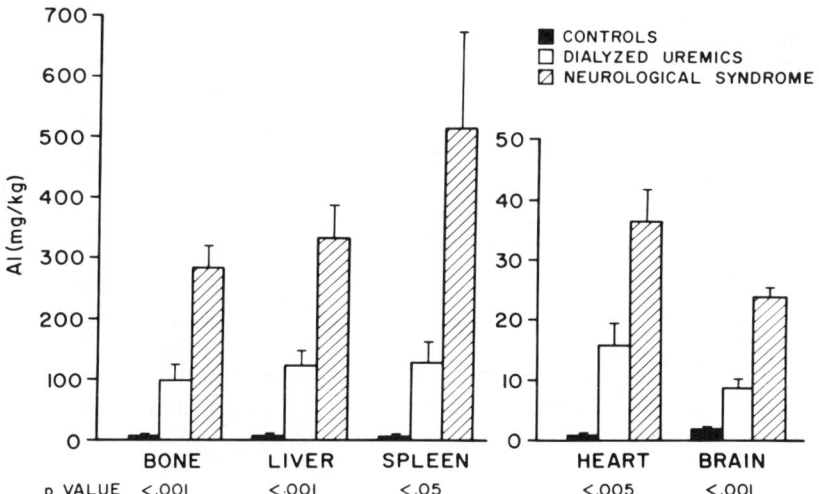

FIG. 2. Tissue aluminum concentrations in controls and dialyzed uremic patients dying of a neurological syndrome (dialysis encephalopathy) and other causes. (From Ref. A2.)

and Caryophyllacae) which are considered to be aluminum accumulators (C5). One accumulator, Lycopodiaceae, yields as much as 70% Al_2O_3 in ash (H13). Aluminum is probably necessary for the best development of some plants. The beneficial effects of aluminum consist of promoting germination of seeds (L6, S7), assisting in the uptake of water (H12), and protecting citrus cuttings from damage by excess copper (L5). Other studies would suggest that under certain conditions aluminum may also be toxic to plants. Aluminum has been found to inhibit cell extension and division and to interfere with plant mineral nutrition (M2). Since aluminum binds to DNA in plant cell nuclei (M1) it could limit template activity, thus explaining the deleterious effects of aluminum on plant metabolism.

Besides having apparent plant toxicity, aluminum is also toxic to fish and other aquatic biota (D7). It has been suggested that labile inorganic aluminum in acidified lake water may affect trout survival. Furthermore, it has been shown by complexing aluminum with fluoride that much of its toxicity for brook trout fry is reduced (D7).

It is well established that under some circumstances, excess aluminum can cause a variety of physiological and biological effects in humans. These effects of aluminum can be divided into three major categories: (1) the effect of aluminum compounds in the gastrointestinal tract, (2) the effect of inhaled aluminum compound, and (3) the systemic toxicity of absorbed aluminum compounds.

Orally administered aluminum compounds have several effects on the absorption of a number of other elements and compounds. The orally administered aluminum compounds have been shown to reduce the absorption of iron (H1), fluoride (S10, S11), phosphorus (L9), strontium (S9), and to a lesser extent, calcium (S9). Because of this property, aluminum compounds have been administered therapeutically as a method of reducing phosphorus absorption in uremic patients. However, when aluminum compounds are given in excessive amounts they can lead to a state of severe phosphate depletion, resulting in osteomalacia (L9) and in interference with phosphorylation processes in tissues (O2). Recently, it has been shown that cholesterol absorption can also be decreased by aluminum compounds (N1). It has been suggested that aluminum combines with pectin to bind fats to nondigestible vegetable fibers, thus preventing the absorption of fats.

Another interesting property of aluminum compounds is their effect on gastrointestinal tract mobility. Aluminum ions inhibit acetylcholine-induced contractions of rodent and human gastric smooth muscle (H4). In addition, free aluminum in the stomach has been shown to markedly reduce gastric emptying (H11). A well-recognized complication of oral aluminum compounds is chronic constipation, which probably results from this altered mobility of the gastrointestinal tract.

Pulmonary disease resulting from the inhalation of aluminum powder was first described in 1934 (F1). Fourteen years later, Schaver (S2) described an occupational pneumoconiosis in 11 patients resulting from the inhalation of bauxite ($Al_2O_3 \cdot 3H_2O$). Another equally toxic aluminum compound is "pyro," which is a finely divided powder (M14). The inhalation of these compounds produces pulmonary fibrosis, resulting in restrictive and obstructive airway disease. More recent studies show that animals exposed to high concentrations of aluminum chlorhydrate, as found in aerosol antiperspirants, develop granulomatous lesions in the respiratory bronchioles (D6).

Most recent interest in regard to aluminum toxicity has been directed at its role in the pathogenesis of multiple neurologic disease. Experimentally, aluminum has been known to be a neurotoxin since 1942 (K5). Kopeloff first described the epileptogenic effect of aluminum hydroxide when it was applied subdurally to animals (K5). In 1965 it was discovered that following aluminum phosphate injection in the brain, animals developed severe convulsions and striking neuronal changes characterized by neurofibrillary degeneration (K3). In 1967, DeBoni and associates were the first to show that aluminum lactate or tartrate injected subcutaneously would induce similar changes and concluded that aluminum could cross the blood–brain barrier (D2). More recently, Wisniewski et al. (W4) have developed a chronic model of aluminum-induced neurofibrillary changes in the rabbit. Additional evidence for aluminum causing the overproduction of neurofilament was obtained from rabbits given aluminum; they showed twice as much incorporation of [^{3}H]leucine into their neuronal perikarya as normal animals (E4). It is of further interest that whereas aluminum is neurotoxic in rabbits and cats, other animals such as rats and mice are resistant to this toxicity.

In vitro experiments by Miller and Levine (M13) showed that cultured neuroblastoma cells exposed to aluminum had a proliferation of neurofilaments, a reduction in cellular rRNA, an increased protein content, an elevated rate of [^{3}H]leucine incorporation, and reduced acetylcholine activity. Similarly, DeBoni et al. (D1) have shown that human neurons, when exposed to a culture media containing aluminum, undergo neuronal death with an overproduction of neurofilaments.

In view of the experimental evidence for the neurotoxicity of aluminum, it has recently been suggested that two human diseases characterized by neurofibrillary degeneration of the brain, Alzheimer's and Guam and Kii peninsula ALS, might result from aluminum intoxication (M8, Y1). In support of this is the finding of increased brain aluminum levels in these conditions (C7, Y1). Furthermore, in Alzheimer's disease the aluminum is present intranuclearly in chromatin only in the neurons with neurofibrillary changes (P3). It would appear that aluminum is increased only in the brain and that other tissue aluminum levels as well as total body aluminum are normal, as suggested by normal plasma aluminum levels in this condition (S4). This increased brain aluminum could result from either a secondary deposit of aluminum in diseased tissue or from a disturbance in the blood–brain barrier in this disorder. However, Wisniewski et al. (W3) have recently challenged the hypothesis that aluminum is responsible for this neurological disease. This challenge is based on the fact that brain aluminum concentration appears to increase with advanced aging and that these investigators found no differences between brain aluminum in patients with Alzheimer's disease and age-matched controls without this neurological illness (M6). In addition, these authors have further suggested that the neurofibrillary changes induced by aluminum in experimental animals are markedly different from those found in Alzheimer's disease. Finally, central cholinergic activity, which is markedly reduced in Alzheimer's disease, is normal in rabbits with aluminum-induced neurofibrillary changes. Obviously, additional studies are required to settle this controversy.

Although it has not been clearly established that Alzheimer's disease is a result of aluminum intoxication, there is considerable evidence that aluminum has neurotoxicity for man. Two cases of supposedly aluminum-induced progressive dementing encephalopathy, one in a ball-millroom worker in an aluminum powder factory, have been reported (L2, M9). In both of these cases, brain aluminum, as well as other tissue stores of aluminum, was increased. However, more convincing evidence of the human neurotoxicity resulting from aluminum is the dementing disease dialysis encephalopathy, which occurs in uremic patients (A1, A3). It is of further interest that neither patients with dialysis encephalopathy nor the other two reported cases of aluminum encephalopathy developed the intraneuronal lesions of neurofibrillary degeneration (B8, L2, M9). Since there are well-known species differences in animals developing neurofibrillary degeneration from aluminum administration, this finding may be explained on this basis. Furthermore, unlike Alzheimer's, where the aluminum is deposited in the nuclear chromatin of only the neurons with neurofibrillary degeneration (C8), in dialysis encephalopathy the aluminum is not only present in greater concentration but is also in the cytoplasm (C8) and possibly the lysosomes (G1) of histologically normal-appearing neurons throughout the central gray matter. That dialysis encephalopathy truly represents a form of human aluminum intoxication is based on strong biochemical and epidemiological data. First, several different investigators have consistently shown brain aluminum levels to be markedly elevated and significantly higher in patients dying of dialysis encephalopathy than in other uremic patients (A2, A8, C3, C8, M5) (Table 5). Even two groups that disagree on increased aluminum levels in Alzheimer's disease have both found high aluminum levels in patients

TABLE 5
ALUMINUM VALUES[a]

Normal	Dialysis patients	Dialysis encephalopathy		References
2.2 ± 1.3[b]	8.5 ± 3.3 (21)[c]	24.5 ± 9.9	(34)	Alfrey (A2)
1.7 ± 0.6	4.7 ± 1.8 (11)	21.0 ± 18.5	(6)	McDermott et al. (M5)
—	—	33.0 ± 9.6	(8)	Cartier et al. (C3)
0.9 ± 0.9	3.8 ± 1.8 (5)	12.4 ± 9.7	(4)	Arieff et al. (A8)
1.9 ± 0.7	—	10–27	(5)	Crapper et al. (C8)

[a] Al mg/kg dry weight.
[b] Data are given as mean ± 1 SD.
[c] Numbers in parentheses are number of patients studied.

dying of dialysis encephalopathy (C8, M5) (Table 5). Second, other tissue stores of aluminum (bone, heart, muscle, spleen, and liver) are also significantly higher in patients dying of dialysis encephalopathy than in dialysis patients dying of other causes (Fig. 2). Thus, unlike Alzheimer's disease, total body aluminum is markedly increased in patients dying of dialysis encephalopathy.

It was initially found that whereas 35 to 50% of patients in some dialysis units died of dialysis encephalopathy, this disease was rarely if ever seen in other dialysis units. It was subsequently shown that all units having a large number of cases of this disease also had large amounts of aluminum (90 to 500 µg/liter) in the water used to prepare the dialysate (A1, A3, A6, A8, B2, B8, C3, C8, D8, F2, G1, M5, R2, S3). Aluminum present in the dialysate was found to be transferred to the patients during dialysis (K1). Furthermore, by removing the aluminum from the water by water treatment, the disease could be completely eradicated in all of the affected units (A1, A3, A6, A8, B2, B8, C3, C8, D8, F2, G1, M5, R2, S3).

Increasing evidence has been presented suggesting that aluminum-contaminated dialysate may also be responsible for a disabling bone disease that occurs in some dialysis patients. This disease is characterized by severe osteomalacia which is unresponsive to vitamin D therapy and clinically manifested by bone pain and pathological fractures (E2, H5, H9, P1, P4, P5). Two epidemiological surveys showed that this bone disease occurred in association with dialysis encephalopathy in areas where dialysis was performed with aluminum-contaminated dialysate (P1, P5). In addition, it has been shown by removing the aluminum from the dialysate that the bone disease will spontaneously heal (H10, P4). Bone aluminum levels have also been shown to be markedly increased in patients with this disease (A5, E2).

Although the toxicity of aluminum administered through the dialysate has been reasonably well established, it is unknown what if any toxicity results from the aluminum retained from the orally administered aluminum-containing phosphate-binding gels commonly given to uremic patients. Some evidence has been presented which suggests that these compounds may not be totally safe. As stated earlier, tissue aluminum levels are increased in nondialyzed uremic patients receiving these agents (A2, A6). Dialysis encephalopathy has been described in nondialyzed uremic patients (B1, M10), in patients receiving peritoneal dialysis (S6), and in dialyzed uremic patients receiving dialysis with aluminum-free dialysate (M7). Tissue aluminum levels were found to be markedly increased in a number of these patients, supporting a source of aluminum exposure other than the dialysate. In addition, it has recently been shown that uremic patients with vitamin D-resistant osteomalacia, who have no apparent dialysate aluminum exposure, have significantly higher bone aluminum levels than uremic patients with other types of bone disease (H5). Similarly, we have also shown a highly significant correlation between bone aluminum concentration and the amount of noncalcified osteoid in these patients (H5). Although this would suggest that aluminum may be responsible for the failure of osteoid to calcify, it is equally possible that noncalcified osteoid accumulates aluminum. Direct testing will be required to resolve these possibilities. Additional evidence has been presented suggesting that aluminum can induce bone disease. Ellis *et al.* (E2) were able to induce a mineralization defect in rats with normal renal function by the parenteral injection of aluminum.

Since the frequency of bone disease and dialysis encephalopathy is much lower in uremic patients maintained on aluminum-free dialysate than in patients receiving dialysis with aluminum-contaminated dialysate, it appears that for unknown reasons, some patients, but not all, may absorb and/or retain aluminum excessively. This may be especially true in children (B1).

The mechanism by which aluminum could manifest these toxic effects has not been defined. It has been suggested that inhibitions of glycolysis and phosphorylation are the most toxic reactions to aluminum-containing compounds (S8). Aluminum has been shown to inhibit in a dose–response manner the enzymic conversion of isocitric acid to α-ketoglutamate in the presence of nicotinamide adenine dinucleotide phosphate (K8). Three independent investigative groups have shown that hexokinase, a magnesium-dependent phosphorylating enzyme, is inhibited by aluminum (H3, T3, V3). It has been shown that aluminum displaces magnesium from ATP, with the resultant stabilization of ATP preventing the phosphate transfer (H3, T3). It has been further suggested that possibly all phosphate-transferring systems involving ATP and magnesium may be biological targets for excess aluminum (S8). Thus, aluminum excess may in part explain the abnormal ATP metabolism described in uremic patients (M15).

Aluminum has been shown to inhibit, although less effectively than cadmium and manganese, synaptosomal Na–K-ATPase and Mg-ATPase as well as synaptosomal choline uptake (L1).

Aluminum is also selectively taken up by liver nuclei where it may combine with DNA (K9). In association with this, it has been shown that aluminum inhibits δ-aminolevulinic acid dehydratase activity (M11).

Aluminum has also been shown to accelerate the rate of oxygen uptake by the succinodehydrogenase cytochrome system (H6) and to increase acetylcholine hydrolysis and acetylcholinesterase (P2).

Finally, it seems possible that aluminum toxicity could result from aluminum displacing copper from some of its metalloenzymes, rendering them inactive. It is of interest that copper is necessary for brain cytochrome oxidase activity (H7) and bone lysyl oxidase activity (C4, H2). Furthermore, two organ systems primarily affected in copper deficiency, the central nervous system and bone, are also effected by aluminum excess. In addition, aluminum has been shown to bind to conalbumin (chicken ovotransferrin), displacing copper from its binding site (D5). Aluminum has also been shown to inhibit ferroxidase (ceruloplasmin) activity (H8). Additional evidence for an interrelationship between these two elements is the finding that aluminum protects citrus cuttings from damage by excess copper (L5).

References

A1. Alfrey, A. C., Dialysis encephalopathy syndrome. *Annu. Rev. Med.* **29**, 93–98 (1978).
A2. Alfrey, A. C., Aluminum metabolism in uremia. *Neurotoxicology* **1**, 43–53 (1980).
A3. Alfrey, A. C., Mishell, M. M., Burks, J., Contiguglia, S. R., Rudolph, H., Lewin, E., and Holmes, J. H., Syndrome of dyspraxia and multifocal seizures associated with chronic hemodialysis. *Trans. Am. Soc. Artif. Intern. Organs* **18**, 257–261 (1972).
A4. Alfrey, A. C., LeGendre, G. R., and Kaehny, W. D., The dialysis encephalopathy syndrome. Possible aluminum intoxication. *N. Engl. J. Med.* **294**, 184–188 (1976).
A5. Alfrey, A. C., Hegg, A., Miller, N., Berl, T., and Berns, A., Interrelationship between calcium and aluminum metabolism in dialyzed uremic patients. *Miner. Electrolyte Metab.* **2**, 81–87 (1979).
A6. Alfrey, A. C., Hegg, A., and Craswell, P., Metabolism and toxicity of aluminum in renal failure. *Am. J. Clin. Nutr.* **33**, 1509–1516 (1980).
A7. Allain, P., Thebaud, H.-E., Dupouet, L., Coville, P., Pisant, M., Speisser, J., and Alquier, P., Etude des taux sanguins de quelques métaux (Al, Mn, Cd, Pb, Cu, Zn) chez les hémodialyses chroniques avant et après dialyse. *Nouv. Presse Med.* **7**, 92–96 (1978).
A8. Arieff, A. I., Cooper, J. D., Armstrong, D., and Lazarowitz, V. C., Dementia, renal failure, and brain aluminum. *Ann. Intern. Med.* **90**, 741–747 (1979).
B1. Baluarte, J. H., Gruskin, A. B., Kiner, L. B., Foley, C. M., and Grover, W. D., Encephalopathy in children with chronic renal failure. *Proc.—Clin. Dial. Transplant Forum* **7**, 95–98 (1977).
B2. Berkseth, R., Mahowald, M., Anderson, D., and Shapiro, F., Dialysis encephalopathy (DE): Diagnostic criteria and epidemiology of 39 patients. *Am. Soc. Nephrol.* **11**, 36A (1978).
B3. Berlyne, G. M., Pest, D., Ben-Ari, J., Weinberger, J., Stern, M., Gilmore, G. R., and

Levine, R., Hyperaluminaemia from aluminum resins in renal failure. *Lancet* **2**, 494–496 (1970).

B4. Boukari, M., Rottembourg, J., Jaudon, M. C., Galli, A., and Legrain, M., Influence of prolonged ingestion of phosphate-binding aluminum gels on serum aluminum levels in patients with chronic renal failure. *Kidney Int.* **12**, 373–376 (1977).

B5. Boukari, M., Jaudon, M. C., Rottembourg, P. F. J., Luciani, J., Legrain, M., and Galli, A., Kinetics of serum and urinary aluminum after renal transplantation. *Lancet* **2**, 1044 (1978).

B6. Boukari, M., Rottembourg, J., Laudon, M.-C., Clavel, J.-P., Legrain, M., and Galli, A., Influence de la prise prolongée de gels d'alumine sur les taux sériques d'aluminum chez les patients atteints d'insuffisance renale chronique. *Nouv. Presse Med.* **7**, 85–88 (1978).

B7. Buge, A., Poisson, M., Masson, S., Bleibel, J.-M., Lafforgue, B., Raymond, P., and Jaudon, M.-C., Encéphalopathie prolongée et reversible chez un dialyse chronique responsabilité probable des sels d'aluminum. *Nouv. Presse Med.* **7**, 2053–2059 (1978).

B8. Burks, J. S., Alfrey, A. C., Huddlestone, J., Norenberg, M. D., and Lewin, E., A fatal encephalopathy in chronic haemodialysis patients. *Lancet* **1**, 764 (1976).

B9. Butt, E. M., Nusbaum, R. E., and Gilmour, T. C., Trace metal levels in human serum and blood. *Arch. Environ. Health* **8**, 52–57 (1964).

C1. Cam, J. M., Luch, V. A., Eastwood, J. B., and deWardner, H. E., The effect of aluminum hydroxide orally on calcium, phosphorus and aluminum metabolism in normal subjects. *Clin. Sci. Mol. Med.* **51**, 407–414 (1976).

C2. Campbell, I. R., Cass, J. S., Cholak, J., and Kehoe, R. A., Aluminum in the environment of man. *Arch. Ind. Health* **15**, 359–448 (1957).

C3. Cartier, F., Allain, P., Gary, J., Chatel, M., Menault, F., and Pecker, S., Encéphalopathie myclonique progressive des dialyses. Rôle de l'eau utilisée pour l'hémodialyse. *Nouv. Presse Med.* **7**, 97–102 (1978).

C4. Chavapil, M., and Misiorowski, R., *In vivo* inhibition of lysyl oxidase by high dose of zinc (40836). *Proc. Soc. Exp. Biol. Med.* **164**, 137–141 (1980).

C5. Chenery, E. M., Aluminum in the plant world. *Bull. Kew Gardens* **1**, 1973 (1948).

C6. Clarkson, E. M., Luck, V. A., Hynson, W. V., Bailey, R. R., Eastwood, J. B., Woodhead, J. S., Clements, V. R., O'Riordan, J. L. H., and deWardner, H. E., The effect of aluminum hydroxide on calcium, phosphorus and aluminum balances, the serum parathyroid hormone concentration and the aluminum content of bone in patients with chronic renal failure. *Clin. Sci.* **43**, 519–531 (1972).

C7. Crapper, D. R., Krishnan, S. S., and Dalton, A. J., Brain aluminum distribution in Alzheimer's disease and experimental neurofibrillary degeneration. *Science* **180**, 511–513 (1973).

C8. Crapper, D. R., Quittkat, S., Krishnan, S. S., Dalton, A. J., and DeBoni, U., Intranuclear aluminum content in Alzheimer's disease, dialysis encephalopathy, and experimental aluminum encephalopathy. *Acta Neuropathol.* **50**, 19–24 (1980).

D1. DeBoni, U., Scott, J. W., and Crapper, D. R., Intracellular aluminum binding: A histochemical study. *Histochemistry* **40**, 31–37 (1974).

D2. DeBoni, U., Otvos, A., Scott, J. W., and Crapper, D. R., Neurofibrillary degeneration induced by systemic aluminum. *Acta Neuropathol.* **35**, 285–294 (1976).

D3. Delaney, J. F., Spinal fluid aluminum levels in patients with Alzheimer disease. *Ann. Neurol.* **5**, 580–581 (1979).

D4. Diehl, H., "Calcein, Calmagite, and 1,1-Dihydroxyazobenzene: Titrimetric, Colorimetric and Fluorometric Reagents for Calcium and Magnesium," pp. 1–4. G. Frederick Smith Chemical Co., Columbus, Ohio, 1964.

D5. Donovan, J. W., and Ross, K. D., Nonequivalence of the metal binding sites of conalbumin. Colorimetric and spectrophotometric studies of aluminum binding. *J. Biol. Chem.* **250**, 6022–6025 (1975).

D6. Drew, R. T., Gupta, R. N., Bend, J. R., and Hook, G. E. R., Inhalation studies with a glycol complex of aluminum-chloride-hydroxide. *Arch. Environ. Health* **28**, 321–326 (1974).

D7. Driscoll, C. T., Jr., Gaker, J. P., Bisogni, J. J., Jr., and Schofield, C. L., Effect of aluminum speciation on fish in dilute acidified waters. *Nature (London)* **284**, 162–164 (1980).

D8. Dunea, G., Mahuraker, S. D., Mamdani, B., and Smith, E. C., Role of aluminum in dialysis dementia. *Ann. Intern. Med.* **88**, 502–504 (1978).

E1. Elliott, H. L., Macdougall, A. I., and Fell, G. S., Aluminum toxicity syndrome. *Lancet* **2**, 1203 (1978).

E2. Ellis, H. A., McCarthy, J. H., and Herrington, J., Bone aluminum in haemodialysed patients and in rats injected with aluminum chloride: Relationship to impaired bone metabolism. *J. Clin. Pathol.* **32**, 832–844 (1979).

E3. Emmons, W. H., "Geology: Principles and Processes," 5th ed., p. 60. McGraw-Hill, New York, 1960.

E4. Exss, R. E., and Summer, G. K., Basic proteins in neurons containing fibrillary deposits. *Brain Res.* **49**, 151–164 (1973).

F1. Filipo, D., Influenza delle polveri di allumino sulla pathologia delle vie respiratorie. *Rass. Med. Appl. Lav. Ind.* **5**, 128–144 (1934).

F2. Flendrig, J. A., Kruis, H., and Das, H. A., Aluminum intoxication: The cause of dialysis dementia? *Proc. Eur. Dial. Transplant Assoc.* **13**, 355–361 (1976).

F3. Fuchs, C., Brasche, M., Paschen, K., Nordbeck, H., Quellhorst, E., and Peek, U., Aluminum-Bestimmung im Serum mit flammenloser Atomabsorption. *Clin. Chim. Acta* **52**, 71–80 (1974).

G1. Galle, P., Chatel, M., Berry, J. P., and Menault, F., Encéphalopathie myoclonique progressive des dialyses: Présence d'aluminum et forte concentration dans les lysosomes des cellules cérébrales. *Nouv. Presse Med.* **8**, 4091–4094 (1979).

G2. Gorsky, J. E., and Dietz, A. A., Determination of aluminum in biological samples by atomic absorption spectrophotometry with a graphite furnace. *Clin. Chem. (Winston-Salem, N.C.)* **24**, 1485–1490 (1978).

G3. Gorsky, J. E., Dietz, A. A., Spencer, H., and Osis, D., Metabolic balance of aluminum studied in six men. *Clin. Chem. (Winston-Salem, N.C.)* **25**, 1739–1743 (1979).

H1. Hall, G. J. L., and Davis, A. D., Inhibition of iron absorption by magnesium trisilicate. *Med. J. Aust.* **2**, 95 (1969).

H2. Harris, E. D., Copper-induced activation of aortic lysyl oxidase in vivo. *Proc. Natl. Acad. Sci. U.S.A.* **73**, 371–380 (1976).

H3. Harrison, W. H., Codd, E., and Gray, R. M., Aluminum inhibition of hexokinase. *Lancet* **2**, 277 (1972).

H4. Hava, M., and Hurwitz, A., The relaxins effect of aluminum and lanthanum on rats and human gastric smooth muscle in vitro. *Eur. J. Pharmacol.* **22**, 156–161 (1973).

H5. Hodsman, A. B., Sherrard, D. J., Brickman, A. S., Alfrey, A. C., Goodman, W. G., Maloney, N., Lee, D. B. N., and Coburn, J. W., Bone aluminum in osteomalacic renal osteodystrophy correlation with excess osteoid. *Am. Soc. Nephrol.* **13**, 20A (1980).

H6. Horecker, B. L., Statz, E., and Hogness, T., The promoting effect of Al, Cr, and rare earths in the succinic dehydrogenase cytochrome system. *J. Biol. Chem.* **128**, 251–256 (1939).

H7. Howell, J. M., and Davison, A. N., The copper content and cytochrome oxidase activity of tissues from normal and swayback lambs. *Biochem. J.* **72**, 365–367 (1959).

H8. Huber, C. T., and Frieden, E., The inhibition of ferroxidase by trivalent and other metal ions. *J. Biol. Chem.* **245,** 3979–3984 (1970).

H9. Hudson, G. A., Milne, F. J., Oliver, N. J., Reis, P., Murray, J., and Meyers, A. M., Bone disease in patients on maintenance haemodialysis using softened or de-ionized water. *S. Afr. Med. J.* **56,** 439–443 (1979).

H10. Hudson, G. A., Milne, F. J., Meyers, A. M., and Reis, P., Treatment of dialysis fracturing bone disease. *Kidney Int.* **17,** 532 (1980).

H11. Hurwitz, A., Robinson, R. G., Vats, T. S., Whittier, F. C., and Herrin, W. F., Effects of antacids on gastric emptying. *Gastroenterology* **71,** 268–273 (1976).

H12. Hutchinson, G. E., Aluminum in soils, plants and animals. *Soil Sci.* **60,** 29–40 (1945).

H13. Hutchinson, G. E., and Wollack, A., Biological accumulators of aluminum. *Trans. Conn. Acad. Arts Sci.* **35,** 73–128 (1943).

K1. Kaehny, W. D., Alfrey, A. C., Holman, R. E., and Shorr, W. J., Aluminum transfer during hemodialysis. *Kidney Int.* **12,** 361–365 (1977).

K2. Kaehny, W. D., Hegg, A. P., and Alfrey, A. C., Gastrointestinal absorption of aluminum from aluminum-containing antacids. *N. Engl. J. Med.* **296,** 1389–1390 (1977).

K3. Klatzo, I., Wisniewski, H., and Streicher, E., Experimental production of neurofibrillary degeneration. I. Light microscopic observations. *J. Neuropathol. Exp. Neurol.* **24,** 187–199 (1965).

K4. Koo, S. I., Fullmer, C. S., and Wasserman, R. H., Intestinal absorption and retention of ^{109}Cd: Effects of cholecalciferol, calcium status and other variables. *J. Nutr.* **108,** 1812–1822 (1978).

K5. Kopeloff, L. M., Barrern, S. W., and Kopelott, N., Recurrent convulsive seizures in animals produced by immunologic and chemical means. *Am. J. Psychiatry* **98,** 881–902 (1942).

K6. Kopp, J. F., Occurrence of trace element in water. *Trace Subst. Environ. Health* **3,** 59 (1970).

K7. Kovalchik, M. T., Kaehny, W. D., Jackson, T., and Alfrey, A. C., Aluminum kinetics during hemodialysis. *J. Lab. Clin. Med.* **92,** 712–716 (1978).

K8. Kratochvil, B., Boyer, S. L., and Hicks, G. P., Effects of metals on the activation and inhibition of isocitric dehydrogenase. *Anal. Chem.* **39,** 45 (1967).

K9. Kushelevsky, A., Yagil, R., Alfasi, Z., and Berlyne, G. M., Uptake of aluminum ion by the liver. *Biomedicine* **25,** 59–60 (1976).

L1. Lai, J. C. K., Guest, J. F., Leung, T. K. C., Lim, L., and Davison, A. N., The effects of cadmium, manganese and aluminum on sodium-potassium-activated and magnesium-activated adenosine triphosphatase activity and choline uptake in rat brain synaptosomes. *Biochem. Pharmacol.* **29,** 141–146 (1980).

L2. Lapresle, J., Duckett, S., Galle, P., and Cartier, L., A case of aluminum encephalopathy in man. *C.R. Seances Soc. Biol. Ses Fil.* **169,** 282–285 (1975).

L3. LeGendre, G. R., and Alfrey, A. C., Measuring picogram amounts of aluminum in biological tissue by flameless atomic absorption analysis of a chelate. *Clin. Chem. (Winston-Salem, N.C.)* **22,** 53–56 (1976).

L4. Lichte, F. E., Hopper, S., and Osborn, T. W., Determination of silicon and aluminum in biological matrices by inductively coupled plasma emission spectrometry. *Anal. Chem.* **52,** 120–124 (1980).

L5. Liebig, G. F., Vanslow, A. P., and Chapman, H. D., Effects of aluminum on copper toxicity as revealed by solution—culture and spectrographic studies of citrus. *Soil Sci.* **53,** 341–352 (1942).

L6. Lipman, C. B., Importance of silicon, aluminum and chlorine for higher plants. *Soil Sci.* **45,** 189–187 (1938).

L7. Liu, J. H., and Huber, C. O., Aluminum determination by atomic emission spectrometry with calcium atomization inhibition titration. *Anal. Chem.* **50**, 1253–1256 (1978).
L8. Long, J. F., Nagode, L. A., Kindig, O., and Liss, L., Axonal swellings of purkinje cells in chickens associated with high intake of $1,25(OH)_2D_3$ and elevated dietary aluminum including x-ray microanalysis. *Neurotoxicology* **1**, 111–120 (1980).
L9. Lotz, M., Zisman, E., and Bartter, F. C., Evidence for phosphorus-depletion syndrome in man. *N. Engl. J. Med.* **278**, 409–415 (1968).
M1. Matsumoto, H., Morimura, S., and Takahashi, E., Binding of aluminum to DNA of DNP in pea root nuclei. *Plant Cell Physiol.* **18**, 987–993 (1977).
M2. Matsumoto, H., Morimura, S., and Takahashi, E., Less involvement of pectin in the precipitation of aluminum in a pea root. *Plant Cell Physiol.* **18**, 325–335 (1977).
M3. Mayor, G. H., Keiser, J. A., Makdoni, D., and Ku, P., Aluminum absorption and distribution: Effect of parathyroid hormone. *Science* **197**, 1187–1189 (1977).
M4. Mayor, G. H., Sprague, S. M., Hourani, M. R., and Sanchez, T. V., Parathyroid hormone-mediated aluminum deposition and egress in the rat. *Kidney Int.* **17**, 40–44 (1980).
M5. McDermott, J. R., Smith, A. I., Ward, M. K., Fawcett, R. W. P., and Kerr, D. N. S., Brain-aluminum concentration in dialysis encephalopathy. *Lancet* **1**, 901–903 (1978).
M6. McDermott, J. R., Smith, A. I., Iqbal, K., and Wisniewski, H. M., Brain aluminum in aging and Alzheimer disease. *Neurology* **29**, 809–814 (1979).
M7. McKinney, T. D., Dewberry, F. L., Stone, W. J., and Alfrey, A. C., Dialysis dementia at the Nashville Veterans Administration Hospital (NVAH). *Abstr. Am. Soc. Nephrol.* **11**, 46A (1978).
M8. McLachlan, D. R. C., and DeBoni, U., Aluminum in human brain disease—an overview. *Neurotoxicology* **1**, 3–16 (1980).
M9. McLaughlin, A. I. G., Kazantzis, G., King, E., Teare, D., Porter, R. J., and Owen, R., Pulmonary fibrosis and encephalopathy associated with the inhalation of aluminum dust. *J. Ind. Med.* **19**, 253–263 (1962).
M10. Mehta, R. P., Encephalopathy in chronic renal failure appearing before the start of dialysis. *Can. Med. Assoc. J.* **120**, 1112–1114 (1979).
M11. Meredith, P. A., Moore, M. R., and Goldberg, A., Effects of aluminum, lead and zinc on deltaaminolaevulinic acid dehydratase. *Enzyme* **22**, 22–27 (1977).
M12. Meredith, P. A., Elliott, H. L., Campbell, B. C., and Moore, M. R., Changes in serum aluminum, blood zinc, blood lead and erythrocyte δ-aminolaevulinic acid dehydratase activity during haemodialysis. *Toxicol. Lett.* **4**, 419–424 (1979).
M13. Miller, C. A., and Levine, E. M., Effects of aluminum salts on cultured neuroblastoma cells. *Neurochemistry* **22**, 751–758 (1974).
M14. Mitchell, J., Pulmonary fibrosis in workers exposed to finely powdered aluminum. *Br. J. Ind. Med.* **18**, 10–20 (1961).
M15. Moss, A. H., Solomon, C. C., and Alfrey, A. C., Elevated plasma adenine nucleotide levels in chronic renal failure and their possible significance. *Proc.—Clin. Dial. Transplant Forum* **9**, 184–188 (1979).
N1. Nagyvary, J., and Bradbury, E. L., Hypochlosterolemic effects of Al^{3+} complexes. *Biochem. Biophys. Res. Commun.* **2**, 592–598 (1977).
N2. Navia, J. M., Effects of minerals on dental caries. *Adv. Chem. Ser.* **94**, 123–160 (1970).
O1. Ondreicka, R., Ginter, E., and Kortus, J., Chronic toxicity of aluminum in rats and mice and its effect on phosphorus metabolism. *Br. J. Ind. Med.* **23**, 305–312 (1966).
O2. Ondreicka, R., Kortus, J., and Ginter, E., Aluminum, its absorption, distribution, and effects on phosphorus metabolism. *In* "Intestinal Absorption of Metal Ions, Trace Ele-

ments and Radionuclides" (S. C. Skoryna and D. Waldron, eds.), p. 293. Pergamon, Oxford, 1971.
P1. Parkinson, I. S., Ward, M. K., Feest, T. G., Fawcett, R. W. P., and Kerr, D. N. S., Fracturing dialysis osteodystrophy and dialysis encephalopathy. An epidemiological survey. *Lancet* 1, 406–409 (1979).
P2. Patocka, J., The influence of Al on cholinesterase and acethycholinesterase activity. *Acta Biol. Med. Ger.* 26, 845–846 (1971).
P3. Perl, D. P., and Brody, A. R., Alzheimer's disease: X-ray spectrometric evidence of aluminum accumulation in neurofibrillary tangle-bearing neurons. *Science* 208, 297–299 (1980).
P4. Pierides, A. M., Edwards, W. G., Jr., Cullum, U. X., Jr., McCall, J. R., and Ellis, H. A., Hemodialysis encephalopathy with osteomalacic fractures and muscle weakness. *Kidney Int.* 18, 115–124 (1980).
P5. Platts, M. M., Goode, G. C., and Hislop, J. S., Composition of the domestic water supply and the incidence of fractures and encephalopathy in patients on home dialysis. *Br. Med. J.* 2, 657–660 (1977).
R1. Recker, R. R., Blotcky, A. J., Leffler, J. A., and Rack, E. P., Evidence for aluminum absorption from the gastrointestinal tract and bone deposition by aluminum carbonate ingestion with normal renal function. *J. Lab. Clin. Med.* 90, 810–815 (1977).
R2. Rozas, V. V., Port, K. F., and Rutt, W. M., Progressive dialysis encephalopathy from dialysate aluminum. *Arch. Intern. Med.* 138, 1375–1377 (1978).
S1. Salvadeo, A., Minoia, C., Segagni, S., and Villa, G., Trace metal changes in dialysis fluid and blood of patients on hemodialysis. *Int. J. Artif. Organs* 2, 17–21 (1979).
S2. Schaver, C. G., Pulmonary changes encountered in employees engaged in the manufacture of aluminum abrasives: Clinical and roentgenologic aspects. *Occup. Med.* 5, 718–728 (1948).
S3. Schreeder, M. T., Dialysis encephalopathy. *Arch. Intern. Med.* 139, 510–511 (1979).
S4. Shore, D., King, S. W., Kaye, W., Forrey, E. F., Winfrey, H. J., Potkins, S. G., Weinberger, D. R., Savory, J., Wills, M. R., and Wyatt, R. H., Serum and cerebrospinal fluid aluminum and circulating parathyroid hormone in primary degenerative (senile) dementia. *Neurotoxicology* 1, 55–63 (1980).
S5. Smith, C. M., DeLuca, H. F., Tanaka, Y., and Mahaffee, K. R., Stimulation of lead absorption by vitamin D administration. *J. Nutr.* 108, 843–847 (1978).
S6. Smith, D. B., Lewis, J. A., Burks, J. S., and Alfrey, A. C., Dialysis encephalopathy in peritoneal dialysis. *JAMA, J. Am. Med. Assoc.* 244, 365–366 (1980).
S7. Sommer, A. L., Studies concerning the essential nature of aluminum and silicon for plant growth. *Chem. Abstr.* 21, 2917 (1927).
S8. Sorenson, J. R. J., Campbell, I. R., Tepper, L. B., and Lingg, R. D., Aluminum in the environment and human health. *Environ. Health Perspect.* 8, 3–95 (1974).
S9. Spencer, H., Lewin, I., Belcher, M. J., and Scamachson, J., Inhibition of radiostrontium absorption. *Int. J. Appl. Radiat. Isot.* 20, 507–516 (1969).
S10. Spencer, H., Wiatrowski, E., Osis, D., and Norris, C., Effect of aluminum containing antacids on fluoride metabolism in man. *J. Dent. Res.* 56, B131 (1977).
S11. Spencer, H., Kramer, L., Norris, C., and Wiatrowski, E., Effect of aluminum hydroxide on fluoride metabolism. *Clin. Pharmacol. Ther.* 28, 529–535 (1980).
S12. Stone, C. J., McLaurin, D. A., Steinhagen, W. H., Cavender, F. L., and Haseman, J. K., Tissue deposition patterns after chronic inhalation exposures of rats and guinea pigs to aluminum chlorhydrate. *Toxicol. Appl. Pharmacol.* 49, 71–76 (1979).
T1. Tipton, I. H., Stewart, P. L., and Martin, P. G., Trace elements in diets and excreta. *Health Phys.* 12, 1683–1689 (1966).

T2. Tipton, I. H., Stewart, P. L., and Dickerson, J., Patterns of elemental excretion in long term balance studies. *Health Phys.* **16**, 455–462 (1969).
T3. Trapp, G. A., Studies of aluminum interaction with enzymes and proteins—the inhibition of hexokinase. *Neurotoxicology* **1**, 89–100 (1980).
T4. Tsukamoto, Y., Iwanami, S., and Marumo, F., Disturbances of trace element concentrations in plasma of patients with chronic renal failure. *Nephron* **26**, 174–176 (1980).
V1. Valkovic, V., Is aluminum an essential element for life. *Origins Life* **10**, 301–305 (1980).
V2. Versieck, J., and Cornelis, R., Measuring aluminum levels. *N. Engl. J. Med.* **302**, 468 (1980).
V3. Viola, R. E., Morrison, J. F., and Cleland, W. W., Interaction of metal (III)-adenosine 5′-triphosphate complexes with yeast hexokinase. *Biochemistry* **19**, 3131–3137 (1980).
W1. Ward, M. K., Ellis, H. A., Feest, T. G., Parkinson, I. S., Kerr, C. N. S., Herrington, J., and Goode, G. L., Osteomalacic dialysis osteodystrophy: Evidence for a water-borne aetiological agent, probably aluminum. *Lancet* **1**, 841–845 (1978).
W2. Wasserman, R. H., Studies on vitamin D_3 and the intestinal absorption of calcium and other ions in the rachitic chick. *J. Nutr.* **77**, 69–80 (1962).
W3. Wisniewski, H. M., and Iqbal, K., Aluminum-induced neurofibrillary changes: Its relationship to senile dementia of the Alzheimer's type. *Neurotoxicology* **1**, 121–124 (1980).
W4. Wisniewski, H. M., Sturman, J. H., and Shek, J. W., Aluminum chloride induced neurofibrillary changes in the developing rabbit: A chronic animal model. *Ann. Neurol.* **8**, 479–490 (1980).
W5. Wolf, A., Graf, H., Pinggera, W. F., Stummvoll, H. K., and Meisinger, V., Serum aluminum and continuous ambulatory peritoneal dialysis. *Ann. Intern. Med.* **92**, 130–131 (1980).
W6. Worker, N. A., and Migicovsky, B. B., Effect of vitamin D on the utilization of beryllium, magnesium, calcium, strontium, and barium in the chick. *J. Nutr.* **74**, 490–494 (1961).
Y1. Yase, Y., the Role of aluminum in CNS degeneration with interaction of calcium. *Neurotoxicology* **1**, 101–109 (1980).
Z1. Zumkley, H., Bertram, H. P., Lison, A., and Ernst, M., Oral administration of magnesium and aluminum during renal insufficiency. *Trace Subst. Environ. Health* **12**, 248 (1978).
Z2. Zumkley, H., Bertram, H. P., Lison, A., Knoll, O., and Losse, H., Aluminum, zinc, and copper concentrations in plasma in chronic renal insufficiency. *Clin. Nephrol.* **12**, 18–21 (1979).

CLINICAL CHEMISTRY OF THIAMIN

Richard E. Davis and Graham C. Icke

Department of Haematology,
Royal Perth Hospital,
Perth, Western Australia

1. Introduction 93
   1.1. History 93
   1.2. Nomenclature 94
2. Chemistry and Biochemistry 95
   2.1. Chemistry 95
   2.2. Biochemistry 100
3. Methods for the Assessment of Thiamin Status 107
   3.1. Methods for the Direct Measurement of Thiamin 107
   3.2. Indirect Measurement of Thiamin 111
4. Clinical Chemistry 112
   4.1. Thiamin and Alcoholism 112
   4.2. Wernicke–Korsakoff Syndrome 114
   4.3. Subacute Necrotizing Encephalomyelopathy (Leigh's Disease) 117
   4.4. Cerebrocortical Necrosis in Cattle and Sheep 118
   4.5. Megaloblastic Anemia 119
   4.6. Thiamin-Responsive Maple Syrup Urine Disease 121
   4.7. Beriberi 121
   4.8. Other Clinical States Which May Be Associated with a Change in Thiamin Status 123
   References 130

1. Introduction

1.1. History

Thiamin is an essential nutrient for both plants and animals. In plants, it is synthesized in the leaves and transported to the roots, where it controls growth and appears to have a hormone-like action.

In animals, thiamin has an essential role in carbohydrate metabolism, and there is some evidence of it also being concerned with protein (C1) and lipid (V1) biosynthesis.

The vitamin was first isolated from rice polishings in 1926 by Jansen and Donath (J1). Thiamin has been and remains of special interest among the deficiency diseases. Early work on beriberi linked it to a deficiency state, and when the vitamin was eventually isolated it became clear that a deficiency of thiamin was responsible. This deficiency syndrome had been recognized as early as 1611 by the then Governor-General of the Dutch East Indies who wrote about beriberi as a disease that caused paralysis of the hands and feet (J2). However, it was not until 1885 that Takaki, a surgeon general in the Japanese Navy, recognized that the disease could be eliminated by the addition of fish, meat, and vegetables to the rice diet (G1). Until this time beriberi was a relatively uncommon disease, but with the introduction of highly milled rice, the incidence increased alarmingly and remains a problem in some countries where rice is the staple diet.

The chemical structure of thiamin was determined by Windaus in 1931 (W1) and this was followed by synthesis of the vitamin in 1936 by Williams and Cline (W2), which led to commercial production on a large scale. Like pyridoxine, its production has constantly increased while the cost of manufacture has steadily decreased (W3). Much has been written on the history of thiamin and this is referred to in some of the many reviews that have been published over the years (F1, L1, S1, V2, W3).

1.2. Nomenclature

In the past, vitamins of the B group have been given a number; for example, thiamin was designated B_1. This resulted in some confusion and it is becoming more usual now to ignore the letter–symbol terminology and to refer to the vitamin as thiamin. The spelling of "thiamin" without an "e" is recommended by the International Union of Nutrition Sciences Committee on Nomenclature and the Committee on Nomenclature of the American Institute of Nutrition, as adopted by the *Journal of Nutrition* in 1976 (A1). These bodies also recommended that the designations "vitamin B_1" and "aneurin(e)" no longer be used. However, it may be difficult to abandon the long-held concept of B-group vitamins.

In man, the coenzyme forms of the vitamin are phosphorylated and are present as monophosphates, pyrophosphates (cocarboxylase), and triphosphates. The vitamin is available as a pharmaceutical preparation, usually in the form of thiamin hydrochloride. Thiamin is water and alcohol soluble and in recent years a number of lipotropic forms have been synthesized (e.g., thiamin tetrahydrofurfuryl disulfide), and these will be referred to later.

Since the isolation of the vitamin more than 50 years ago, it has interested biochemists, physicians, and the general public to the extent that a large number of references appear each year in the scientific literature. Like some other B-group vitamins, the biological function of thiamin is widespread and it is easy to implicate it in a variety of disease states in both animals and man. Because the vitamin is cheap to produce and appears to have no harmful effects, it is found in a wide range of multivitamin preparations and so-called tonics. Until recently, the vitamin has been difficult to measure in biological fluids and this has made it difficult to assess some of the claims made for its therapeutic effectiveness. This article attempts to synthesize the most recent literature on the subject and to collate the information into a form which will stimulate both clinical chemists and physicians to take a greater interest in this important field.

2. Chemistry and Biochemistry

2.1. CHEMISTRY

The chemistry and biochemistry of thiamin have been the subject of a number of reviews (J3, L1, N1, V2) and will not be dealt with in depth here. Although thiamin is not synthesized by higher animals, it is produced by plants and some microorganisms. Synthesis by microorganisms is interesting because some organisms, such as *Absidia repens*, are autotrophic, while *Mucor ramannianus* requires the presence of the thiazole ring. *Rhodotorula rubra* requires only the pyrimidine moiety for synthesis, while *Phycomyces* requires both the thiazole ring and pyrimidine (S2).

Although it is clear that plants make a significant contribution to thiamin nutrition in man, the significance of bacterial synthesis of the vitamin has not received much attention. Wostmann *et al.* (W4) calculated that in the rat, bacterial production of thiamin in the cecum amounted to between 6 and 10 ng/day. Thiamin synthesized by bacteria in the large bowel is unlikely to be available for absorption except in animals practicing coprophagy. However, in ruminants, bacterial synthesis of thiamin is well able to meet the animals' requirement.

Thiamin is composed of pyrimidine and thiazole moieties which are joined by a methylene bridge. Its structure was established in the early 1930s as 3-(2'-methyl-4'-amino-5'-pyrimidylmethyl)-5-(2-hydroxyethyl)-4-methylthiazole (W1, W2) (Fig. 1a).

In the original synthesis of the vitamin (W2), the pyrimidine and thiazole moieties were formed separately followed by the coupling of the two fractions. The 4-hydroxypyrimidine was synthesized in a reaction between acetamidine and a derivative of ethyl-2-ethoxyl-1-formylpropionate. The 4-hydroxypyrimidine was converted in a series of steps to 2-methyl-4-amino-5-bromomethylpyrimidine. The thiazole moiety was synthesized by reacting

a

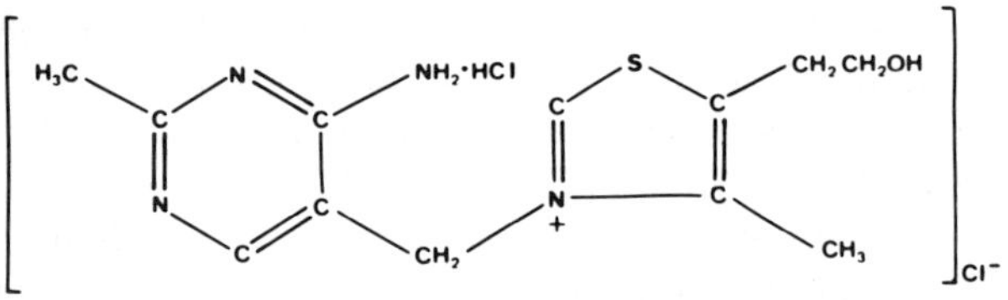

THIAMIN HYDROCHLORIDE

b

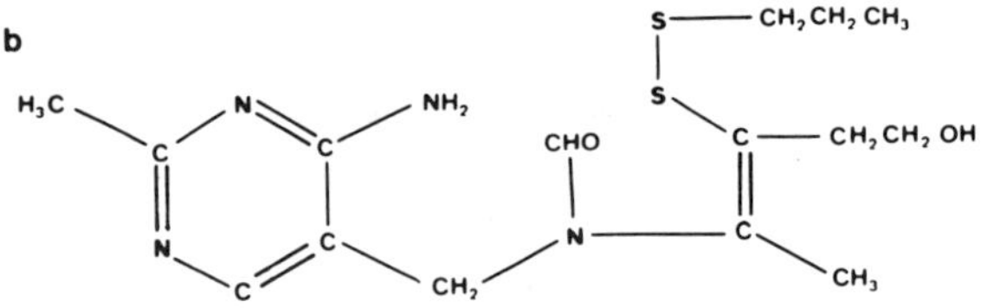

THIAMIN PROPYL DISULFIDE

c

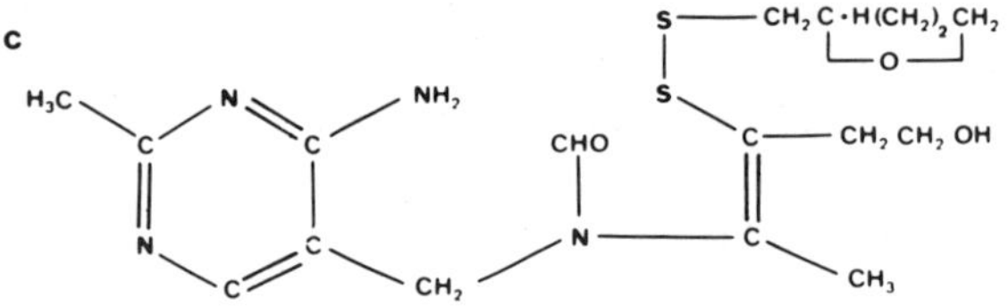

THIAMIN TETRAHYDROFURFURYL DISULFIDE

FIG. 1. Chemical structure of thiamin hydrochloride and two allithiamins.

3-acetyl-3-chloropropanol with thioformamide. An alternative approach was described by Todd and Bergel (T1), who synthesized the pyrimidine with a suitable side chain which could be reacted with a second compound to form the thiazole ring. These workers condensed ethyl-1-ethoxylmethylene-1-cyanoacetate with acetamide to form 5-cyano-4-hydroxy-2-methylpyrimidine. This in turn was converted to 2-methyl-4-amino-5-aminomethylpyrimidine, which was reacted with potassium dithioformate to yield 2-methyl-4-amino-5-thioformamide methylpyrimidine. Condensation of this compound with 3-acetyl-3-chloropropanol yielded thiamin.

Both of these reactions gave worthwhile amounts of thiamin and were used for commercial production.

2.1.1. *Thiamin Antagonists*

These can be divided into three categories: (1) those compounds that are able to displace thiamin but are in themselves without coenzyme activity, (2) the thiaminases which have the ability to cleave the thiamin molecule, and (3) agents with direct antithiamin activity, such as tannic acid.

There are a considerable number of agents that are able to displace thiamin, and only a few examples will be dealt with here. They are not naturally occurring.

Pyrithiamin is a potent thiamin analog, first synthesized in 1941 (T2), and animal experiments provided convincing evidence of its antivitamin activity (W5); other analogs are oxythiamin, 2-methylthiamin, 2'-butylthiamin, 2'-butylpyrithiamin, oxypyrithiamin, and 2'-ethylthiamin. All of these analogs contain the hydroxyethyl group present in thiamin but differ in other features. The biological activity of thiamin is dependent on the 2-position of the 2-methenyl group remaining unsubstituted (R1), and the analogs take advantage of this.

Thiaminase is a naturally occurring substance found in some shellfish, in certain parts of raw freshwater and saltwater fish, in certain plants, mainly ferns, and in three species of bacteria, *Bacillus thiaminolyticus*, *Bacillus aneurinolyticus*, and *Clostridium thiaminolyticum* (F2).

There are two forms of this enzyme, thiaminase I [thiamin:base 2-methyl-4-aminopyrimidine-5-methenyltransferase (EC 2.5.1.2)] and thiaminase II [thiamin hydrolase (EC 3.5.99.2)] (F2, L1). Thiaminases act by cleaving the pyrimidine from the thiazole ring. Thiaminase I behaves differently in that it mediates in a base-exchange reaction which requires a basic cosubstrate. Thiaminase II is found only in certain organisms such as *B. aneurinolyticus*.

Thiaminases have been shown to be responsible for cerebrocortical necrosis of cattle and sheep, a disease of the central nervous system (E1, E2). Untreated affected animals die in 2–6 days following the onset of clinical symptoms. The disease is caused by a deficiency of thiamin brought about by a thiamin-destroying enzyme in the rumen of affected animals (L2). Bacterial thiaminase has been found in the intestine of man, and a study in the Japanese district of Nilgata indicated that about 3% of the population had thiaminase disease (F2). However, in man, thiaminases have seldom been proved to be the cause of significant clinical thiamin deficiency. The classic example of a thiamin deficiency caused by thiaminase is that of Chastek paralysis, a polyneuritic disease of foxes that are fed on carp (F2). The disease also affects mink when they are fed a diet of uncooked fish. A study of these animals has shown that fish comprises only about one-third of their natural diet (H1). When commercial mink-ranch diets contain fresh fish on only 2 or 3 days/week, the animals appear able to obtain adequate thiamin from the diet on the days when it does not contain the thiaminase-rich fish (L3).

Bracken fern (*Pteridium aquilinum*) contains a powerful thiaminase I and is the cause of the disease "bracken staggers" in horses. Bracken poisoning has also occurred in pigs. Rats fed a complete diet mixed with dried bracken-frond powder were found to develop thiamin deficiency (E1). Thiaminase I has also been found in the rock fern "Nardoo" (*Marsilea drummondii*). This plant is found in many of the river watercourses in Eastern Australia (E3).

Tannic acid appears to have an appreciable antithiamin effect and Hilker *et al.* (H2) noted that tea, which may be high in tannic acid, depending to some extent on the method of preparation, had antithiamin properties. In communities where fermented tea leaves are chewed, there was found to be a reduction in available thiamin. Betel nuts also contain a significant concentration of tannic acid, and people who chew them have been shown to have a biochemical deficiency of thiamin in the presence of an adequate dietary intake of the vitamin. Abstention from betel nut chewing resulted in a significant improvement in biochemical thiamin status (V3). It has not been shown that the tannic acid in betel nuts is responsible for its antithiamin activity, but results of the study on the effect of tannic acid on thiamin make this a plausible explanation (K1).

Studies on the nature of the tannic acid effect have shown that following incubation for 30 minutes with thiamin, structural changes take place. The reaction was biphasic, and the first phase of the reaction occurred very quickly and was oxygen independent; during this phase more than 30% of the vitamin disappeared. The second phase was relatively slow and was oxygen dependent, and at the end of 3 hours only 8% of the original free thiamin was detected. Ascorbic acid was found to inhibit the reaction and "protected" the thiamin. This protective effect was not due to changes in pH (R2).

2.1.2. *Allithiamins*

These compounds were first discovered by Fujiwara and Watanabe (F3). They observed that extracts from certain plants, particularly the juice from garlic (*Allium sativum* L), when heated with thiamin at 60°C, resulted in a loss of the characteristic thiochrome reaction, suggesting that the thiamin had in some way become inactivated. However, the mixture when given to thiamin-deficient animals was found to exert full biological activity. Thiamin treated in this manner is changed to an alkyl disulfide, and this may also be effected by using the juice from plants such as onion (*Allium cepa* L), garden shallot (*Allium Bakeri regal*), Japanese leek (*Allium tuberosum rottl*), and many others. Prior treatment of the extracts with allinose from Cruciferae such as raddish or cabbage resulted in a good yield of thiamin methyl disulfide (F4). Following this work, the allithiamins have been produced synthetically and the range of products includes not only the alkyl and propyl disulfides, but also S-acyl (M1), O,S-dibenzoyl, O,5-diacetyl (M1), and cyclocarbothiamin derivatives (M2).

These compounds are lipotropic and less soluble in water than thiamin HCl. They are rapidly absorbed from the intestine and appear to bypass the absorption–control mechanism (B1). When given orally, urinary excretion in the form of free thiamin and not allithiamin occurs. There is some evidence that thiamin propyl disulfide is retained in the body better than the conventional forms of the vitamin (K2) and that higher concentrations of it can be obtained within the red blood cells. In the case of thiamin tetrahydrofurfuryl disulfide, following absorption, it enters the red cells and is then reduced catalytically in the presence of glutathione; the free thiamin is then slowly released into the circulation (M3). It has been shown that the propyl disulfide compound readily penetrates nerve tissue, CSF, and eyeball (L1) and also exerts a protective effect against cyanide poisoning. Mice protected with thiamin propyl disulfide and then injected with 150 mg of potassium cyanide per 10 g of body weight had an 11.7% mortality, compared with 70.6% of control mice not pretreated with the allithiamin. The protective action was thought to be due to the S—S bond in the molecule. Although other thiamin derivatives such as S-acyl thiamin were found to be absorbed across the gut mucosa, they did not readily penetrate into erythrocytes like the propyl disulfide. Neither did they have any demonstrable activity in protecting mice from the effects of cyanide.

Only two allithiamin derivatives have received significant clinical trials, thiamin propyl disulfide and thiamin tetrahydrofurfuryl disulfide. These compounds are formed by destroying the thiazolium ring structure of the thiamin molecule and replacing it with a disulfide with a different prosthetic group attached (Fig. 1b and c).

The propyl disulfide appears to be as effective clinically as the tetrahydrofurfuryl disulfide. It has been claimed that the tetrahydrofurfuryl disulfide is preferred because the propyl disulfide imparts a strong garlic odor to the breath of patients (L1).

Orally administered allithiamins have been recommended for the prophylaxis and treatment of thiamin deficiency. They have essentially the same biological properties as parenterally administered water-soluble thiamin. They appear to be more efficiently utilized and have not produced any recognizably untoward effects following long-term administration (B1, T3).

2.2. BIOCHEMISTRY

Thiamin is phosphorylated to the pyrophosphate coenzyme form which is essential to the oxidative decarboxylation of α-keto acids to aldehydes using pyruvate and α-ketoglutarate as substrates. Pyruvate is metabolized to acetyl coenzyme A, and α-ketoglutarate to succinyl coenzyme A. Coenzyme A is condensed with oxaloacetate and is then catalyzed by citrate synthase to form citrate, the first step in the citric acid cycle.

Thiamin has a major role as the cofactor for transketolation reactions in the pentose phosphate cycle, which is necessary for the provision of pentose phosphate for nucleotide synthesis and for the reduced NADP required for various synthetic pathways. The measurement of erythrocyte transketolase is one of the laboratory tests for thiamin deficiency.

Apart from its coenzyme function, thiamin has an important role in the nervous system, where it is present in the form of a triphosphate as well as the diphosphate. The vitamin is essential for normal brain metabolism, but the precise molecular events underlying the neurological changes associated with thiamin deficiency are still poorly understood. Spector has examined the transport of thiamin into the cerebrospinal fluid, brain, and choroid plexus in rabbits (S3). He found that with a normal total plasma thiamin concentration, less than 5% of total thiamin entered these tissues by means of simple diffusion. Using ^{35}S-labeled thiamin, he calculated that the relative turnover of total thiamin in the choroid plexus, whole brain, and cerebrospinal fluid was 5, 2, and 14%/hour, respectively. This was based on the penetration of ^{35}S-labeled thiamin into the tissues following an intravenous injection. Spector (S3) concluded that the entry of total thiamin into the brain and cerebrospinal fluid is regulated by a saturable transport mechanism. He suggested that the locus of this system may be, in part, in the choroid plexus.

Barchi and Viale (B2) studied the activation of membrane-associated thiamin triphosphatase from rat brain and found that it required a divalent cation such as Mg^{2+}. An excess of free Mg^{2+} inhibited thiamin triphosphatase activity, but excess thiamin triphosphate had no effect on the membrane-associated enzyme. Barchi and Braun (B3) had earlier shown in rats that membrane-associated thiamin triphosphatase was different from the less active soluble thiamin triphosphatase.

In another study (I1) on the action of chlorpromazine on rat brain thiamin phosphatase, it was shown that chlorpromazine at concentrations of 0.25–1.0 mM caused a marked decrease in the activity of microsomal and soluble thiamin triphosphate and a marked increase of microsomal thiamin diphosphate activity.

Seltzer and McDougal (S4) measured thiamin pyrophosphate and α-ketoglutarate during the development of pyrithiamin-induced thiamin deficiency in four brain regions of the mouse. Although the animals lost weight, they showed no signs of neurological involvement. Thiamin pyrophosphate fell to 10% of normal in all four brain regions. The animals became ataxic during days 7 to 8. Of the four mouse brain regions studied, it was thought that two would be relatively resistant to thiamin depletion. These two regions showed little further decrease in thiamin pyrophosphate levels, but the susceptible regions decreased to 5% of normal. The α-ketoglutarate showed a marked increase in the susceptible regions (i.e., vestibular nucleus and thalmus) compared with the resistant areas, the cerebral cortex and cerebellar hemispheres. These workers concluded that the areas studied had large reserves of coenzyme and that resistant areas have a small but important advantage in their capacity to retain thiamin pyrophosphate during the last stages of depletion.

Thiamin has been thought to have a role in lipid elaboration. Geel and Dreyfus (G2) studied the brain lipid composition of young rats exposed to dietary thiamin deficiency. These rats developed neurological signs and marked diminution of growth, while controls showed no such changes. No changes in brain lipids of either whole brain or selected brain areas were observed. As part of the same study, undernutrition produced a highly significant depression of all brain lipids. From the results of their work on the regulation of thiamin diphosphatase activity in rat brain microsomes by lipids, Baba et al. (B4) concluded that the phospholipids in microsomes provide effective protection for thiamin diphosphatase against heat inactivation. Studies on cultured glial cells have shown that when these cells are deprived of thiamin, their rate of fatty acid synthesis was only 13% of the rate found in thiamin-supplemented cells. The levels of fatty acid synthetase and acetyl coenzyme A carboxylase, two important enzymes in the pathway, were also reduced. There was also a reduction in cholesterol biosynthesis in the thiamin-depleted cells. The disturbance of fatty acid synthesis could be reversed with as little as 0.01 μg/ml of thiamin. In contrast, cultured neuronal cells showed only a slight reduction in lipid synthesis when deprived of thiamin (V1).

The effect of thiamin-deficiency encephalopathy on the cholinergic neurotransmitter function has been studied in rat brain by Vorhees et al. (V4). They found no significant changes in acetylcholine levels in thiamin-deficient animals. However, they did find some regional changes, with a decrease in acetylcholine utilization in the midbrain, medulla–pons, diencephalon, and cortex. The reason for this reduced utilization was uncertain, but it was thought that thiamin may participate in neuronal membrane function and in synaptosomal acetylcholine release. Considerable work has been done in this area and there is increasing evidence that thiamin has a function in membrane transport. However, much remains to be done before there is a clear understanding of the part played by thiamin in mammalian neurochemistry.

2.2.1. Absorption

The absorption of thiamin appears to be controlled by a rate-limiting process (T4), and it has been shown that an increase in the oral intake of the vitamin above 10 mg does not significantly increase the concentration appearing in the blood or the quantity excreted in the urine. There remains some uncertainty about the amount of thiamin that can be absorbed from a single oral dose. In a series of studies, Thomson and Leevy found that there was a maximum absorption of approximately 8.3 mg following a single dose of thiamin hydrochloride (T5). The absorption tests were based on the use of an oral dose of [^{35}S]thiamin hydrochloride ranging from 1.0 to 20.0 mg, and this was given with an intravenous flushing dose of 200 mg of nonradioactive thiamin. However, in an earlier study Thomson *et al.* reported a maximum absorption of 4.77 mg in normal subjects following a 20-mg oral dose (T4).

Unlike other B-group vitamins, an oral dose of 10 mg or less did not significantly change the serum pattern of radioactivity in the presence of a 200-mg flushing dose. Using microbiological assay techniques, only a small rise in the serum thiamin concentration was observed following a 10-mg oral load of the vitamin (Fig. 2).

In the rat, absorption of thiamin takes place mainly in the duodenum and proximal jejunum; little thiamin is absorbed in the remainder of the small intestine (S5). In man, only limited studies have been possible, but in segments of small bowel which were obtained at autopsy, absorption was found to occur mainly in the jejunum and ileum (personal observation).

It has been suggested that in addition to an active transport system required for the absorption of small quantities of thiamin, large doses may be absorbed by passive diffusion (R3). However, Morrison and Campbell (M4) found that in normal male subjects receiving adequate diets, the percentage of thiamin excreted in the urine following an oral dose decreased markedly when the dose exceeded 2.5 mg. Increasing the dose from 2.5 to 20 mg increased the amount of thiamin excreted in the urine by only 0.2 mg. This was in contrast to other B-group vitamins such as riboflavin, the excretion of which remained a constant percentage of the oral dose.

A number of studies on absorption control have been made, but these have failed to provide any clear information on the mechanism involved. Lazarov (L4) found that in adrenalectomized rats, absorption of [^{35}S]thiamin was significantly reduced. Corticosteroids appear to play some part in thiamin absorption. Injection of prednisolone restored absorption, but not to the same extent as that observed in nonadrenalectomized control rats.

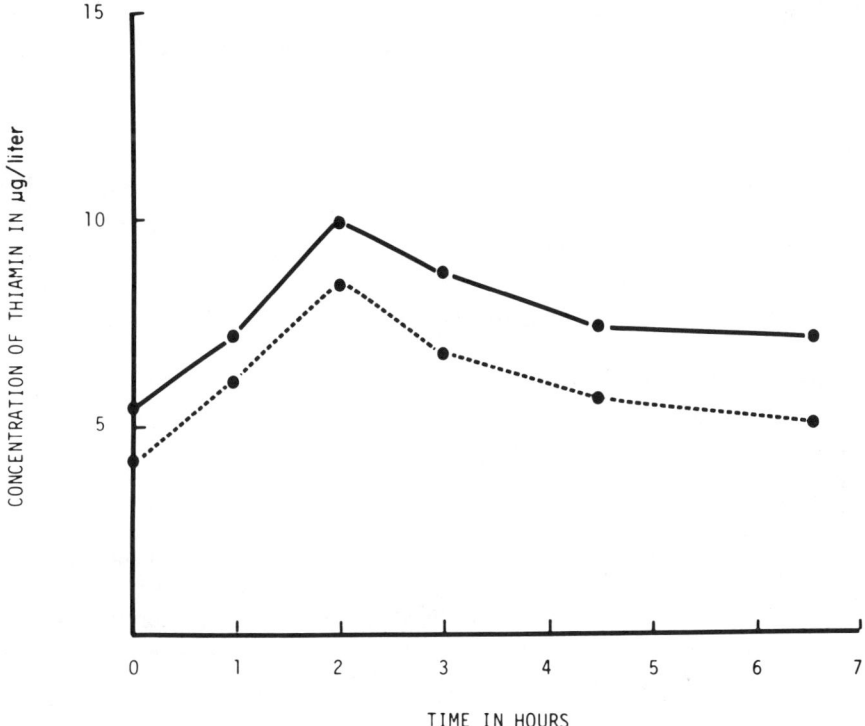

FIG. 2. Rise in serum thiamin concentration in two subjects following a 10-mg oral load.

Lazarov (L4) also found that the concentrations of thiamin mono- and diphosphates in the intestinal mucosa and liver were lower in the adrenalectomized animals. The concentration of thiamin triphosphate was higher in the intestinal mucosa following adrenalectomy, but remained unchanged in the liver. This work suggests that the adrenal glands play a role in the phosphorylation of thiamin and may provide some clues with regard to the nature of the active transport mechanism.

Rindi and Ventura (R3) concluded that thiamin entry into the mucosal cells is linked with a carrier-mediated mechanism dependent either on phosphorylation–dephosphorylation coupling or, alternatively, on a system requiring metabolic energy and Na^+. The concept of phosphorylation–dephosphorylation was not supported by Komai et al. (K3); they found that chloroethylthiamin, a thiamin analog which has no effect on pyrophosphokinase, inhibited thiamin transport in the small intestine of rats. Moreover, these workers scarcely detected thiamin pyrophosphokinase in microvillous membranes.

Although there is agreement that absorption of thiamin is rate limited and that this process is saturable at 0.5 to 1.0 μM (D1, S5), the precise method of control is not known. Absorption of large oral doses of thiamin by passive diffusion appears to contribute only marginally to body stores.

2.2.2. Transport

Most of the vitamin present in serum appears to be bound to protein, predominately albumin. Therapeutic doses of thiamin result in rapid saturation of the binding sites and as with most other B-group vitamins, the excess free form is excreted in the urine. The affinity of the various esters for particular protein binders has received little attention. The transport of the vitamin across cell membranes has been the subject of intensive study and appears to be based on specific membrane-bound phosphatases. This permits certain cells which are particularly sensitive to thiamin depletion to effect a priority over less vital areas (B3, S3, S4).

2.2.3. Daily Allowance

Thiamin has an important role in the metabolism of glucose and the daily requirement for the vitamin is related to energy need, particularly that which is derived from carbohydrate. It has been found that 0.33 mg of thiamin is required for each 4400 kJ of energy requirement (F5). The Food and Nutrition Board of the National Research Council therefore recommends a thiamin intake of 0.5 mg/4400 kJ and considers that this will maintain a satisfactory vitamin/carbohydrate balance (N2). The recommendation pertains to adults and children of all ages, including infants, and provides for some excess of the vitamin, particularly in infants and very young children. However, little of the vitamin in excess of that actually required is likely to be absorbed in the normal infants because of the rate-limiting absorption control.

It has also been recommended that because elderly people may use thiamin less efficiently, their supplementary intake should be 1 mg/day regardless of their dietary intake. In pregnancy, an additional 0.2 mg/day has been recommended (N2).

It has been claimed, following a nutritional study of over 1000 dentists and their wives, that approximately 9 mg of thiamin may be designated as the "ideal" daily allowance. The investigators also claimed that the higher the thiamin intake, the healthier the individual (C2). From a general nutrition point of view, it is very unlikely that this goal could be achieved without thiamin supplementation. Such a recommendation would be valid only if thiamin played a major role in a number of metabolic pathways in addition to those already known.

2.2.4. Increased Demand

Conditions resulting in an increased demand for thiamin are rare. A well-known example is the unusual thiamin-responsive form of maple syrup urine disease (see Section 4.6).

2.2.5. Hyperalimentation

Despite the widespread use of intravenous hyperalimentation, there have been few studies to determine the micronutrient requirements. Kishi et al. (K4) recommended a vitamin combination providing 5 mg/day of thiamin. They found this to be a sufficient and safe level, and recommended that it be used in place of a standard combination of vitamins used for this purpose which provided 55 mg/day of thiamin.

2.2.6. Thiamin Content of Foodstuff

There is continuing concern about the apparent lack of thiamin in the average "western" diet, and in an attempt to overcome this, certain foodstuffs are artificially supplemented with the vitamin (for example, breakfast cereals, infant milk formulas, and some flours used in bread manufacture) (H3).

In trying to estimate the dietary intake of the vitamin, it is important to determine its concentration in the food as served, as losses during processing may be substantial.

In a major study of the thiamin content of food served to members of the United States Armed Forces, 174 food items served for breakfast, lunch, dinner, and brunch, representing all the meals served over a period of 14 consecutive days, were analyzed (D2). Twenty-two cereal-based items had a mean thiamin content of 3.34 mg/kg (range 0.58–8.98). Twenty-five meat, poultry, and fish items had a mean concentration of 2.94 mg/kg (range 0.37–13.30) and 23 vegetable items had a mean of 1.14 mg/kg (range 0.21–4.65). Cereal-based foods appeared to provide a substantial part of the daily thiamin requirement. They were found to have a higher average concentration than other foodstuffs and they are usually consumed in larger quantities. It was not stated whether any of the cereal foods were vitamin enriched.

In these studies, a microbiological assay using *Lactobacillus viridescens* as the test organism was used to assay the thiamin. Ang and Moseley (A2) have recently described a method using high-pressure liquid chromatography to measure thiamin in meat and meat products. Only a limited number of results were available using this method, but the mean content of five meat items was 1.58 mg/kg, which was below that found by Dong et al. (D2) in their much larger study.

Certain procedures used in food processing may lead to the loss of the vitamin; for example, canned Californian garbanzo beans show an improvement in color following a soaking in a solution of sodium bisulfite. However, if the concentration of bisulfite was raised by as little as 0.05%, there was an increase in the loss of the vitamin (L5). Ang *et al.* (A3) studied the loss of thiamin in frozen beef–soy patties and frozen–fried breaded chicken parts heated by various methods. It was found that chicken parts reheated by convection and held hot for 3 hours retained only 74% of their thiamin. Other methods of heating resulted in smaller losses of between 4 and 19%.

There are substantial differences in the thiamin content of various foodstuffs and also between different samples of the same food. Heating and various manufacturing procedures may result in a considerable loss of the vitamin.

2.2.7. *Light and Heat Stability of Thiamin*

The stability of the vitamin when heated appears to be largely pH dependent. In a series of experiments, Beadle *et al.* (B5) observed that heating for 1 hour in a boiling water bath in the pH range of 3 to 7 resulted in complete destruction of thiamin. At pH 7, there was 100% destruction of thiamin when heated in the presence of borate or acetate buffers, whereas phosphate buffer appeared to have some protective effect and only 40% of the vitamin was destroyed.

Ten percent of thiamin may be lost from milk during pasteurization and 30% or more may be lost by heat sterilization (K5). Light appears to have little effect on the vitamin; bottles of cow's milk left exposed to direct sunlight for up to 8 hours showed no significant loss of the vitamin (F6).

2.2.8. *Toxicity*

Thiamin is thought to be nontoxic in humans unless it is administered in doses which are many thousands of times larger than those required for optimal nutrition. This has resulted in very large doses of the vitamin being administered on the basis that as it is harmless, a large dose is likely to do more good. Nevertheless, under normal physiological conditions, it is impossible to substantially raise the serum level of the vitamin by oral administration because of a rate-limiting absorption mechanism. Thiamin is unique among the B-group vitamins in this respect, and this suggests that there may be good reason for avoiding significantly raised serum levels of the sort found following large parenteral doses of the vitamin. To date, there is no evidence except possibly in dependency syndromes that a 100-mg parenteral dose is any more effective than a 10-mg dose.

Thiamin given by injection has been responsible for reactions resembling anaphylactic shock in man. In animals, large intravenous doses of thiamin result in death from depression of the respiratory center. The lethal dose of the vitamin administered intravenously has been reported as 250 mg/kg in rats, 125 mg/kg in mice, 300 mg/kg in rabbits, and 350 mg/kg in dogs (N1). In a series of studies on dogs, Smith et al. (S6) observed that if the animals were given artificial respiration they could survive a thiamin load of up to 125 mg/kg, which produced blood concentrations of up to 36.9 mg/100 ml. Without such support, doses resulting in blood levels between 7.2 and 10.0 mg/100 ml were invariably fatal.

Further studies are required to determine the effect of large parenteral doses of thiamin in man and whether, in fact, large doses offer any additional benefit.

3. Methods for the Assessment of Thiamin Status

There are numerous methods available for the separation and assay of thiamin and its esters in biological fluids and tissues, few of which have found widespread acceptance.

The majority of techniques for assessing the vitamin fall into the following categories: (1) direct measurement of blood levels, (2) measurement of the excretion rate of the vitamin, (3) measurement of the abnormal metabolic products resulting from a deficient state, and (4) the measurement of some other product dependent on the concentration of the vitamin in the body. All of these approaches to the assessment of thiamin have significant problems, either technical or physiological, and interpretation of results should not be made without an understanding of the technique employed. Due to the limitation of space, it is only intended here to review a few of the wide range of methods available and to discuss some of their individual advantages and disadvantages.

3.1. Methods for the Direct Measurement of Thiamin

3.1.1. *Biological Assays*

Biological assays using laboratory animals were the first methods to be developed, and these were mainly used before pure thiamin became available. One method adopted was to observe the curative effect of food extracts on neuropathies associated with thiamin deficiency in pigeons (K6). Other workers used methods based on the growth response of thiamin-deficient rats (F7). Young rats received a basal diet until their weight declined. A single dose of the material under study was administered orally and this caused a temporary gain in weight. The length of time the effect lasted was proportional to the amount of thiamin present in the test material. A sensitivity of as little as 1 µg has been claimed for these rat growth techniques. Other assay methods involving animals included the bradycardia test, in which the decrease in the heart rate of polyneuritic rats was measured. The duration of the temporary increase in heart rate after a single dose of thiamin was proportional to the amount of thiamin administered (E4).

Biological assays using animals are expensive and inconvenient and are now seldom used.

Several microbiological methods have been developed for the measurement of thiamin, the first of which used one of the lower fungi, *Phycomyces blakesleeanus,* to determine the thiamin content of foodstuffs (S7). The organism required a long incubation period of up to 2 weeks and this has made it unsatisfactory for use with clinical material.

Numerous other organisms have been used for the assay of the vitamin, including *Streptococcus salivarius* (N3), *Kloeckera brevis* (H4), *Lactobacillus viridescens* (D3), and *Ochromonas danica* (B6). Assays using these organisms have been found to be particularly useful for determining the thiamin content of foodstuffs, but they have not found widespread use in the assay of clinical material. Some have not been sufficiently sensitive, while others have required special technical skills which resulted in the assay functioning satisfactorily only in a few special laboratories. A method using *Lactobacillus fermenti* as the test organism was first described in 1944 for assaying the thiamin content of foodstuffs (S8). This method was modified by Fang and Butts (F8), and although it incorporated a number of improvements over earlier methods, it had the disadvantage of requiring prolonged heating of the samples at high temperatures; it had been reported earlier that thiamin was heat labile (B5, R4).

In 1980, Icke (I2) described an automated method for the assay of thiamin in serum and red cells using a streptomycin-resistant mutant strain of *L. fermenti* as the test organism. Streptomycin was added to the assay medium, and this avoided the need for sterilization or aseptic technique and permitted automation of the procedure. This automated method had a set-up capacity of 160 assays/hour, and following 16–18 hours incubation at 37°C, results could be read at a rate of 160/hour. The test was sensitive to 0.5 µg/liter of thiamin. The test organism was found to respond to both free and phosphorylated forms of the vitamin. However, the growth response of the organism to thiamin pyrophosphate was approximately 30% greater than that obtained with thiamin hydrochloride.

Results using the automated microbiological assay were found to be between 21 and 28% higher than those obtained by a modified thiochrome method (S9). It was subsequently shown that up to 30% of thiamin pyrophosphate activity was lost after heating at 100°C for 30 minutes, a prerequisite of the thiochrome method. No demonstrable loss of thiamin activity occurred using the automated assay, as heating was minimal for erythrocytes (90°C for 5 minutes), and for the assay of serum, no heating at all was required.

3.1.2. Chemical Methods

Hennessy and Cerecedo (H5) described a thiochrome method for measuring the vitamin based on the ferricyanide oxidation of thiamin to thiochrome, which exhibited an intense blue fluorescence in ultraviolet light. A thiochrome-interfering fluorescent substance called F2 (n-methylnicotinamide) has been described by some workers (M5, N4). Other extraneous substances in urine which may contribute to the fluorescence in the thiochrome test have also been reported (C3). Numerous modifications of this method have been proposed (D4, L1, P1, S9) despite the fact that the adequacy of the method in detecting thiamin deficiency has been questioned by nutritional-survey teams (A4). The main problem in using the thiochrome technique or one of its modifications for the assay of the vitamin in biological materials is that the low concentrations present have previously been beyond the sensitivity of the method. However, a recent modification of the thiochrome technique for the determination of endogenous thiamin and its phosphate esters in biological material has been proposed in which the ion-exchange chromatography is replaced by counter ion-exchange chromatography (T6, W6). This method was found suitable for measuring nanogram quantities of thiamin in biological material.

A colorimetric method based on the coupling of p-aminoacetophenone with thiamin to produce an insoluble purple-red compound was one of the earliest chemical methods described (M6, P2). However, uric acid and ascorbic acid, both of which are difficult to remove, interfere with the development of the color, and the method was not sufficiently sensitive for use with material having a low concentration of the vitamin.

Other direct biochemical methods for assessing thiamin deficiency include the formaldehyde–diazotized sulfanilic method (K7) and the bromothymol blue method (G3).

3.1.3. Urinary Loading Tests

A number of thiamin loading tests have been described (J4, L7, P3, S10). The most common procedure used is to measure the urinary excretion of thiamin following a test load of the vitamin.

Deficient subjects will usually excrete less than 20 µg in the 4 hours following the test dose. Considerable variation has been noted in the results from individuals tested under similar conditions. Baker and Frank (B7) suggested that divergencies could be partly explained by one or all of the following: (1) differences in absorption in the gastrointestinal tract, (2) variations in the renal threshold, (3) differences in the rate and volume of urine excreted, (4) inadequate emptying of the bladder, and (5) variations in the test dose administered.

Urinary load tests have been shown to be of little value in grading the severity of thiamin deficiency (K8).

3.1.4. Chromatographic Techniques

A number of chromatographic techniques have been used for the separation and determination of thiamin and its esters in animal tissue and in commercial vitamin preparations using thin-layer chromatography (B8) and classic ion-exchange chromatography (R5, S11, S12). Few have been found suitable for use as routine assay systems and none has found a place in the routine measurement of blood levels.

Recently, a method for the separation of thiamin phosphoric esters in tissue using a Sephadex ion-exchange column has been described (P4). The aim of this method was to obtain sharp separation between thiamin triphosphate and thiamin pyrophosphate. Using an acetate buffer (pH 3.8), it was demonstrated by electrophoresis that thiamin triphosphate, under the conditions of the test, had a negative charge, thiamin pyrophosphate was electroneutral or weakly positively charged, and thiamin monophosphate was positively charged. Hence, thiamin triphosphate was not affected by the negatively charged ionogen group of Sephadex and emerged first from the column, while thiamin pyrophosphate was only slightly affected and eluted next. Thiamin monophosphate and thiamin HCl combine strongly with the Sephadex and could only be eluted by using a high-ionic-strength buffer. Unfortunately, the thiamin triphosphate fraction was contaminated with interfering substances which hindered its quantitative determination by the thiochrome methods unless an additional purification on the ion-exchange resins was used.

3.1.5. High-Performance Liquid Chromatography (HPLC)

A method for the analysis of commercial vitamin preparations using HPLC was introduced in 1974 (C4). The system utilized an ultraviolet monitor which permitted measurements at either 254 or 280 nm. With this technique, an assay based on a single run could be completed in 10–15 minutes. Although this method was found to be accurate and reproducible for the milligram quantities of thiamin present in commercial vitamin preparations, it is doubtful as to whether the method would be sufficiently sensitive for clinical use.

Roser *et al.* (R6) combined a modified thiochrome method with high-performance liquid chromatography in order to determine urinary thiamin. The method was shown to be sensitive, reproducible, and specific for urinary thiamin.

A recent description of a method using high-performance liquid chromatography to detect thiamin and its phosphate esters suggested that quantities as low as 1 pmol (I3) could be detected after the conversion of the vitamin and its esters to the corresponding fluorescent thiochrome derivatives. Another report claimed a sensitivity of 0.05 pmol (K9), which suggests that these methods might be satisfactorily adapted for the measurement of serum thiamin levels.

3.1.6. *High-Voltage Paper Electrophoresis*

Because of the sensitivity of the thiochrome method to interfering substances, the use of high-voltage paper electrophoresis as an alternative method for the separation of thiamin was studied (P5). Its main use appeared to be in the assaying of thiamin in foodstuffs and pharmaceutical preparations.

A modified electrophoretic technique using a citrate buffer containing methanol, ethanol, and propanol was found to improve the elution of thiamin compounds from the electrophoresis paper (P6). For biological material, the proteins were first precipitated with perchloric acid and the sample was neutralized with potassium carbonate; the sample was then lyophilized in order to concentrate the thiamin content of the material.

3.2. INDIRECT MEASUREMENT OF THIAMIN

Horwitt and Kreisler (H6) devised a method for the determination of early thiamin-deficient states by the estimation of blood lactate and pyruvic acids after glucose administration and exercise. Other workers have reported that lactate and pyruvate blood levels were of little diagnostic value in mild thiamin deficiency or liver disease (F9).

The measurement of erythrocyte transketolase has been widely used for the assessment of thiamin deficiency (B9, D5).

Transketolase (EC 2.2.1.1) is an enzyme which occurs in red blood cells, liver, kidney, adrenal cortex, brain, skeletal muscle, and other tissue. In the transketolase reaction, the ketol group of appropriate keto sugars forms an active glycoaldehyde–enzyme intermediate with thiamin pyrophosphate, which is transferred to a suitable acceptor aldehyde. Consequently, transketolase is capable of functioning in a number of donor–acceptance systems. In mammalian tissues, the main action of transketolase is related to ribulose 5-phosphate metabolism (S13).

The technique involves incubating hemolyzed whole blood samples at 38°C in a buffered medium with an excess of ribose 5-phosphate (with and without excess thiamin pyrophosphate), and then determining the amount of ribose 5-phosphate used. Alternatively, the sedoheptulose 7-phosphate or the hexoses produced can be measured. Any increase in enzyme activity after the addition of thiamin pyrophosphate is referred to as the "thiamin pyrophosphate effect."

In a study of 20 normal individuals aged 20–40 years, Williams (W7) found that this effect ranged from 0 to 25%. In a similar study, Bayoumi and Rosalki (B10) reported a normal range of up to 23%. An increase in enzyme activity above these levels is accepted as indicating thiamin deficiency. This is still the most widely used test by clinical biochemists for the assessment of thiamin status.

Like so many indirect tests of vitamin nutrition, results can be difficult to interpret; transketolase activity may be inhibited in alcoholics and the addition of thiamin may not result in increased enzyme activity in these patients (L8).

The direct methods now available such as the automated microbiological assay provide a simple, accurate, and highly sensitive test. Results correlate well with the patients' clinical assessment, and for these reasons may be more acceptable to clinicians and nutritionists.

4. Clinical Chemistry

4.1. Thiamin and Alcoholism

Vitamin deficiency is a common finding in alcoholics for a variety of reasons. Alcoholics may eat nothing for days, and when they do eat, their diet is often high in carbohydrates and low in vitamins (V5). The alcohol provides even more carbohydrate, and since thiamin is actively involved in the metabolism of sugars, the stage is set for thiamin deficiency.

Since there are also metabolic pathways for ethanol in the tissues which require thiamin as a cofactor, it is probable that long-term ethanol ingestion would induce severe deficiency of the vitamin. Acetaldehyde, an intermediate in the metabolism of ethanol, is much more toxic than ethanol to transketolase (A5), and it may be that high levels of acetaldehyde play an important role in producing thiamin deficiency.

Alcoholics may also develop vitamin depletion syndromes because of defects in absorption. Intestinal malabsorption and pancreatic dysfunction were initially demonstrated in patients with alcoholic cirrhosis of the liver (B11, M7, S14). Similar abnormalities were later found in patients with chronic alcoholism who had minimal or no liver disease (S15, T7, T8). Hoyumpa et al. (H7) studied the effect of chronic ethanol administration on intestinal transport in rats. Their studies confirmed the inhibitory effect of ethanol on thiamin transport and showed that inhibition is dependent more on the ethanol concentration than on the duration of exposure to ethanol and that thiamin malabsorption in the rat may be intermittent. This has largely been confirmed by the work of Shaw et al. (S16), who found that in baboons, chronic alcohol feeding for up to 3 years had no effect on blood levels of thiamin or on urinary thiamin excretion.

The fact that the effects of ethanol on the small intestine are accompanied by functional changes has been observed by Hillman (H8), who also noted alterations in mucosal enzymes and in the subsequent absorption of many nutrients.

In addition to the direct effect of ethanol on thiamin transport, other nutritional factors related to ethanol consumption may also interfere with thiamin transport. Folate deficiency is a common occurrence in alcoholics (D6) and is frequently encountered in these cases in association with thiamin deficiency (I2). It has been reported that folate-deficient rats absorb thiamin less efficiently than control rats. Folate repletion reversed the thiamin malabsorption (H9).

Other workers have suggested that there may be impairment of tissue thiamin storage in the presence of alcohol (A5). Abe and Itokawa (A5) measured the thiamin content of brain and liver of ethanol-fed rats and found it to be lower than in control rats and they suggested the possibility that ethanol denatures the apoenzyme of transketolase. After treatment with thiamin, the urinary excretion of the vitamin was higher than in the controls, which may indicate that the tissue's ability to retain thiamin following ethanol administration is impaired.

Heaton (H10) has reported that 10–15% of chronic alcoholics develop cirrhosis of the liver and that up to 30% of hospitalized alcoholics may have cirrhosis. Kinetic studies have shown that alcohol-induced cirrhosis of the liver may impair the hepatic uptake or storage of vitamins. This may be due to the failure of binding sites within the damaged liver cells or to the failure of the liver to elaborate the specific enzymes required for converting the vitamin into an appropriate coenzyme form (L9). The delayed conversion of thiamin to thiamin pyrophosphate has been thought to be responsible for the slow clinical response seen in some patients with Wernicke's encephalopathy (C3), and it may be that this is the result of a reduction in available thiamin phosphorylases.

Malabsorption is found in many patients with cirrhosis (B11), but this may not be severe enough to cause serious impairment of vitamin absorption except in a few patients.

The heart and brain may also undergo pathological change as a result of alcoholism and its associated thiamin deficiency. Specific nutritional heart disease in the alcoholic may occur in a form similar to that of beriberi, but in contrast to beriberi heart disease, alcoholic cardiomyopathy manifests with symptoms of congestive heart failure, low cardiac output, and peripheral vasoconstriction. However, the low urinary and blood thiamin confirm the diagnosis of alcohol-induced cardiomyopathy (S17). The neuroencephalopathies and the thiamin deficiency associated with alcoholism are discussed in Section 4.2.

The effect of acute ethanol intoxication on thiamin metabolism has been studied by Takabe and Itokawa (T9). A temporary decrease in thiamin and transketolase levels was produced, which reverted to normal after 3 days. The decrease in thiamin following ethanol ingestion was much greater than after the administration of an equivalent quantity of glucose. The ratios of free thiamin to total thiamin and of plasma thiamin to whole blood thiamin were significantly decreased after ethanol ingestion.

Thiamin depletion in alcoholics has usually been assessed on the basis of the results of the erythrocyte transketolase activation test (W8). Some workers have reported that there was no correlation between blood thiamin levels and red cell transketolase activity in a group of hospitalized alcoholics (L8). Baker (B6) also noted some discrepancies between erythrocyte transketolase activity and blood thiamin levels, especially among patients with liver disease. The suggestion that some patients with Wernicke–Korsakoff syndrome have inborn abnormalities of transketolase (B12) also contraindicates the use of the transketolase activation test for the investigation of possible thiamin deficiency in those patients with suspected Wernicke–Korsakoff syndrome.

The high incidence of thiamin deficiency found in alcoholics has led some workers to recommend that alcoholic beverages be fortified with thiamin (P7). Centrewall and Criqui (C5) compared the annual cost of institutionalizing 1200 patients with Korsakoff's psychosis in the United States with the cost of fortifying alcoholic beverages with thiamin. The relative cost of $3 million for a fortification program, when compared to the $17 million cost of institutionalization, makes fortification look like an attractive proposition. However, a reduced intake of the vitamin is only one of the causes of thiamin deficiency in alcoholics. Problems of malabsorption, liver pathology, storage, and transportation of the vitamin all play major roles in causing deficiencies.

The classic studies of Sullivan and Herbert (S18) have provided convincing evidence that folate is not utilized in the presence of alcohol. So far, there is no convincing evidence that the situation is any different with respect to thiamin.

4.2. Wernicke–Korsakoff Syndrome

The concurrence of Wernicke's encephalopathy and Korsakoff's psychosis in alcoholic polyneuropathy has been observed for many years. It is now increasingly recognized that both Wernicke's encephalopathy and Korsakoff's psychosis are manifestations of the same pathological process, albeit at successive stages. Evidence to support this concept comes from a recent study in which 84% of patients hospitalized with Wernicke's disease who survived the acute illness exhibited "a typical Korsakoff's psychosis" (V2).

Following the early deprivation studies in pigeons, the pathological similarities between the thiamin-deprived birds and the pathological changes in Wernicke's encephalopathy were noted, and it was concluded that the symptoms of Wernicke's encephalopathy were due to thiamin deficiency. Wernicke's disease, which presents with a sudden onset, is characterized by paralysis of eye movement, ataxic gait, and mental confusion. Peripheral nerves are also commonly affected; Victor (V5) found peripheral neuropathies in more than 80% of the patients studied. Although many workers have suggested that in the Wernicke–Korsakoff syndrome there is selective damage to the hypothalamus, including the mammilary bodies, it has been reported by others that lesions of the mammilary bodies are not essential for the development of amnesia (V6).

Few reports have been published on the incidence of Wernicke's disease. Harper (H11) made an extensive study of 2891 necropsies over a 4-year period and noted that 51 cases (1.7%), 45 of which were alcoholics, had evidence of Wernicke's disease. Only seven had been diagnosed during life.

Of 186 patients with Wernicke's encephalopathy studied by Victor and Silby (V2), 157 (84%) exhibited Korsakoff's psychosis, which is essentially a disorder of mental function in which memory is impaired out of all proportion to other cognitive functions. Loss of past memory (retrograde amnesia) and impaired ability to acquire new permanent memory (anterograde amnesia) are common. Confabulation, which occurs in many patients, may be due to glimpses from past memory which are inserted into the gaps in present memory without regard for the time relationship.

The concept of a "biochemical lesion," first suggested by Peters (P8) in relation to thiamin deficiency and the nervous system, has been the basis for subsequent biochemical studies. Seltzer and McDougal (S4) contended that the regional distribution of lesions in thiamin-deficient mice could be accounted for by a regional difference in the ability of the brain to retain the last 10% of thiamin diphosphate. Although they found such differences, they did not suggest a mechanism to account for the failure of Korsakoff's psychosis to respond to thiamin therapy with the same rapidity as Wernicke's encephalopathy.

Ultrastructural studies on brain-stem lesions found in many animal species, including man, indicate that the earliest changes produced by thiamin deficiency can be detected in the membranous structures. Lesions in thiamin-deficient rat brain stem in the early stages of deficiency show a predominance of intracellular edema involving glial cells (R7). In the more advanced stages of the disease process, edema appears to involve myelin sheaths and the extravascular compartments. Dreyfus (D7) noted metabolic differences and variations in enzymic activity in various areas of the brain and suggested that these differences might be due to differences in vasculature such as the density of neurons, glial cells, and nerve fibers and in the degree of myelination. Of particular interest is the work of Pincus and Grove (P9), who reported a decrease in the concentration of thiamin diphosphate in brain tissue when rats were fed a diet low in thiamin.

In the pons of symptomatic animals, the thiamin diphosphate decreased to 26% of that of control levels. The decrease was less marked in other regions of the brain. The absolute concentration of thiamin triphosphate did not change during depletion, but when expressed as a percentage of total thiamin, there was a fourfold rise in thiamin triphosphate in the midbrain and pons, suggesting that in thiamin deficiency, parts of the brain are able to retain thiamin triphosphate.

Although the Wernicke–Korsakoff syndrome occurs mainly in alcoholics, a role for genetic factors in its pathogenesis has been suggested (L10). The Wernicke–Korsakoff syndrome occurs only in a small minority of alcoholics and other malnourished individuals, and it appears to occur more frequently in Europeans than non-Europeans ingesting diets having a similar thiamin content. Of particular interest were the findings that some alcoholic patients with Wernicke–Korsakoff syndrome may have a transketolase (a thiamin-dependent enzyme) abnormality (B12). The abnormality found in the enzyme appeared to be genetic rather than a consequence of alcohol abuse, since it persisted in fibroblasts through more than 20 generations of cultures in a medium containing excess thiamin and no ethanol, suggesting that individuals with this abnormality may have a predilection to thiamin deficiency. Two other thiamin-dependent enzymes, pyruvate dehydrogenase and α-ketoglutarate dehydrogenase, were normal. Patients with this enzyme abnormality were able to grow and thrive on the usual intake of the vitamin, but required extraordinarily high doses of the vitamin when they drank heavily.

Even though a few milligrams of thiamin is usually sufficient to correct the occular signs of Wernicke's encephalopathy, it has been recommended by some workers that large intravenous doses (500 mg) be used (I4). Problems in management may arise once the patient has recovered from the Wernicke's encephalopathy and the Korsakoff's psychosis becomes more prominent. The recovery of mental function may proceed very slowly over several months, and in some patients recovery is incomplete. This may reflect the extent of the damage to the brain and the slowness of the damaged brain tissue to recover.

4.3. SUBACUTE NECROTIZING ENCEPHALOMYELOPATHY (LEIGH'S DISEASE)

This is a rare degenerative disease of childhood, first described by Leigh in 1951 (L11), and up to 1976, little more than 100 cases had been reported in the literature (D8). The disease is characterized by weakness, anorexia, speech difficulty, irregular eye movements, and failure to thrive. The age at onset ranges between 1 week and 11 years, with a mean of 21 months. Affected children appear to develop normally and then rather suddenly deteriorate; for example, they may cease talking, become incontinent, and may have difficulty in walking (G4). The disease has a high familial preponderance and is probably inherited as an autosomal recessive gene.

Biochemical changes include elevated blood pyruvate and lactate levels, suggesting defective decarboxylation of pyruvate. Increased blood concentrations of α-ketoglutarate have also been found, suggesting a generalized defect in the oxidation of α-keto acids (V2).

Although the precise defect in necrotizing encephalomyelopathy is not known, it has been claimed that it has neuropathological and clinical signs similar to Wernicke's encephalopathy, and this has led to the study of the thiamin metabolism and thiamin-dependent enzymes in Leigh's disease. Greenhouse and Schneck (G5) reported normal red cell transketolase activity and normal thiamin pyrophosphate enhancement effect (4%) in a 3-month-old girl. The patient was treated with thiamin hydrochloride, 10 mg intramuscularly daily, but without effect. Pincus et al. (P10) measured the thiamin triphosphate levels in various brain regions in 13 patients with subacute necrotizing encephalomyelopathy and in 11 controls. They observed that a consistent reduction in thiamin triphosphate levels in the brain stem correlated well with the regularity with which lesions were found in this region. In an earlier study, Cooper et al. had demonstrated a substance in urine, blood, and cerebrospinal fluid from patients with necrotizing encephalomyelopathy which inhibited the phosphotransferase enzyme that is required for the conversion of thiamin pyrophosphate to thiamin triphosphate (C6). They also found thiamin to be absent from the brain of a child who had a subacute form of the disease, although there were normal concentrations in both liver and kidney. The inhibitor has now been found in most autopsy-proven cases of necrotizing encephalomyelopathy (M8, P11), but there remains some uncertainty about its significance.

Although successful treatment with doses up to 2 g orally per day has been claimed (D9), in view of the limited absorption of large doses of the vitamin, it seems unlikely that much of this dose was in fact absorbed. Thiamin tetrahydrofurfuryl disulfide has also been tried, but without discernible effect. This allithiamin was supplied in the form of sugar-coated tablets which the child had chewed because she was unable to swallow them. This resulted in her breath and the whole house smelling of burning rubber! The smell is not a problem when the tablets are swallowed (G4). Other workers have had limited success with this material (D10, P12).

Biotin is a cofactor in pyruvate carboxylase activity, and this vitamin together with lipoic acid, which is involved in the oxidation of pyruvate, has been used with some success (G4). However, Grover *et al.* (G6) and Gruskin *et al.* (G7) found it to be without effect.

It seems likely because of the wide range of signs, symptoms, and responses to various treatments found in patients that subacute necrotizing encephalomyelopathy may not be a single disease entity or, if it is, it presents in a variable manner. There is evidence that the most common form of this rare disease is associated with a failure to elaborate thiamin triphosphate in the brain.

4.4. Cerebrocortical Necrosis in Cattle and Sheep

This disease of cattle and sheep has affected animals in the United States, Australia, Canada, Europe, and South Africa. While the term cerebrocortical necrosis has been generally used to describe the disease, in America, a similar condition has been called polioencephalomalacia (E5). Although there are some clinical differences between the two conditions and the mortality rate of polioencephalomalacia is reported to be lower, they appear to be the same disease and both are characterized by a focal necrosis which is disseminated throughout the cerebral cortex.

The disease was first recognized as a distinct entity by Terlecki and Markson (T10) and is the result of thiamin depletion. Treatment with this vitamin cures the clinical disease, although it cannot repair the damage already done and the underlying cause may remain. The deficiency has been associated with a high-activity thiaminase type I found in the rumen of affected animals (B13). Boyd and Walton isolated thiaminase-producing *Bacillus thiaminolyticus* and *Clostridium sporogenes* from the rumen contents of affected cattle and sheep, but they concluded that neither of these organisms was the real source of rumen thiaminase (B13). Linklater and Dyson studied fecal thiaminase in three flocks of sheep in which one or more of the animals had developed cerebrocortical necrosis. They found that up to one-third of clinically normal animals were excreting thiaminase, and it was also noted that in two flocks, multiple cases of the disease appeared following treatment with the antihelmintics levamisole HCl and thiabendazole (L2). Although acute thiamin deficiency has been produced in lambs and calves by feeding them a thiamin-free diet, this has little relevance to the thiamin deficiency found in adult animals, since mature ruminants do not require an exogenous source of the vitamin. However, the disease resulting from this feeding experiment was similar to the cerebrocortical necrosis seen in adult animals (E2). There seems to be little doubt that cerebrocortical necrosis is the result of an acute thiamin deficiency brought about by the breakdown of the vitamin by a thiaminase (E5, E6, S19). This is supported by changes in pyruvate kinase, pyruvate, lactate, and transketolase levels in affected animals (E7).

The final diagnosis of cerebrocortical necrosis rests on the characteristic histopathology of the brain, i.e., necrosis of the cerebral cortex, and necrosis in the lateral geniculate bodies of the thalamus and the posterior coliculi of the midbrain in many cases (E6). The essential criterion for the diagnosis is the necrosis in the cerebral cortex.

4.5. MEGALOBLASTIC ANEMIA

Only four patients with a thiamin-responsive megaloblastic anemia have so far been reported in the literature (E8, R8, V7). The first was an 11-year-old caucasian girl with diabetes mellitus and complete bilateral sensorineural deafness. She presented with a hemoglobin of 5.8 g/dl, many macrocytes in the blood film, and a bone marrow which showed erythroid hyperplasia with numerous megaloblasts. She was treated with a multivitamin preparation containing 10 mg thiamin HCl, 10 mg riboflavin, 20 mg pyridoxine, 80 mg niacinamide, 20 mg pantothenic acid, 0.2 mg biotin, and 100 mg ascorbic acid. A reticulocyte response of 19% was reached on the eighth day of treatment. The patient's treatment was then changed to a multivitamin preparation containing only 1.2 mg thiamin. Three months later she relapsed, but on this occasion sequential treatment with various vitamins was given. Only treatment with 50 mg/day thiamin resulted in a significant reticulocytosis (13.5%). There was no response to cobalamin or folate, the vitamins most commonly associated with megaloblastic anemia.

It appears that this patient had an increased requirement for thiamin, and although the megaloblastosis suggested that there was a defect in the DNA pathway, there is no evidence to implicate thiamin in DNA synthesis.

A second patient with a thiamin-responsive megaloblastic anemia was reported in 1978 (V7). This was a 6-year-old girl with a history of anemia, convulsions, and bilateral deafness. She was the third child of a consanguineous marriage; her parents were first cousins.

The patient was admitted with a hemoglobin of 7.2 g/dl. Her blood film showed macrocytes, anisocytosis, and a mild degree of hypersegmentation of the neutrophils. A bone marrow biopsy revealed megaloblastic hyperplasia. Radiological examination showed the spleen, stomach, and heart to be in the right half of the abdomen, while the liver was in the left. At this time, the patient had been taking 100 mg pyridoxine plus an oral preparation containing 5 mg folate and 15 μg cobalamin. Urinary examination for orotic acid was negative and serum levels of folate and cobalamin were normal. There was a rapid hematological response to the administration of thiamin alone and a hematological relapse when thiamin supplementation was ceased. This was repeated twice, and on each occasion there was a hematological relapse when thiamin was stopped.

The two most recent patients were reported by Evans et al. (E8). They were a brother and sister of Asian origin. Both had deafness, diabetes mellitus, and an anemia with megaloblastic and sideroblastic changes. Electron microscopy showed many iron-laden mitochondria within a high proportion of the erythroblasts. The deoxyuridine-suppression test indicated that the megaloblastic changes were not caused by any impairment of the methylation of deoxyuridylate. In both patients, the anemia responded to thiamin and was unresponsive to treatment with folate or cobalamin. The deafness and diabetes mellitus were unaffected by treatment with thiamin.

In all of these children, there appeared to be an increased demand for thiamin which was restricted to one particular pathway. There was no other evidence of thiamin deficiency such as beriberi, nor was there any evidence of an abnormality of thiamin-dependent enzymes such as erythrocyte transketolase, leukocyte pyruvate dehydrogenase, or leukocyte α-ketoglutarate dehydrogenase. All four patients appeared to have multiple systemic defects.

4.5.1. Megaloblastic Anemia Due to Cobalamin and Folate Deficiency

Megaloblastic anemia due to cobalamin deficiency is associated with an elevated level of erythrocyte transketolase activity, in contrast to that due to a deficiency of folate where the enzyme level remains normal. However, the wide range of transketolase activity found in patients with folate-deficient megaloblastic anemia precludes its use as a diagnostic test. Fasting blood pyruvate levels are also frequently elevated in both cobalamin and folate deficiency, but no correlation has been found between hemoglobin concentration, transketolase activity, thiamin pyrophosphate effect, or serum and erythrocyte folate activity (W9).

Thiamin deficiency was found to be more common in folate-deficient patients than in those with cobalamin deficiency, in whom it was rare. This is not surprising, as folate deficiency is frequently associated with a defect in dietary intake.

4.6. Thiamin-Responsive Maple Syrup Urine Disease

Maple syrup urine disease is a rare inborn error of metabolism characterized by the accumulation of leucine, isoleucine, valine, and the corresponding keto acids in the body fluids and urine. The urine of patients has an odor resembling maple syrup. The frequency of the disorder is approximately 1/250,000 live births (S20).

Several variants of the disease have been described (W10), and these range from a severe form, having a poor prognosis and in which there is almost complete absence of decarboxylase activity for the branched-chain keto acids in leukocytes and skin fibroblasts, to a mild form exhibiting none of the symptoms observed in the classic cases, but which nevertheless shows the biochemical abnormalities associated with the disease. Physical signs range from irritability, refusal to feed, and absent tendon reflexes in severely affected babies, to delayed milestones in those who are mildly affected (S21).

A thiamin-responsive variant of the disease was described by Scriver *et al.* (S20). The patient had an unusually benign clinical history, but her urine contained an excess of leucine, isoleucine, and valine, and her plasma contained five times the normal level of these amino acids, plus an abnormal amount of alloisoleucine.

Because thiamin pyrophosphate is the coenzyme for the first step in oxidative decarboxylation of branched-chain keto acids, the patient was started on 10 mg of thiamin hydrochloride daily and this resulted in a return to normal of the abnormal plasma amino acid levels. Withdrawal of the thiamin caused the plasma concentrations of leucine, isoleucine, valine, and alloisoleucine to rise to pretreatment levels. Treatment of classic maple syrup urine disease with thiamin has been reported to be without effect on the abnormal amino acid levels (W10).

4.7. Beriberi

Much has been written on beriberi (F1, K10, T11, V2), and the purpose here is to give only a brief account of the subject.

The term beriberi is used to describe a clinical state resulting from a diet-induced deficiency of thiamin, and although a deficiency of other vitamins may also be present, there is no doubt that it is the lack of thiamin that is responsible for a major part of the pathology. Beriberi was, and still is to a lesser extent, found mainly in communities where rice formed the staple diet, and its appearance in epidemic proportions coincided with the introduction of highly milled rice. This process removed a large part of the thiamin content and also probably removed a number of other vitamins and essential nutrients. The result was a diet high in carbohydrate and low in thiamin.

Wheat is also refined to a point where it loses a major portion of its thiamin content, but this is overcome by a program of fortification in many countries. It has been calculated that if the intake of thiamin is reduced to less than 0.4 mg/day over a long period of time, beriberi results (W11). The disease may therefore also be found among the poor, the aged, and some alcoholics.

Clinical symptoms have been grouped under three headings, the so-called wet, dry, and cardiac types. The wet form is characterized by effusions and peripheral edema. Polyneuropathy is associated with the dry form and acute congestive heart failure is a feature of the cardiac form. Cardiac changes are, however, observed at some stage of the disease in nearly all patients, and in some, these may appear suddenly and be followed by death (S13). Patients presenting with a variety of these symptoms are not uncommon. Laryngeal paralysis and deafness have sometimes been found in patients with severe beriberi (F1). Symptoms are generally variable and reflect the normal variation in human susceptibility. It is perhaps more appropriate to consider the symptoms in terms of the severity of the disease rather than to try to place patients into one of the arbitrary categories of the disease (P13).

4.7.1. *Infantile Beriberi*

This is seen in breast-fed babies whose mothers are severely depleted of thiamin. These infants appear well fed but are unable to effectively utilize carbohydrates, and the ingestion of these foods causes a grave depletion of the vitamin in vital areas of the body. Symptoms may include vomiting, abdominal pain, diarrhea, abdominal distention, convulsions, cyanosis, and tachycardia. Death may be sudden and has been reported to be similar to that seen in the sudden infant death syndrome (R9). When there is a chronic but less than fatal depletion, infants may develop edema, oliguria, constipation, enlargement of heart and liver, and retardation of growth (L1).

The disease is not frequently found where there is a shortage of food, since the restriction of carbohydrates reduces the thiamin requirement (L1).

Symptoms of beriberi may occur in the rare instance where a patient has an abnormality of thiamin metabolism. Pestel reported the case of a 12-year-old boy who developed cardiac failure with increased cardiac output, polyneuritis, and ankle edema while on what was described as a perfectly normal diet. He was found to have a low blood thiamin level and recovered fully with thiamin therapy. However, he was subsequently found to have a persistently low level of the phosphorylated form of the vitamin, suggesting that he may have had an enzymic defect (P14).

4.8. Other Clinical States Which May Be Associated with a Change in Thiamin Status

4.8.1. Thiamin Status of Hospital Patients

The study of vitamin nutrition among hospital inpatients is attractive because of the easy accessibility of the target population. Lemoine et al. (L12) examined the thiamin, riboflavin, pyridoxal, and ascorbate status of 656 hospital inpatients over a period of 2 years. They found that 57% of the patients had a low daily intake of thiamin, and that in 19%, the intake was less than half the recommended dietary allowance (N2). On the basis of the erythrocyte transketolase activation test, 25% of the patients studied showed a biochemical deficiency of the vitamin. As would be expected, 27 of 57 patients with alcoholic cirrhosis of the liver showed evidence of thiamin deficiency. In these studies, classic signs of severe vitamin deficiency were seldom observed; however, there was an increase in nonspecific symptoms.

In a smaller study (T12), the red cell transketolase activity was measured in 86 hospital patients selected because it was thought that they might have been deficient in thiamin. Those showing evidence of vitamin depletion included those with Wernicke's encephalopathy (5), peripheral neuropathy (6), cardiomyopathy (6), and alcoholism (9). Four of the patients with Wernicke's encephalopathy were alcoholics.

Although the measurement of red cell transketolase is a useful test for assessing thiamin status, it appears to be less sensitive than direct microbiological assays. It is therefore likely that thiamin deficiency is more common than that suggested by the results of current biochemical methods.

4.8.2. Thiamin Status of the Aged

In a study of 98 long-term geriatric patients, MacLennan et al. (M9) concluded that thiamin deficiency was comparatively rare. It appeared that in the prolonged-stay hospital used for the study, patients were offered a regular, balanced diet which was sufficient to not only supply their daily need, but was also able to make up for any initial deficiency on admission. In a study of 93 acute geriatric hospital admissions, none had a completely normal nutritional profile. However, there was little evidence of thiamin malnutrition (M10). On the other hand, Hoorn et al. (H12) found 35 of 153 (22.9%) geriatric patients to have a biochemical deficiency of thiamin, and in a similar study, Griffiths et al. (G8) found that 40% of the elderly patients admitted to the hospital were deficient in thiamin. In a study of 196 institutionalized and noninstitutionalized elderly people, 14 (7.1%) were found to be consuming less than two-thirds of the recommended daily intake of thiamin. While 17.6% of males and 12.5% of females had evidence of a biochemical deficiency, no characteristic clinical features of thiamin deficiency were seen.

It appears from these and other major studies (B14, H13) that elderly people living at home or in a sheltered situation have a high preponderance of thiamin and other vitamin deficiencies. These may not be severe enough to result in marked clinical changes, but it seems likely that they contribute to a general lowering of health standards.

The level of vitamin nutrition in hospitals and institutions for the aged appears variable, and probably depends not only on the standard of food, but also on the dedication of the nursing staff. It has been suggested that elderly people utilize thiamin less efficiently and that their daily allowance should be 1 mg/day (N2). There remains a need, however, for a study of vitamin utilization in the aged.

4.8.3. Effect of Drugs

Only a few drugs have so far been found to have an adverse effect on thiamin metabolism. Furosemide is a widely used diuretic agent, and its long-term use in patients with congestive cardiac failure has been found to be associated with a thiamin deficiency which is characterized by an increased thiamin pyrophosphate effect on erythrocyte transketolase activity (Y1). In rats, an intraperitoneal injection of furosemide (20 mg/kg) resulted in a significant increase in the urinary excretion of thiamin. Increasing the rats' consumption of water also increased the urinary output of thiamin, but did not increase it to the same extent as the diuretic (Y2).

Patients with malignant disease may have a poor intake of nutrients. However, there is some evidence to suggest that a particular deficiency may be linked to a specific type of tumor (D11) or to the use of a particular cytotoxic agent. Treatment with drug combinations which include 5-fluorouracil has been associated with the development of thiamin deficiency, while patients receiving drug combinations which did not include 5-fluorouracil showed no such deficiency (A6, B15). The drug did not cause any side effects such as nausea or vomiting and the appetite of both groups was similar. It therefore seemed unlikely that the deficiency reflected a reduced dietary intake. However, Soukop and Calman (S22) were of the opinion that 5-fluorouracil served only to exaggerate an already existing biochemical deficiency. Nine of their ten patients with an initial biochemical deficiency developed an exaggerated deficiency following 5-fluorouracil therapy. It has been suggested that the drug prevents phosphorylation of the vitamin (B15), but further work is required to confirm this. Patients having a cytotoxic drug-related thiamin deficiency have been reported to respond adequately to 100 mg of the vitamin. However, while replacement therapy may improve the signs and symptoms of the deficiency state, such treatment has been found to coincide with rapid tumor progression (S22).

4.8.4. Oral Contraceptive Steroids and Thiamin Metabolism

Women taking oral contraceptive steroids have a reduced serum concentration of folate and cobalamin compared with women not taking these agents (D12). Tryptophan metabolism is also altered in the presence of oral contraceptives, and an increase in the urinary excretion of a number of intermediates of the tryptophan–kynurenine pathway occurs following a tryptophan load (W12). However, although the excretion of these intermediates suggested a deficiency of pyridoxal, no reduction in the serum level of this vitamin was found (D12, D13). Nevertheless, substantial supplementation with pyridoxine is required to correct the abnormalities of tryptophan metabolism.

There is no convincing evidence that oral contraceptive agents have any adverse effect on thiamin metabolism. Joshi *et al.* (J5) measured both the erythrocyte transketolase level and the effect of thiamin pyrophosphate stimulation on this enzyme in 10 women taking Ovral and in 8 women taking Norgestrel. No significant changes were observed. Vir and Love (V8) found no significant difference between the transketolase measurements in a group of women taking oral contraceptives and those in controls. However, they did note that 30% of women taking oral contraceptives and 25% of the control group had results indicative of thiamin deficiency. Ahmed and Bamji (A7) found that the thiamin level, reflected by transketolase, decreased in women taking oral contraceptives and recommended supplementation with 3 mg of the vitamin daily. Briggs and Briggs (B16), in a study of 20 women taking oral contraceptives, also noted that erythrocyte transketolase became more unsaturated by its cofactor, indicating a mild hypovitaminosis.

No frank clinical signs of thiamin deficiency associated with the taking of oral contraceptive steroids have been observed.

4.8.5. Pregnancy

There is an increased requirement for thiamin in pregnancy which may result in a deficiency in the mother. This is, of course, more likely to occur in communities where the dietary intake of the vitamin is borderline. It has been recommended that the thiamin intake be raised by 0.2 mg/day in pregnancy (N2) and that in underprivileged communities, the diet should be increased by 17% over the nonpregnant intake in order to satisfy the increased need for energy and nutrients (A8).

Until recently, there have been methodological problems in trying to study thiamin metabolism in pregnancy. While direct assay methods for measuring serum and red cell folate and cobalamin have been available for many years, similar methods for measuring thiamin have only rarely been available in clinical laboratories. The assessment of the vitamin status in pregnancy has largely depended on measurement of the urinary excretion of the vitamin or its metabolites or on the transketolase activation test. Thanangkul and Amatayakul (T13) found that 20% of the pregnant Thai women studied excreted less than 40 µg of the vitamin in 24 hours, which was indicative of a biochemical deficiency, although clinical signs of a deficiency were found in only 7%. In a study of 500 pregnant women, 25–30% were considered to be thiamin deficient; this was based on the results obtained using the erythrocyte transketolase activation test and comparisons with 300 male and female blood donors who served as controls. By the last month of pregnancy, 30–80% of mothers were considered biochemically deficient. No correlation between thiamin status and the course or outcome of the pregnancy could be demonstrated (H14). These findings were confirmed in a study of 300 pregnant women of Mexican descent in which 22% were considered to be low or deficient in thiamin (J6), and in a Dutch study in which it was found that 25–30% of pregnant women were in the biochemically deficient range at term (V9). It has been suggested that a deficiency of thiamin is established early in pregnancy and then remains constant until delivery (H14). However, it seems more likely that it follows a pattern similar to that observed with respect to cobalamin and folate, in which the blood level of both vitamins decreases progressively as pregnancy advances (D14, L13). A decline in blood concentration of these vitamins probably reflects a redistribution rather than a deficiency, because not only are vitamins for which there is only a relatively small body store affected, but the cobalamins, stores of which are normally sufficient for several years, are also affected (M11).

Following parturition, the maternal serum vitamin levels rapidly increase. Nevertheless, when vitamin levels are found to be substantially below the lower limit of the reference range, appropriate supplementation should be given.

Vir and her colleagues (V10) found that the occurrence of thiamin deficiency increased with the number of previous pregnancies. However, this requires further examination because the group studied appeared to have a very high prevalence of thiamin deficiency. Based on the measurement of the transketolase–pyrophosphate effect, 30% of nonpregnant and 28–39% of 60 pregnant women in the second or third trimester were considered to be deficient. In a study using a direct microbiological assay, only 2% of 104 pregnant women were found to have a low red cell thiamin concentration. Over a period of 11–15 weeks, the red cell thiamin concentration in these women fell by an average of 28%, but in the majority of women it remained within the reference range (12). A vitamin profile of 41 mothers and their newborn infants at parturition showed the mothers to have a mean blood thiamin level of 19 µg/liter, and their infants a mean level of 78 µg/liter, a mother to neonate thiamin ratio of 1:4. In a group of diabetic mothers, this ratio appeared reversed (B17). This was based on a small series of tests, and further studies are required.

Thiamin deficiency has been implicated as one of the factors associated with toxemia of pregnancy. Chaudhuri (C7) studied 1210 pregnancies and concluded that there was a definite relationship between malnutrition and toxemia of pregnancy, and of the nutrients involved, thiamin, calcium, and iron were the most important.

4.8.6. Lactation

In common with other water-soluble vitamins, the concentration of thiamin in human breast milk is related to the maternal intake of the vitamin. Thiamin supplementation during lactation resulted in an increased concentration of thiamin in the milk of malnourished women (H15), but once an optimal intake had been established and body stores replenished, the concentration in the milk tended to plateau (H15, V11). The thiamin content of human milk is very variable, and in a study of healthy Filipino women, the concentration of the vitamin increased from 18.6 µg/liter on the second to third day of lactation to 113.6 µg/liter after 4 weeks of lactation (G9). No correlation was found between the number of infants breast fed and the vitamin content of the maternal milk. In a group of healthy, middle-class American women, the mean milk thiamin concentration was found to be 138 µg/liter at 5 to 7 days lactation, rising to 220 µg/liter between days 43 and 45 (N5). A second group of women received thiamin supplements, but this was found to have no effect on the concentration of the vitamin in the milk over a 6-week period. It was concluded that at least in well-nourished middle-class American women, thiamin supplementation during lactation was unnecessary.

In thiamin-deficient mothers, the breast-fed infant will rapidly show signs of thiamin deficiency. Trostler et al. (T14) fed pregnant rats a diet low in thiamin from the tenth day of their pregnancy through lactation. Dams showed signs of thiamin deficiency after 30 days on the diet, whereas the pups displayed signs of thiamin deficiency from the fourteenth postpartum day. Brain transketolase activity was depressed in day-old pups from thiamin-deficient dams and brain pyruvate levels were elevated by the seventh postpartum day. At 14 days postpartum, the brain of thiamin-deficient pups contained less phospholipids, cerebrosides, and cholesterol than found in control animals. An attempt was made to rehabilitate some pups with thiamin at day 19, but after 24 days, the brain cerebroside content was still lower than that found in the controls. This work suggests that early thiamin deprivation may effect changes that cannot respond to later therapy.

Cow's milk is widely used in a modified form for infant feeding, and it has been shown to contain considerably more thiamin than does human milk. Nobile and Woodhill found it to have a concentration of between 290 and 670 µg/liter (N6), while in a limited study in this laboratory, a range of 255–684 µg/liter was obtained for pasteurized cow's milk and 360–896 µg/liter for fresh cow's milk. The difference between the results with the pasteurized and fresh milk is probably due to the effect of heating, as up to 38% of thiamin has been shown to be destroyed during sterilization (F6).

4.8.7. Thiamin and the Sudden Infant Death Syndrome

The sudden infant death syndrome (SIDS) is the major cause of death in infants between the age of 1 week and 1 year. The syndrome is responsible for 2.3 deaths per 1000 live births in the United States, resulting in some 10,000 deaths annually. In other countries, the figures range from 1.2 to 2.5+ per 1000 live births (R9). The immediate cause of death is not clear, although it is a widely accepted working hypothesis that neural regulation of breathing during sleep is involved (W13). In 1978, Read, in his review of the etiology of the sudden infant death syndrome, suggested that it may be linked to deranged thiamin neurochemistry (R9). Petersen *et al.* examined this hypothesis by measuring the erythrocyte transketolase in SIDS infants, but found no difference between the group studied and the controls (P15).

Davis *et al.* (D15) examined thiamin levels in SIDS infants using an automated microbiological assay with *Lactobacillus fermenti* as the test organism. They measured the serum thiamin concentration in 121 SIDS infants and in 26 control infants dying of explicable causes. The thiamin concentration in most of the SIDS infants was exceptionally high, with a range of 1.0–>500 µg/liter (mean 98.7) compared with <1.0–64.0 µg/liter (mean 21.4) for the controls. In a further study (D16), these authors confirmed that SIDS infants had very high levels of serum thiamin (Fig. 3). The measurement of transketolase will not give information on elevated red cell thiamin concentrations and provides no direct information on serum levels. They suggested that these high levels may be due to excessive absorption of the vitamin through the failure of the rate-limiting absorption control mechanims. However, this simple explanation does not take into account the rapid renal excretion of the vitamin which normally occurs. It is possible that the final answer may involve both absorption and excretory mechanisms.

High serum levels of thiamin in animals cause death by depression of the respiratory center. However, the serum thiamin concentrations observed in SIDS infants were considerably less than that which is required to cause death by respiratory paralysis in rabbits and mice (N1). Some larger animals

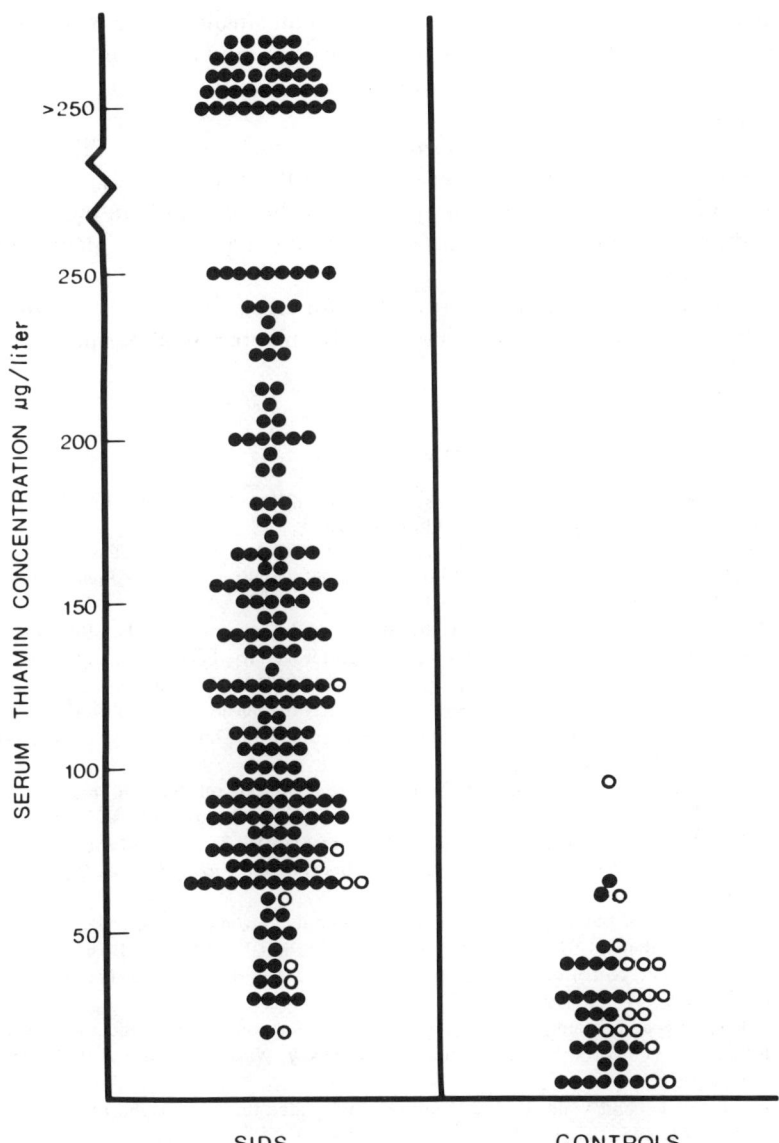

FIG. 3. Serum thiamin concentrations in 233 infants dying from SIDS compared with those from 46 control infants dying of explicable causes. Solid circles denote infants less than 12 months old; open circles, infants 12–18 months. [Reproduced from Davis et al. (D16), with permission.]

appear to be more sensitive to the vitamin than smaller animals, but this is variable and it is not known just how sensitive human infants might be. Moreover, the thiamin concentration in SIDS infants would be expected to build up over a period of weeks or months.

Infants fed on artificial milk formulas may be exposed to thiamin concentrations as high as 2100 µg/liter because of the practice of some manufacturers of adding several times the quantity of the vitamin than indicated on the product label. Bottle-fed babies were found to have serum thiamin levels similar to those who were breast fed.

It is not suggested that high serum thiamin levels are directly responsible for SIDS, and at present they represent only an apparent abnormality which requires further study.

References

A1. Anonymous, Nomenclature policy: Generic descriptions and trivial names for vitamins and related compounds. *J. Nutr.* **106**, 8–14 (1976).
A2. Ang, C. Y. W., and Moseley, F. A., Determination of thiamin and riboflavin in meat and meat products by high-pressure liquid chromatography. *J. Agric. Food Chem.* **28**, 483–486 (1980).
A3. Ang, C. Y. W., Basillo, L. A., Cato, B. A., and Livingston, G. E., Riboflavin and thiamine retention in frozen beef-soy patties and frozen fried chicken heated by methods used in food service operations. *J. Food Sci.* **43**, 1024–1027 (1978).
A4. Anonymous, Interdepartmental Committee on Nutrition for National Defence, Nutrition Survey. The Kingdom of Thailand. Department of Defence, Washington, D.C., 1962.
A5. Abe, T., and Itokawa, Y., Effect of ethanol administration on thiamine metabolism and transketolase activity in rats. *Int. J. Vitam. Nutr. Res.* **47**, 307–314 (1977).
A6. Aksoy, M., Basu, T. K., Brient, J., and Dickerson, J. W. T., Thiamin status of patients treated with drug combinations containing 5-fluorouracil. *Eur. J. Cancer* **16**, 1041–1045 (1980).
A7. Ahmed, F., and Bamji, M. S., Vitamin supplements to women using oral contraceptives (studies of vitamins B1, B2, B6 and A). *Contraception* **14**, 309–318 (1976).
A8. Arroyave, G., Nutrition in pregnancy in Central America and Panama. *Am. J. Dis. Child.* **129**, 427–430 (1975).
B1. Baker, H., and Frank, O., Absorption, utilisation and clinical effectiveness of al-lithiamines compared to water-soluble thiamines. *J. Nutr. Sci. Vitaminol.* **22**, Suppl., 63–68 (1976).
B2. Barchi, R. L., and Viale, R. O., Membrane associated thiamine triphosphatase. *J. Biol. Chem.* **251**, 193–197 (1976).
B3. Barchi, R. L., and Braun, P. E., A membrane-associated thiamine triphosphatase from rat brain. *J. Biol. Chem.* **247**, 7668–7673 (1972).
B4. Baba, A., Matsuda, T., and Iwata, H., Possible regulation of thiamine diphosphatase activity in rat brain microsomes by lipids. *Biochim. Biophys. Acta* **482**, 71–78 (1977).
B5. Beadle, B. W., Greenwood, D. A., and Kraybill, H. R., Stability of thiamine to heat. I. Effect of pH and buffer salts in aqueous solutions. *J. Biol. Chem.* **149**, 339–347 (1943).
B6. Baker, H., Pasher, I., Frank, O., Hutner, S., Aronson, S., and Sobotka, H., Assay of thiamine in biological fluids. *Clin. Chem. (Winston-Salem, N.C.)* **5**(1), 13–17 (1959).

B7. Baker, H., and Frank, O., *In* "Modern Nutrition in Health and Disease" (R. S. Goodhart and M. E. Shils, eds.), 5th ed., Chapter 17, p. 523. Lea & Febiger, Philadelphia, Pennsylvania, 1973.

B8. Bican-Fister, T., and Drazin, V., Quantative analysis of water soluble vitamins in multicomponent pharmaceutical forms. *J. Chromatogr.* **77**, 389–395 (1973).

B9. Brin, M., Tai, M., Ostashever, A. S., and Kalinsky, H., The effect of thiamine deficiency on the activity of erythrocyte hemolysate transketolase. *J. Nutr.* **71**, 273–281 (1960).

B10. Bayoumi, R. A., and Rosalki, S. B., Evaluation of methods of co-enzyme activation of erythrocyte enzymes for detection of deficiencies of vitamins B1, B2 and B6. *Clin. Chem. (Winston-Salem, N.C.)* **22**, 327–335 (1976).

B11. Baraona, E., Orrego, H., Fernandez, O., Amenabar, E., Maldonado, E., Tag, F., and Salinas, A., Absorptive function of the small intestine in liver cirrhosis. *Am. J. Dig. Dis.* **7**, 318–329 (1962).

B12. Blass, J. P., and Gibson, G. E., Abnormality of a thiamine requiring enzyme in patients with Wernicke-Korsakoff syndrome. *N. Engl. J. Med.* **297**, 1367–1370 (1977).

B13. Boyd, J.W., and Walton, J. R., Cerebrocortical necrosis in ruminants; an attempt to identify the source of thiaminase in affected animals. *J. Comp. Pathol.* **87**, 581–589 (1977).

B14. Baker, H., Frank, O., Thind, I. S., Jaslow, S. P., and Louria, D. B., Vitamin profiles in elderly persons living at home or in nursing homes versus profile in healthy young subjects. *J. Am. Geriatr. Soc.* **27**, 444–450 (1979).

B15. Basu, T. K., Dickerson, J. W. T., Raven, R. W., and Williams, D. C., The vitamin status of patients with cancer as determined by red cell transketolase activity. *Int. J. Vitam. Nutr. Res.* **44**, 53–58 (1974).

B16. Briggs, M. H., and Briggs, M., Thiamin status and oral contraceptives. *Contraception* **11**, 151–154 (1975).

B17. Baker, H., Frank, O., Thomson, A. D., Langer, A., Munves, E. D., De Angelis, B., and Kaminetzky, H. A., Vitamin profile of 174 mothers and newborns at parturition. *Am. J. Clin. Nutr.* **28**, 59–65 (1975).

C1. Chakrabarti, C. H., and Pandit, V. I., Influence of thiamine on protein biosynthesis. *J. Vitaminol.* **15**, 211–214 (1969).

C2. Cheraskin, E., Ringsdorf, W. M., Medford, F. H., and Hicks, B. S., The "ideal" daily vitamin B1 intake, *J. Oral Med.* **33**, 77–79 (1978).

C3. Cole, M., Turner, A., Frank, O., Baker, H., and Leevy, C., Extraocular palsy and thiamine therapy in Wernicke's encephalopathy. *Am. J. Clin. Nutr.* **22**, 44–51 (1969).

C4. Callmer, K., and Davies, L., Separation and determination of vitamin B1, B2, B6 and nicotinamide in commercial vitamin preparations using high performance cation-exchange chromatography. *Chromatographia* **7**, 644–640 (1974).

C5. Centrewall, B. S., and Criqui, M. H., Prevention of the Wernicke-Korsakoff syndrome—A cost benefit analysis. *N. Engl. J. Med.* **299**, 285–289 (1978).

C6. Cooper, J. R., Itokawa, Y., and Pincus, J. H., Thiamine triphosphate deficiency in subacute necrotising encephalomyelopathy. *Science* **164**, 74–75 (1969).

C7. Chaudhuri, S. K., Role of nutrition in the etiology of toxemia of pregnancy. *Am. J. Obstet. Gynecol.* **110**, 46–48 (1971).

D1. Dakshinamurti, K., B vitamins and nervous system function. *Nutr. Brain* **1**, 249–318 (1977).

D2. Dong, M. H., McGown, E. L., Schwenneker, B. W., and Sauberlich, H. E., Thiamin, riboflavin and vitamin B6 contents of selected foods as served. *J. Am. Diet. Assoc.* **76**, 156–160 (1980).

D3. Deibel, R. H., Evans, J. B., and Niven, C. F., Jr., Microbiological assay for thiamine using *Lactobacillus viridescens*. *J. Bacteriol.* **74**, 818–821 (1957).

D4. Dong, M. H., Green, M. D., and Sauberlich, H. E., Determination of urinary thiamine by the thiochrome method. *Clin. Biochem.* **14**, 16–18 (1981).
D5. Dreyfus, P. M., Clinical application of blood transketolase determinations. *N. Engl. J. Med.* **267**, 596–598 (1962).
D6. Davis, R. E., and Smith, B. K., Pyridoxal and folate deficiency in alcoholics. *Med. J. Aust.* **2**, 357–360 (1974).
D7. Dreyfus, P. M., Thiamine deficiency encephalopathy: Thoughts on its pathogenesis. *In* "Thiamine" (C. J. Gubler, M. Fujiwara, and P. M. Dreyfus, eds.), pp. 229–243. Wiley, New York, 1976.
D8. David, R. B., Mamunes, P., and Rosenblum, W. I., Necrotizing encephalomyelopathy (Leigh). *In* "Handbook of Clinical Neurology" (P. J. Vinken and G. W. Bruyn, eds.), Vol. 28, pp. 349–363. Elsevier, Amsterdam, 1976.
D9. De Groot, C. J., and Hommes, F. A., Further speculation on the pathogenesis of Leigh's encephalomyelopathy. *J. Pediatr.* **82**, 541–542 (1973).
D10. Dunn, H. G., and Dolman, C. L., Necrotizing encephalomyelopathy. *Eur. Neurol.* **7**, 34–55 (1972).
D11. Dickerson, J. W. T., and Basu, T. K., Specific vitamin deficiencies and their significance in patients with cancer and receiving chemotherapy. *Curr. Concepts Nutr.* **6**, 95 (1977).
D12. Davis, R. E., and Smith, B. K., Pyridoxal, vitamin B12 and folate metabolism in women taking oral contraceptive agents. *S. Afr. Med. J.* **48**, 1937–1940 (1974).
D13. Davis, R. E., Serum vitamin levels and human nutrition. *Proc. Nutr. Soc. Aust.* **4**, 45–52 (1979).
D14. Davis, R. E., Stenhouse, W. S., and Woodliff, H. J., Serum folate levels in pregnancy. *Med. J. Aust.* **1**, 52–53 (1969).
D15. Davis, R. E., Icke, G. C., and Hilton, J. M. N., High thiamine levels in sudden-infant death syndrome. *N. Engl. J. Med.* **303**, 462 (1980).
D16. Davis, R. E., Icke, G. C., and Hilton, J. M. N., High serum thiamine and the sudden infant death syndrome. *Clin. Chim. Acta* **123**, 321–328 (1982).
E1. Evans, W. C., Thiaminases and their effects on animals. *Vitam. Horm. (N.Y.)* **33**, 467–504 (1975).
E2. Evans, W. C., Evans, I. A., Humphreys, D. J., Lewin, B., Davies, W. E. J., and Axford, R. F. E., Indications of thiamine deficiency in sheep, with lesions similar to those of cerebrocortical necrosis. *J. Comp. Pathol.* **85**, 253–267 (1975).
E3. Everett, M. R., "Medical Biochemistry," 2nd ed. New York Press, 1946.
E4. Edwin, E. E., and Lewis, G. J., Reviews of the progress of dairy science. *Dairy Res.* **38**, 79–90 (1971).
E5. Edwin, E. E., Spence, J. B., and Woods, A. J., Thiaminases and cerebrocortical necrosis. *Vet. Rec.* **83**, 417 (1968).
E6. Edwin, E. E., Markson, L. M., Shreeve, J., Jackman, R., and Carroll, P. J., Diagnostic aspects of cerebrocortical necrosis. *Vet. Rec.* **104**, 4–8 (1979).
E7. Edwin, E. E., Plasma enzyme and metabolite concentrations in cerebrocortical necrosis. *Vet. Rec.* **87**, 396–398 (1970).
E8. Evans, D. I. K., Haworth, C., and Wickramasinghe, S., Thiamine responsive anaemia: Two further cases. *Abstr. Int. Soc. Haematol., Eur. Afr. Div., 6th Meet.* p. 164 (1981).
F1. Farmer, T. W., Vitamin B1 deficiency. *In* "Metabolic and Deficiency Diseases of the Nervous System" (P. J. Vinken and G. W. Bruyn, eds.), Part II, pp. 49–57. North-Holland Publ., Amsterdam, 1976.
F2. Fujita, A., Thiaminase. *Adv. Enzymol.* **15**, 389–421 (1954).
F3. Fujiwara, M., and Watanabe, H., Allithiamine, a newly found compound of vitamin B1. *Proc. Jpn. Acad.* **28**, 156–158 (1952).

F4. Fujiwara, M., Tsuno, S., and Yoshimura, M., Jujikaka shokubutsu Ni yoru allithiamine no dozokutai no seisei tsuitte. *Bitamin* **10**, 506–507 (1956).

F5. Food and Agriculture Organization World Health Organization of the United Nations, Requirements of vitamin A, thiamine, riboflavin and niacin. *FAO Nutr. Meet. Rep. Ser.* **41**, 32 (1967).

F6. Ford, J. E., The influence of the dissolved oxygen in milk on the stability of some vitamins towards heating and during subsequent exposure to sunlight. *J. Dairy Res.* **34**, 239–247 (1967).

F7. Franke, K. W., and Franke, W. R., A metabolism cage for rats. *J. Lab. Clin. Med.* **19**, 669–671 (1934).

F8. Fang, S. C., and Butts, J. S., Microbiological assay method for thiamine using *Lactobacillus fermenti* 36 as the test organism. *Proc. Soc. Exp. Biol. Med.* **78**, 463–466 (1951).

F9. Fennelly, J., Frank, O., Baker, H., and Leevy, C., Peripheral neuropathy of the alcoholic. 1. Aetiological role of aneurin and other B-complex vitamins. *Br. Med. J.* **2**, 1290–1292 (1964).

G1. Goodhart, R. S., and Shils, M. E., "Modern Nutrition in Health and Disease," 6th ed. Lea & Febiger, Philadelphia, Pennsylvania, 1980.

G2. Geel, S. E., and Dreyfus, P. M., Brain lipid composition of immature thiamine-deficient and undernourished rats. *J. Neurochem.* **24**, 353–360 (1975).

G3. Gupta, V. D., and Cadwallader, D. E., Acid dye method for the analysis of thiamine. *J. Pharm. Sci.* **57**, 112–116 (1968).

G4. Gordon, W., Marsden, H. B., and Lewis, D. M., Subacute necrotising encephalomyelopathy in three siblings. *Dev. Med. Child Neurol.* **16**, 64–78 (1974).

G5. Greenhouse, A. H., and Schneck, S. A., Subacute necrotising encephalomyelopathy. *Neurology* **18**, 1–8 (1968).

G6. Grover, W. D., Auerbach, V. H., and Patel, M. S., Biochemical studies and therapy in subacute necrotising encephalomyelopathy (Leigh's syndrome). *J. Pediatr.* **81**, 39–44 (1972).

G7. Gruskin, A. B., Patel, M. S., Linshaw, M., Ettenger, R., Huff, D., and Grover, W., Renal function studies and kidney pyruvate carboxylase in subacute necrotising encephalomyelopathy. *Pediatr. Res.* **7**, 832–841 (1973).

G8. Griffiths, L. L., Brocklehurst, J. G., Scott, D. L., Marks, J., and Blackley, J., Thiamine and ascorbic acid levels in the elderly. *Gerontol. Clin.* **9**, 1–10 (1969).

G9. Greogry, M. E., Water-soluble vitamins in milk and milk products. *J. Dairy Res.* **42**, 197–216 (1975).

H1. Høie, J., Experiments with feeding mink little fresh food (meat and fish). *Nor. Pelsdyrbl.* **21**, 2–16 (1954).

H2. Hilker, D. M., Chan, K. C., Chen, R., and Smith, R. L., Antithiamin effects of tea. I. Temperature and pH dependence. *Nutr. Rep. Int.* **4**, 223–227 (1971).

H3. Howarter, K. B., and Klein, B. P., Thiamine content and palatability of quick breads made with soy-egg flours. *J. Food Sci.* **43**, 1010–1011 (1978).

H4. Hoff-Jorgensen, E., and Hansen, B., A microbiological assay of vitamin B1. *Acta Med. Scand.* **9**, 562–566 (1955).

H5. Hennessy, D. J., and Cerecedo, L. R., The determination of free and phosphorylated thiamin by a modified thiochrome assay. *J. Am. Chem. Soc.* **61**, 179–183 (1939).

H6. Horwitt, M. K., and Kreisler, O., The determination of early thiamine deficient state by estimation of blood lactic and pyruvic acids after glucose administration and exercise. *J. Nutr.* **37**, 411–427 (1949).

H7. Hoyumpa, A. M., Jr., Nichols, S., Henderson, G. I., and Schenker, S., Intestinal thiamin transport: Effect of chronic ethanol administration in rats. *Am. J. Clin. Nutr.* **31**, 938–945 (1978).

H8. Hillman, R. W., Alcoholism and malnutrition. *In* "Biology of Alcoholism" (B. Kissin and H. Gegleiter eds.), Vol. 3, pp. 513–586. Plenum, New York, 1974.

H9. Howard, L., Wagner, C., and Schenker, S., Malabsorption of thiamine in folate deficient rats. *J. Nutr.* **104**, 1024–1028 (1974).

H10. Heaton, K. W., Alcoholic liver disease. *Br. J. Hosp. Med.* **18**, 118–120 (1977).

H11. Harper, C., Wernicke's encephalopathy: A more common disease than realised. *J. Neurol.* **42**, 226–231 (1979).

H12. Hoorn, R. K. J., Flikweert, J. P., and Westerink, D., Vitamin B1, B2 and B6 deficiencies in geriatric patients, measured by coenzyme stimulation of enzyme activities. *Clin. Chim. Acta* **61**, 151–162 (1975).

H13. Harrill, I., and Cervone, N., Vitamin status of older women. *Am. J. Clin. Nutr.* **30**, 431–440 (1977).

H14. Heller, S., Salkeld, R. M., and Körner, W. F., Vitamin B1 status in pregnancy. *Am. J. Clin. Nutr.* **27**, 1221–1224 (1974).

H15. Hytten, F. E., and Thomason, A. M., Nutrition of the lactating woman. *In* "Milk: The Mammary Gland and Its Secretion" (S. K. Kon and A. T. Cowie, eds.), Vol. 2, p. 18. Academic Press, New York, 1961.

I1. Iwata, H., Baba, A., Matsuda, T., and Terashita, Z., Properties of thiamine di- and triphosphatases in rat brain microsomes: Effect of chlorpromazine. *J. Neurol.* **24**, 1209–1213 (1975).

I2. Icke, G. C., The microbiological assay of thiamine and its clinical significance. MSc. Thesis, University of Western Australia, Perth (1980).

I3. Ishi, K., Sarai, K., Sanemori, H., and Kawasaki, T., Analysis of thiamine and its phosphate esters by high-performance liquid chromatography. *Anal. Biochem.* **97**, 191–195 (1979).

I4. Ikonomoff, I. V., Therapie der alkoholtrankheit. *Muench. Med. Wochenschr.* **120**, 905–908 (1978).

J1. Jansen, B. C. P., and Donath, W. F., On the isolation of the anti-beri-beri vitamin. *Proc. R. Acad. Amsterdam* **29**, 1390–1400 (1926).

J2. Jansen, B. C. P., Early nutritional researches on beriberi leading to the discovery of vitamin B1. *Nutr. Abstr. Rev.* **26**, 1–14 (1956).

J3. Jansen, B. C. P., IV. Biochemical systems. *In* "The Vitamins" (W. H. Sebrell and R. S. Harris, eds.), Vol. 3, pp. 425–442. Academic Press, New York, 1954.

J4. Johnson, R. E., Henderson, C., Robinson, P. F., and Consolazio, C. F., Comparative merits of fasting specimens, random specimens and oral loading tests in field nutritional survey. *J. Nutr.* **30**, 89–98 (1945).

J5. Joshi, U. M., Lahiri, A., Kora, S., Dikshit, S. S., and Virkar, K., Short term effect of Ovral and Norgestrel on the vitamin B6 and B1 status of women. *Contraception* **12**, 425–436 (1975).

J6. Jacob, M., Hunt, I. F., Dirige, O., and Swendseid, M. E., Biochemical assessment of the nutritional status of low-income pregnant women of Mexican descent. *Am. J. Clin. Nutr.* **29**, 650–656 (1976).

K1. Kositawattanakul, T., Tosukhowong, P., Vimokesant, S. L., and Panijpan, B., Chemical interactions between thiamin and tannic acid. II. Separation of products. *Am. J. Clin. Nutr.* **30**, 1686–1691 (1977).

K2. Krishnan, M. V., Mahajan, S. N., and Rao, G. R., Colorimetric determination of thiamine propyl disulphide and thiamine disulphide in their respective pharmaceutical preparations. *Analyst* **101**, 601–610 (1976).

K3. Komai, T., Kawai, K., and Shindo, H., Active transport of thiamine from rat small intestine. *J. Nutr. Sci. Vitaminol.* **2**, 163–177 (1974).
K4. Kishi, H., Nishii, S., Ono, T., Yamaji, A., Kasahara, N., Hiraoka, E., Okada, A., Itakura, T., and Takagi, Y., Thiamin and pyridoxal requirements during intravenous hyperalimentation. *Am. J. Clin. Nutr.* **32**, 332–338 (1979).
K5. Kon, S. K., A survey of the place of cows milk in the human diet and of the effects of processing on its nutritive value. *Br. Med. Bull.* **5**, 170–176 (1947).
K6. Kinnersley, H. W., Peters, R. A., and Reader, V., The curative pigeon test: A critique. *Biochem. J.* **22**, 276–291 (1928).
K7. Kinnersley, H. W., and Peters, R. A., Improvements in the use of the formaldehyde azo reaction of vitamin B1. *Biochem. J.* **32**, 1516–1520 (1938).
K8. Krause, R. F., *In* "Modern Nutrition in Health and Disease" (H. Wohl and R. S. Goodhart, eds.), 4th ed., p. 531. Lea & Febiger, Philadelphia, Pennsylvania, 1968.
K9. Kimura, M., Fujita, T., Nishida, S., and Itokawa, Y., Differential fluorometric determination of picogram levels of thiamine, thiamine monophosphate, diphosphate and triphosphate using high-performance liquid chromatography. *J. Chromatogr.* **188**, 417–419 (1980).
K10. Katsura, E., and Oiso, T., Beriberi. *W. H. O. Monogr. Ser.* **62**, 136–145 (1976).
L1. Lonsdale, D., Thiamine metabolism in disease. *CRC, Crit. Rev. Clin. Lab. Sci.* **5**, 289–313 (1975).
L2. Linklater, K. A., Dyson, D. A., and Morgan, K. T., Faecal thiaminase in clinically normal sheep associated with outbreaks of polioencephalomalacia. *Res. Vet. Sci.* **22**, 308–312 (1977).
L3. Leoschke, W. L., Fish in the raising of mink. *In* "Fish as Food" (G. Borgström, ed.), Vol. 2, pp. 435–441. Academic Press, New York, 1962.
L4. Lazarov, J., Effects of adrenalectomy and prednisolone on the absorption and phosphorylation of thiamine in rats. *J. Endocrinol.* **76**, 385–389 (1978).
L5. Luh, B. S., Karbassi, M., and Schweigert, B. S., Thiamine, riboflavin, niacin and color retention in canned small white and garbanzo beans as affected by sulfite treatment. *J. Food Sci.* **43**, 431–434 (1978).
L6. Leveille, G. A., Modified thiochrome procedure for the determination of urinary thiamine. *Am. J. Clin. Nutr.* **25**, 273–274 (1972).
L7. Lossy, F. T., Goldsmith, G. A., and Sarett, H. P., A study of test dose excretion of five B complex vitamins in man. *J. Nutr.* **45**, 213–224 (1951).
L8. Leevy, C. M., Cardi, L., Frank, O., Gellene, R., and Baker, H., Incidence and significance of hypo-vitaminemia in a randomly selected municipal hospital population. *Am. J. Clin. Nutr.* **17**, 259–271 (1965).
L9. Leevy, C. M., and Baker, H., Vitamins and alcoholism. *Am. J. Clin. Nutr.* **21**, 1325–1328 (1968).
L10. Lipton, M. A., Mailman, R. B., and Nemeroff, C. B., Vitamins, Megavitamin therapy and the nervous system. *Nutr. Brain* **3**, 195 (1979).
L11. Leigh, D., Subacute necrotizing encephalomyelopathy in an infant. *J. Neurol., Neurosurg. Psychiatry* **14**, 216–221 (1951).
L12. Lemoine, A., Devehat, C. L., Codaccioni, J. L., Monges, A., Bermond, P., and Salkeld, R. M., Vitamin B1, B2, B6 and C status in hospital inpatients. *Am. J. Clin. Nutr.* **33**, 2595–2600 (1980).
L13. Lowenstein, L., Lalonde, M., Deschenes, E. B., and Shapiro, L., Vitamin B12 in pregnancy and the puerperium. *Am. J. Clin. Nutr.* **8**, 265–275 (1960).
M1. Matsukawa, T., and Kawasaki, H., Studies on vitamin B1 and related compounds. XLV. Thiol-type thiamine derivatives -1. *Yakugaku Zasshi* **73**, 705–708 (1953).

M2. Murakami, M., Takahashi, K., Hirata, Y., and Iwamoto, H., Bitamin B1 to hosugen no hanno. *Bitamin* **34,** 71–77 (1966).

M3. Mitoma, C., Metabolic disposition of thiamine tetrahydrofurfuryl disulphide in dog and man. *Drug Metab. Dispos.* **1,** 698–703 (1973).

M4. Morrison, A. B., and Campbell, J. A., Vitamin absorption studies. I. Factors influencing the excretion of oral test doses of thiamine and riboflavin by human subjects. *J. Nutr.* **72,** 435–444 (1960).

M5. Mickelson, O., Condiff, H., and Keys, A., The determination of thiamine in urine by means of the thiochrome technique. *J. Biol. Chem.* **160,** 361–370 (1945).

M6. Melnick, D., and Field, H., Chemical determination, stability and form of thiamine in urine. *J. Biol. Chem.* **130,** 97–107 (1939).

M7. Marin, G. A., Clark, M. L., and Senior, J. R., Studies of malabsorption occurring in patients with Laënnecs cirrhosis. *Gastroenterology* **56,** 727–736 (1969).

M8. Murphy, J. V., Craig, L., and Glew, R., Leigh's disease: Biochemical nature of the inhibitor. *Pediatr. Res.* **8,** 392 (1974).

M9. MacLennan, W. J., Coombe, N. B., Martin, P., and Mason, B. J., The relationship of laboratory parameters to dietary intake in a long-stay hospital. *Age Ageing* **4,** 189–194 (1975).

M10. Morgan, A. G., Kelleher, J., Walker, B. E., Losowsky, M. S., Droller, H., and Middleton, R. S. W., A nutritional survey in the elderly: Blood and urine vitamin levels. *Int. J. Vitam. Nutr. Res.* **45,** 448–462 (1975).

M11. Metz, J., Festenstein, H., and Welch, P., Effect of folic acid and vitamin B12 supplementation on tests of folate and vitamin B12 nutrition in pregnancy. *Am. J. Clin. Nutr.* **16,** 472–479 (1965).

N1. Neal, R. A., and Sauberlich, H. E., Thiamin. *In* "Modern Nutrition in Health and Disease" (R. S. Goodhart and M. E. Shils, eds.), 5th ed., pp. 191–197. Lea & Febiger, Philadelphia, Pennsylvania, 1980.

N2. National Research Council Food and Nutrition Board, "Recommended Dietary Allowances," 7th Ed., p. 42. Natl. Acad. Sci., Washington, D.C., 1968.

N3. Niven, C. F., and Smiley, K. L., A microbiological assay method for thiamine. *J. Biol. Chem.* **150,** 1–9 (1943).

N4. Najjar, V. A., and Holt, L. E., Jr., Studies on thiamine excretion. *Bull. John Hopkins Hosp.* **67,** 107–124 (1940).

N5. Nail, P. A., Thomas, M. R., and Eakin, R., The effect of thiamin and riboflavin supplementation on the level of those vitamins in human breast milk and urine. *Am. J. Clin. Nutr.* **33,** 198–204 (1980).

N6. Nobile, S., and Woodhill, J. M., A survey of the vitamin content of some 2000 foods as they are consumed by selected groups of the Australian population. *Food Technol. Aust.* **25,** 80–100 (1973).

P1. Pelletier, O., and Madère, R., New automated method for measuring thiamine in urine. *Clin. Chem. (Winston-Salem, N.C.)* **18,** 1937–1942 (1972).

P2. Prebluda, H. J., and McCollum, E. V., A chemical reagent for thiamine. *J. Biol. Chem.* **127,** 495–503 (1939).

P3. Pearson, W. N., Biochemical appraisal of the vitamin nutritional status in man. *JAMA, J. Am. Med. Assoc.* **180,** 49–55 (1962).

P4. Parkhomenko, J. M., Rybina, A. A., and Khalmuradov, A. G., Separation of thiamine phosphoric esters on Sephadex cation exchanger. *In* "Methods in Enzymology" (D. B. McCormick and L. D. Wright, eds.), Vol. 62, p. 59. Academic Press, New York, 1979.

P5. Panjipan, B., and Detkriangkraikun, P., High voltage paper electrophoresis as an alternative method for thiamin determination in the presence of substances capable of interfering with thiochrome formation. *Am. J. Clin. Nutr.* **32,** 723–725 (1979).

P6. Penttinen, H. K., Electrophoretic separation of thiamine and its mono, di and triphosphate esters. In "Methods in Enzymology" (D. M. McCormick and L. D. Wright, eds.), Vol. 62, p. 68. Academic Press, New York, 1979.
P7. Price, J., and Theodoros, M. T., The supplementation of alcoholic beverages with thiamine—a necessary preventive measure in Queensland. Aust. N. Z. J. Psychiatry 13, 315–320 (1979).
P8. Peters, R. A., The biochemical lesion in vitamin B1 deficiency. Lancet 1, 1161–1165 (1936).
P9. Pincus, J. H., and Grove, I., Distribution of thiamine phosphate esters in normal and thiamine deficient brains. Exp. Neurol. 28, 477–483 (1970).
P10. Pincus, J. H., Solitare, G. B., and Cooper, J. R., Thiamine triphosphate levels and histopathology. Arch. Neurol. (Chicago) 33, 759–763 (1976).
P11. Pincus, J. H., Subacute necrotising encephalomyelopathy (Leigh's disease): A consideration of clinical features and etiology. Dev. Med. Child Neurol. 14, 87–101 (1972).
P12. Pincus, J. H., Cooper, J. R., Murphy, J. V., Rabe, E. F., Lonsdale, D., and Dunn, H. G., Thiamine derivatives in subacute necrotising encephalomyelopathy. A preliminary report. Pediatrics 51, 716–721 (1973).
P13. Platt, B. S., Thiamine deficiency in human beriberi and in Wernicke's encephalopathy. In "Thiamine Deficiency" (G. E. W. Wolstenholme and M. O'Connor, eds.). Churchill, London, 1967.
P14. Pestel, M., Sebastien, P., Rouffy, J., Tellement, J. P., Singlas, E., and Koin, G., Béribéri aigu par carence en thiamine non phosphorylée. Ann. Med. Interne 125, 507–512 (1974).
P15. Petersen, D. R., Labbe, R. F., Van Belle, G., and Chinn, N. M., Erythrocyte transketolase activity and sudden infant death. Am. J. Clin. Nutr. 34, 65–67 (1981).
R1. Rogers, E. F., Thiamine antagonists. Ann. N. Y. Acad. Sci. 98, 412–429 (1962).
R2. Rungruangsak, K., Tosukhowong, P., Panijpan, B., and Vimokesant, S. L., Chemical interactions between thiamin and tannic acid. I. Kinetics, oxygen dependence and inhibition by ascorbic acid. Am. J. Clin. Nutr. 30, 1680–1685 (1977).
R3. Rindi, G., and Ventura, U., Thiamine intestinal transport. Physiol. Rev. 52, 821–827 (1972).
R4. Rosenberg, H. R., "Chemistry and Physiology of the Vitamins." Wiley (Interscience), New York, 1945.
R5. Rindi, G., and de Giuseppe, L., A new chromatographic method for the determination of thiamine and its mono, di and triphosphates in animal tissues. Biochem. J. 78, 602–606 (1961).
R6. Roser, R. L., Andrist, A. H., Harrington, W. H., Naito, H. K., and Lonsdale, D., Determination of urinary thiamine by high pressure liquid chromatography utilizing the thiochrome fluorescent method. J. Chromatogr. 146, 43–53 (1978).
R7. Robertson, D. J., Wasan, S. M., and Skinner, D. B., Ultrastructural features of early brain stem lesions of thiamine-deficient rats. Am. J. Pathol. 52, 1081–1087 (1968).
R8. Rogers, L. E., Porter, S. P., and Sidbury, J. B., Thiamine-responsive megaloblastic anaemia. Pediatrics 74, 494–504 (1969).
R9. Read, D. J. C., The aetiology of the sudden infant death syndrome: Current ideas on breathing and sleep and possible links to deranged thiamine neurochemistry. Aust. N. Z. J. Med. 8, 322–336 (1978).
S1. Sebrell, W. H., Jr., and Harris, R. S., eds., "The Vitamins," Vol. 3, pp. 404–478. Academic Press, New York, 1954.
S2. Schopfer, W. H., "Plants and Vitamins," pp. 110–114. Chronica Botanica, Walthan, Massachusetts, 1949.

S3. Spector, R., Thiamine transport in the central nervous system. *Am. J. Physiol.* **230**, 1101–1107 (1976).

S4. Seltzer, J. L., and McDougal, D. B., Temporal changes of regional cocarboxylase levels in thiamine-depleted mouse brain. *Am. J. Physiol.* **227**, 714–718 (1974).

S5. Sklan, D., and Trostler, N., Site and extent of thiamin absorption in the rat. *J. Nutr.* **107**, 353–356 (1977).

S6. Smith, J. A., Foa, P. P., and Weinstein, H. R., Some toxic effects of thiamine. *Fed. Proc., Fed. Am. Soc. Exp. Biol.* **6**, 204 (1947).

S7. Schoepfer, W. M., and Jung, A., Un test végétal pour l'aneurine. Méthode critique et résultats. *Congr. Int. Tech. Chim. Ind. Agric., C.R., 5th, 1937* Vol. 1, pp. 22–26 (1937).

S8. Sarett, P., and Cheldelin, V. H., The use of *Lactobacillus fermentum* 36 for thiamine assay. *J. Biol. Chem.* **155**, 153–161 (1944).

S9. Shultz, A. L., and Natelson, S., Studies on the distribution and concentration of thiamine in blood and urine. *Microchem. J.* **17**, 109–118 (1972).

S10. Spector, H., Peterson, M. S., and Freidemann, T. E., "Methods for Evaluation of Nutritional Adequacy and Status," Natl. Acad. Sci.—Natl. Res. Counc., Washington, D. C., 1954.

S11. Saccani, F., and Neri, C., Separazione e determinazione di farmaci mediante resine a scambio ionico. Nota II. Piridossina, tiamina e ciancobalamina. *Boll. Chim. Farm.* **109**, 275–280 (1970).

S12. Saccani, F., and Neri, C., Separazione e determinazione di farmaci mediante resine a scambio ionico. Nota III. Cocarbossilasi, tiamina, piridossina e cianocabalamina. *Boll. Chim. Farm.* **109**, 344–346 (1970).

S13. Sauberlich, H. E., Biochemical alterations in thiamine deficiency—their interpretation. *Am. J. Clin. Nutr.* **20**, 528–542 (1967).

S14. Sun, D. C. H., Albacete, R. A., and Chen, J. J., Malabsorption studies in cirrhosis of the liver. *Arch. Intern. Med.* **119**, 567–572 (1967).

S15. Small, M. A., Longarinni, A., and Zamcheck, N., Disturbances of digestive physiology following acute drinking episodes in "skid-row" alcoholics. *Am. J. Med.* **27**, 575–585 (1959).

S16. Shaw, S., Gorkin, B. D., and Lieber, C. S., Effects of chronic alcohol feeding on thiamin status: Biochemical and behavioural correlates. *Am. J. Clin. Nutr.* **34**, 616 (1981).

S17. Shaw, S., and Lieber, C. S., Nutrition and alcoholism. *In* "Modern Nutrition in Health and Disease" (R. S. Goodhart and M. E. Shils, eds.), 6th ed., pp. 1220–1243. Lea & Febiger, Philadelphia, Pennsylvania, 1980.

S18. Sullivan, L. W., and Herbert, V., Suppression of haematopoiesis by ethanol. *J. Clin. Invest.* **43**, 2048–2062 (1964).

S19. Shreeve, J. E., and Edwin, E. E., Thiaminase-producing strains of *Cl. sporogenes* associated with outbreaks of cerebrocortical necrosis. *Vet. Rec.* **94**, 330 (1974).

S20. Scriver, C. R., MacKenzie, S., Clon, C. L., and Delvin, E., Thiamine-responsive maple-syrup-urine disease. *Lancet* **1**, 310–312 (1971).

S21. Schulman, J. D., Lustberg, T. J., Kennedy, J. L., Museles, M., and Seegmiller, J. E., A new variant of maple syrup urine disease (branched-chain ketoaciduria). *Am. J. Med.* **49**, 118–124 (1970).

S22. Soukop, M., and Calman, K. C., Thiamin status in cancer patients and the effect of 5-fluorouracil therapy. *Br. J. Cancer* **38**, 180 (1978).

T1. Todd, A. R., and Bergel, F., Aneurin. Part VII. A synthesis of aneurin. *J. Chem. Soc.* **1**, 364–367 (1937).

T2. Tracy, A. H., and Elderfield, R. C., Studies in the pyridine series. II. Synthesis of 2-methyl-3-(β-hydroxyethyl) pyridine and of the pyridine analog of thiamine (vitamin B1). *J. Org. Chem.* **6**, 54–62 (1941).

T3. Thomson, A. D., Frank, O., Baker, H., and Leevy, C. M., Thiamine propyl disulphide: Absorption and utilisation. *Ann. Intern. Med.* **74**, 529–534 (1971).

T4. Thomson, A. D., Baker, H., and Leevy, C. M., Patterns of ^{35}S-thiamine hydrochloride absorption in the malnourished alcoholic patient. *J. Lab. Clin. Med.* **76**, 34–45 (1970).

T5. Thomson, A. D., and Leevy, C. M., Observations on the mechanism of thiamine hydrochloride absorption in man. *Clin. Sci.* **43**, 153–163 (1972).

T6. Thornber, E. J., Biochemical studies of thiamine deficiency in the lamb. Ph.D. Thesis, Murdoch University, Western Australia (1979).

T7. Tomasulo, P. A., Kater, R. M., and Liber, F., Impairment of thiamine absorption in alcoholics. *Am. J. Clin. Nutr.* **21**, 1340–1344 (1968).

T8. Thomson, A., Baker, H., and Leevy, C. M., Thiamine absorption in alcoholism. *Am. J. Clin. Nutr.* **21**, 537–538 (1968).

T9. Takabe, M., and Itokawa, Y., An experimental study of thiamine metabolism in acute ethanol intoxication. *Experientia* **36**, 327–328 (1980).

T10. Terlecki, S., and Markson, L. M., Cerebrocortical necrosis. *Vet. Rec.* **71**, 508 (1959).

T11. Tanphaichitr, V., Vimokesant, S. L., Dhanamitta, S., and Valyasevi, A., Clinical and biochemical studies of adult beriberi. *Am. J. Clin. Nutr.* **23**, 1017–1027 (1970).

T12. Truswell, A. S., Konno, T., and Hansen, J. D. L., Thiamine deficiency in adult hospital patients. *S.Afr. Med. J.* **46**, 2079–2082 (1972).

T13. Thanangkul, O., and Amatayakul, K., Nutrition of pregnant women in a developing country—Thailand. *Am. J. Dis. Child.* **129**, 426–427 (1975).

T14. Trostler, N., Guggenheim, K., Havivi, E., and Sklan, D., Effect of thiamine deficiency in pregnant and lactating rats on the brain of their offspring. *Nutr. Metab.* **21**, 294–304 (1977).

V1. Volpe, J. J., and Marasa, J. C., A role for thiamine in the regulation of fatty acid and cholesterol biosynthesis in cultured cells of neural origin. *J. Neurochem.* **3**, 975–981 (1978).

V2. Victor, M., and Silby, H., Thiamine deficiency. *In* "Scientific Approaches to Clinical Neurology" (E. Goldensohn and S. Appel, eds.), Vol. 1, pp. 204–226. Lea & Febiger, Philadelphia, Pennsylvania, 1977.

V3. Vimokesant, S. L., Hilker, D. M., Nakornchai, S., Rungruangsak, K., and Dhanamitta, S., Effects of betel nut and fermented fish on thiamin status of North Eastern Thais. *Am. J. Clin. Nutr.* **28**, 1458–1463 (1975).

V4. Vorhees, C. V., Schmidt, D. E., Barrett, R. J., and Schenker, S., Effects of thiamin deficiency on acetylcholine levels and utilization in vivo in rat brain. *J. Nutr.* **107**, 1902–1908 (1977).

V5. Victor, M., Deficiency diseases of the nervous system secondary to alcoholism. *Postgrad. Med.* **50**, 75–79 (1971).

V6. Victor, M., Adams, R. D., and Collins, G. H., "The Wernicke-Korsakoff Syndrome. A Clinical and Pathological Study of 245 Patients, 82 with Post-mortem Examinations." Davis, Philadelphia, Pennsylvania, 1971.

V7. Viana, M. B., and Carvalho, R. I., Thiamine-responsive megaloblastic anaemia, sensorineural deafness and diabetes mellitus: A new syndrome? *Pediatrics* **93**, 235–238 (1978).

V8. Vir, S. C., and Love, A. H. G., Effect of oral contraceptive agents on thiamin status. *Int. J. Vitam. Nutr. Res.* **49**, 291–295 (1979).

V9. Van Den Berg, H., Schreurs, W. H. P., and Joosten, G. P. A., Evaluation of the vitamin status in pregnancy. *Int. J. Vitam. Nutr. Res.* **48**, 12–21 (1978).

V10. Vir, S. C., Love, A. H. G., and Thompson, W., Thiamin status during pregnancy. *Int. J. Vitam. Nutr. Res.* **50**, 131–140 (1980).

V11. Ventura, U., and Rindi, G., Transport of thiamin by the small intestine in vitro. *Experientia* **21**, 645–646 (1965).
W1. Windaus, A., Darstellung von kristalliziertem antineurischem vitamin aus hefe. *Nachr. Ges. Wiss. Gottingen* (1931).
W2. Williams, R. R., and Cline, J. K., Synthesis of vitamin B1. *J. Am. Chem. Soc.* **58**, 1504 (1936).
W3. Wuest, H. M., The history of thiamine. *Ann. N. Y. Acad. Sci.* **98**, 385–400 (1962).
W4. Wostmann, B. S., Knight, P. L., and Kan, D. F., Thiamine in germfree and conventional animals: Effect of the intestinal microflora on thiamine metabolism of the rat. *Ann. N. Y. Acad. Sci.* **98**, 516–527 (1962).
W5. Woolley, D. W., and White, A. C. G., Production of thiamine deficiency disease by the feeding of a pyridine analog of thiamine. *J. Biol. Chem.* **149**, 285–289 (1943).
W6. Waldenlind, L., An improved method for the determination of endogenous thiamine and its phosphate esters in biological material. *Nutr. Metab.* **23**, 38–41 (1979).
W7. Williams, G. W., Methods for the estimation of three vitamin dependent red cell enzymes. *Clin. Biochem.* **9**, 252–255 (1976).
W8. Wood, B., Breen, K. J., and Penington, D. G., Thiamine status in alcoholism. *Aust. N. Z. J. Med.* **7**, 475–484 (1977).
W9. Wells, D. G., and Marks, V., Anaemia and erythrocyte transketolase activity. *Acta Haematol.* **47**, 217–224 (1972).
W10. Wong, P. W. K., Justice, P., Smith, G. F., and Hsia, D. Y. Y., A case of classical "maple syrup urine disease" thiamine nonresponsive. *Clin. Genet.* **3**, 27–33 (1972).
W11. Walters, J. H., Neurological disease due to malnutrition. *Trans. R. Soc. Trop. Med. Hyg.* **60**, 128–135 (1966).
W12. Wolf, H., Studies on tryptophan metabolism in man: The effects of hormones and vitamin B6 on urinary excretion of metabolites of the kynurenine pathway. *Scand. J. Clin. Lab. Invest.* **33**, Suppl. 136 (1974).
W13. Weitzman, E. D., and Graziani, L. Sleep and the sudden infant death syndrome: A new hypothesis. *Adv. Sleep Res.* **1**, 327 (1974).
Y1. Yui, Y., Fujiwara, H., Mitsui, H., Wakabayashi, A., Kambara, H., Kawai, C., and Itokawa, Y., Furosemide-induced thiamine deficiency. *Jpn. Circ. J.* **42**, 744 (1978).
Y2. Yui, Y., Itokawa, Y., and Kawai, C., Furosemide-induced thiamine deficiency. *Cardiovasc. Res.* **14**, 537–540 (1980).

VITAMIN-RESPONSIVE INBORN ERRORS OF METABOLISM

K. Bartlett

Department of Clinical Biochemistry and Metabolic Medicine,
University of Newcastle upon Tyne,
Royal Victoria Infirmary,
Newcastle upon Tyne, England

1. Introduction 142
2. Biotin (Vitamin H) 144
    2.1. Function and Metabolism 144
    2.2. Propionicacidemia 148
    2.3. 3-Methylcrotonyl-CoA Carboxylase Deficiency 150
    2.4. Pyruvate Carboxylase Deficiency 152
    2.5. Acetyl-CoA Carboxylase Deficiency 153
    2.6. Combined Carboxylase Deficiency 153
3. Cobalamin (Vitamin B_{12}) 158
    3.1. Function and Metabolism 158
    3.2. Deficiency of R-type Cobalamin Binders 161
    3.3. Defects of Intestinal Absorption of Cobalamins 161
    3.4. Transcobalamin II Deficiency 161
    3.5. The Methylmalonicacidurias 162
4. Folate 163
    4.1. Metabolism and Function 163
    4.2. Inherited Folate Malabsorption 166
    4.3. Homocystinuria Due to 5,10-MethyleneTHF Reductase Deficiency 166
    4.4. Dihydrofolate Reductase Deficiency 167
    4.5. Glutamate Formiminotransferase Deficiency 167
5. Pyridoxine (Vitamin B_6) 168
    5.1. Function and Metabolism 168
    5.2. Cystathionine Synthase Deficiency 169
    5.3. Cystathionase Deficiency 172
    5.4. Kynureninase Deficiency 173
    5.5. Glyoxylate:2-Oxoglutarate Carboligase Deficiency 173
    5.6. Ornithine–2-Oxo-acid Aminotransferase 174
6. Riboflavin 174
    6.1. Function and Metabolism 174
    6.2. Congenital Methemoglobinemia 175

6.3.	Pyruvate Kinase Deficiency	175
6.4.	Glutaryl-CoA Dehydrogenase Deficiency	176
6.5.	Defects of β Oxidation	177
7. Thiamin		177
7.1.	Function and Metabolism	177
7.2.	Maple Syrup Urine Disease	179
7.3.	Congenital Lacticacidosis	180
7.4.	Thiamin-Responsive Megaloblastic Anemia	181
8. Conclusions		181
	References	183

1. Introduction

An inborn error of metabolism can be defined as the genetically determined absence of an enzyme activity. Since the pioneering insight of Garrod (G1) it has become apparent that almost any biochemical transformation in man can be affected. Inherited defects of amino acid, fat, carbohydrate, purine, pyrimidine, and complex lipid metabolism have been documented. Even such fundamental events as terminal electron transport are not excluded and cytochrome deficiencies are not, surprisingly, entirely incompatible with life. The absence or partial absence of an enzyme activity does not necessarily imply an inability to synthesize the appropriate protein, but can arise from, for example, defective regulatory mechanisms or cofactor transformations. It has been the need for effective treatment which has provided the impetus to the investigation of these disorders. In a number of inborn errors where the defective enzyme had a cofactor requirement, it was found that the administration of large amounts of the vitamin precursor of the cofactor resulted in a clinical and biochemical improvement. It is this class of disorders, the vitamin-responsive inborn errors of metabolism, which is the subject of this article.

On a theoretical basis, there are a number of sites where the metabolism of a vitamin can be disordered (Fig. 1). Some vitamins are absorbed from the intestinal lumen by complex mechanisms involving a number of proteins. Thereafter, specific carrier proteins in blood may be required for the delivery of the vitamin to tissues. There may be specialized cellular uptake processes followed by a series of intracellular or intraorganelle transformations prior to final activation of an apoenzyme to yield enzymatically active holoenzyme. A defect at any of these points would result in an attenuation of holoenzyme activity. In addition, a mutation which produced an apoprotein with a reduced affinity for the cofactor would result in the formation of holoenzyme at a reduced rate. Whether such a defect would result in a disordered pathway would depend upon the relative rates of synthesis and degradation of both apoenzyme and holoenzyme. For any of these defects to

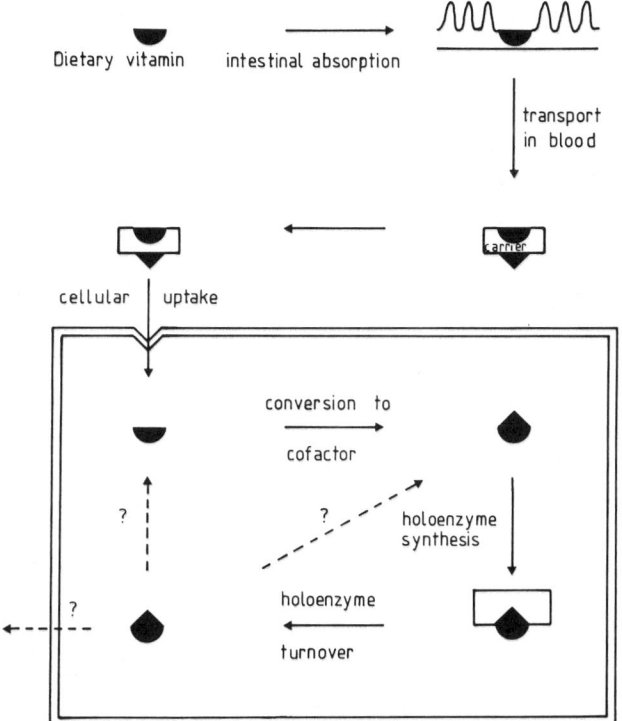

FIG. 1. Generalized scheme of vitamin metabolism showing the possible sites of genetic defects.

be responsive to the administration of large amounts of the appropriate vitamin, it is of course necessary that they be partial or "leaky." The total absence of any of the steps in enzyme activity would not be reversed by large amounts of the substrate (vitamin). From the foregoing, it can be seen that the vitamin-responsive inborn errors of metabolism are potentially an extremely complex group of disorders. This has to a large extent been borne out in practice, and there are examples of most of the theoretically possible defects in the literature.

Generally, vitamins are organic substances which are required for normal metabolism in small amounts which cannot be synthesized endogenously and therefore must be present in the diet. Presumably, at some point in evolution man has lost the ability to synthesize these compounds and instead relies upon their provision in the diet. In some inborn defects of cofactor-requiring enzymes, the *de novo* synthesis of cofactor is affected, and in individuals with these defects, cofactor has to be supplied in the diet. In this situation, the cofactor has become a vitamin as the result of a mutation. If the cofactor in question is present abundantly in the normal diet, then these individuals would not come to our attention as having metabolic defects. An example of such a mutation is the tetrahydrobiopterin-responsive variant of hyperphenylalaninemia. Finally, it is not clear to what extent a vitamin released by the turnover of a holoenzyme can be reutilized. If, for example, biotinidase, the enzyme which cleaves the biocytin released as a result of the degradation of biotin-dependent enzymes, was absent, it is possible that individuals lacking this enzyme would have an increased dietary requirement for biotin (see Section 2.6).

Because of the many advances in this field over the past few years, it is appropriate to review present knowledge. However, an exhaustive review of the literature is not possible here and this article will concentrate on the more recent developments, while making reference to earlier reviews. This article is limited to a consideration of water-soluble vitamins and the inborn errors with which they are associated.

2. Biotin (Vitamin H)

2.1. Function and Metabolism

There are four known mammalian biotin-dependent enzymes (Fig. 2). Three of these, propionyl-CoA carboxylase [PCC; (EC 6.4.1.3)], 3-methylcrotonyl-CoA carboxylase [MCC; (EC 6.4.1.4)], and pyruvate carboxylase [PC; (EC 6.4.1.1)], are located in the mitochondrial compartment of the cell, whereas acetyl-CoA carboxylase [AC; (EC 6.4.1.2)] is cytosolic. PCC is involved in the degradation of the amino acids valine, isoleucine, and methionine, and MCC is part of the degradative pathway of leucine. PC is the rate-determining step of gluconeogenesis from alanine and pyruvate, and AC, which catalyzes the formation of malonyl CoA, is the first committed step in fatty acid biosynthesis. Biotin is unusual in that it is covalently attached to the apocarboxylases via the ϵ-amino group of a lysine residue (Fig. 3), where it functions at the active site as a CO_2 carrier in the carboxylation reactions (M1). The formation of the holocarboxylases from biotin and the apocarboxylases is catalyzed by holocarboxylase synthetase in an ATP-dependent reaction (Fig. 2). There is some evidence that this activating enzyme has a dual subcellular localization (M2). The cytosolic enzyme presumably acts on apoAC, whereas the mitochondrial apocarboxylases are substrates for the mitochondrial holocarboxylase synthetase. It is not clear whether these two enzymes are identical, although there is some evidence that they are not (see Section 2.6).

The intestinal absorption of biotin has not been very well characterized.

1. **Cytosolic**

   Acetyl CoA carboxylase (EC 6.4.1.2)

   $$CH_3CO\text{-}SCoA + ATP \xrightarrow[CO_2]{Mg^{2+}} HOOCCH_2CO\text{-}SCoA + ADP$$

   (+) Citrate, K^+

2. **Mitochondrial**

   (a) Propionyl CoA carboxylase (EC 6.4.1.3)

   $$CH_3CH_2CO\text{-}SCoA + ATP \xrightarrow[CO_2]{Mg^{2+}} HOOC\underset{|}{CH}(CH_3)CO\text{-}SCoA + ADP$$

   (+) K^+

   (b) 3-Methylcrotonyl CoA carboxylase (EC 6.4.1.4)

   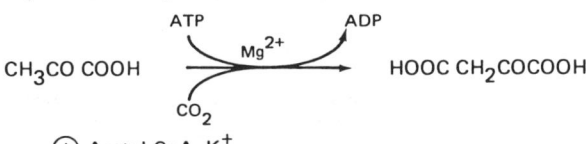

   (+) K^+

   (c) Pyruvate carboxylase (EC 6.4.1.1)

   $$CH_3CO\,COOH + ATP \xrightarrow[CO_2]{Mg^{2+}} HOOC\,CH_2COCOOH + ADP$$

   (+) Acetyl CoA, K^+

   FIG. 2. Mammalian biotin-dependent carboxylases.

There appear to be two species-dependent mechanisms among mammals. Spencer and Brody (S1) demonstrated saturation kinetics for biotin uptake by everted intestinal sacs with an apparent K_m with respect to biotin of 0.06 mM in hamster, mouse, chipmunk, gerbil, and squirrel. In the case of rat, rabbit, guinea pig, and ferret, however, saturation kinetics were not observed and only passive diffusion was demonstrated. In a more recent study, where the uptake of biotin by hamster small intestine mucosal cells was measured, an apparent $K_m = 1$ mM was found (B1). No systematic studies of biotin absorption in man have been reported.

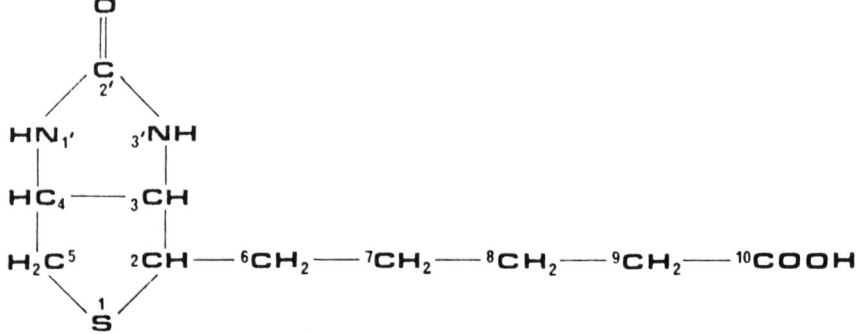

FIG. 3. Structure of biotin (2'-keto-3,4-imidazolido-2-tetrahydrothiophene-n-valeric acid).

The concentration of biotin in human plasma is about 0.5–3.0 nM. The reported ranges are summarized in Table 1. There are two commonly used methods for the estimation of biotin in plasma, a bioassay using either *Lactobacillus plantarum* or *Ochromonas danica* (B2, B3), and a competitive-binding assay using biotin as the biotin binder (H1, H2). The *Lactobacillus*-based bioassay has been criticized for its lack of specificity, although the assay utilizing *O. danica* does not appear to have this problem. The competitive-binding methods for the assay of biotin use either tritiated biotin or an iodinated derivative (H1, H2). The latter method has the advantage of being

TABLE 1
CONCENTRATION OF BIOTIN IN BLOOD AND PLASMA
DETERMINED BY A VARIETY OF METHODS

| Method | Tissue | Concentration (nM) | | | Reference |
		Mean	Range	(n)	
1[a,b]	Blood	1.1	0.5–1.7	25	
2[a,c]	Blood	2.4	0.8–4.1	76	
3[a,d]	Blood	6.0	3.6–11.1	12	B4
4[a,c]	Plasma	1.3	0.9–1.7	12	
5[a,c]	Plasma	1.3	0.6–2.4	31	
6[a,e]	Plasma	6.7	5.3–9.4	12	
7[f]	Plasma	1.26	0.7–3.3	30	H2
8[f]	Plasma	0.7	0.3–1.2	13	B7

[a] Bioassay.
[b] *L. plantarum*, samples acid hydrolyzed.
[c] *O. danica*, samples treated with papain.
[d] *L. plantarum*, samples treated with papain.
[e] *L. casei*, samples acid hydrolyzed.
[f] Avidin competitive-binding assay.

the most sensitive, and biotin can be measured in as little as 10–20 μl of plasma. This consideration is of some importance since investigations of neonates are frequently necessary. A typical calibration curve is shown in Fig. 4. It is theoretically possible to make this assay even more sensitive. If, for example, a biotin derivative of pentalysine was iodinated with Bolton and Hunter reagent instead of hexamethylenediamine, a radioactive conjugate of much higher specific activity would be produced. Since it is the specific activity of the labeled ligand which determines the maximum sensitivity of a competitive-binding assay, much lower amounts of biotin could be measured. However, problems with contamination may place a lower limit on the amounts that could be estimated.

Recently, it has been reported that acid hydrolysis increases the plasma biotin as measured by a competitive-binding assay (S2). This finding is surprising since the binding of biotin by avidin is extremely tight ($K_d = 10^{-15}$ M) and thus incubation of plasma would perhaps be expected to strip biotin from any putative transporting protein. However, the existence of such a

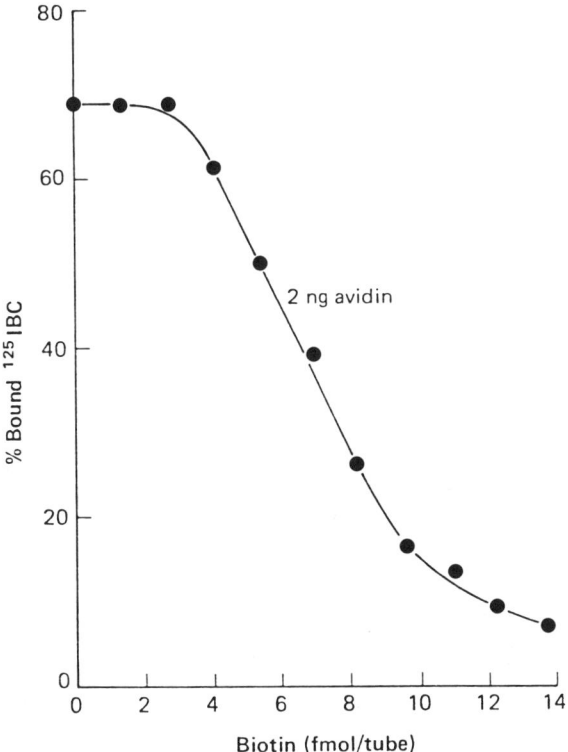

FIG. 4. Competitive-binding assay of biotin.

binder has been demonstrated in hen plasma (M3), and clearly, further studies are required to elucidate this point in man.

The uptake of biotin by yeast has been shown to be carrier mediated and inhibited by p-nitrophenylbiotin. A similar mechanism in cultured human fibroblasts with an apparent $K_m = 80$ nM has been demonstrated (G2). Since the concentration of biotin in plasma is about 1 nM, the rate of uptake of biotin by tissues must be determined by the circulating level of the vitamin. The avidin–biotin complex is taken up by cultured human fibroblasts and HeLa cells by a pinocytotic mechanism. There is, however, no evidence that biotin entering the cell in this way is available for holocarboxylase synthesis (D1). The intracellular metabolism of biotin has received little attention. Biotin can cross the inner mitochondrial membrane and enter the matrix space, but the mechanism of this transport has not been studied (P1).

The normal daily requirement for biotin is probably about 35 µg for neonates, rising to about 150 µg for adults (D2), however, it is likely that the gut flora contributes some biotin, since the urinary plus fecal excretion is in excess of daily intake (B4). Biotinidase is a peptidase found in blood and liver and is specific for ϵ-N-biotinyllysine (biocytin). Biocytin is thought to arise from the turnover of biotin-containing enzymes (P2). It is possible that biotin from this source might be important in the whole-body economy of the vitamin, although the extent to which this occurs is not known.

2.2. Propionicacidemia

Propionyl CoA arises from the degradation of valine, isoleucine, methionine, thymine, the action of some gut flora, and from the oxidation of the side chain of cholesterol. The pathway by which propionyl CoA enters the Krebs cycle via carboxylation to methylmalonyl CoA, racemization, and isomerization to succinyl CoA is shown in Fig. 5, and has recently been reviewed in detail (R1).

The ketotic–hyperglycinemia syndrome, first delineated by Child *et al.* in 1961 (C1), was in reality the first description of propionicacidemia (H3, H4). The patient, a child, presented with coma, dehydration, ketoacidosis, and hyperglycinemia, and died at the age of 7 years after repeated episodes of ketoacidosis precipitated by infections and the ingestion of protein. Subsequently, Hommes *et al.* (H5) described a child with a devastating acidosis who died at the age of 5 days, and who had a markedly elevated plasma propionate concentration of 5.4 mM, about 1000 times normal. Although indirect evidence suggested a primary defect of PCC in these patients (H3, H4), the first direct measurement of the enzyme and the demonstration of its deficiency was provided by Gompertz *et al.* in another patient with severe ketoacidosis and propionicacidemia who died in the neonatal period (G3).

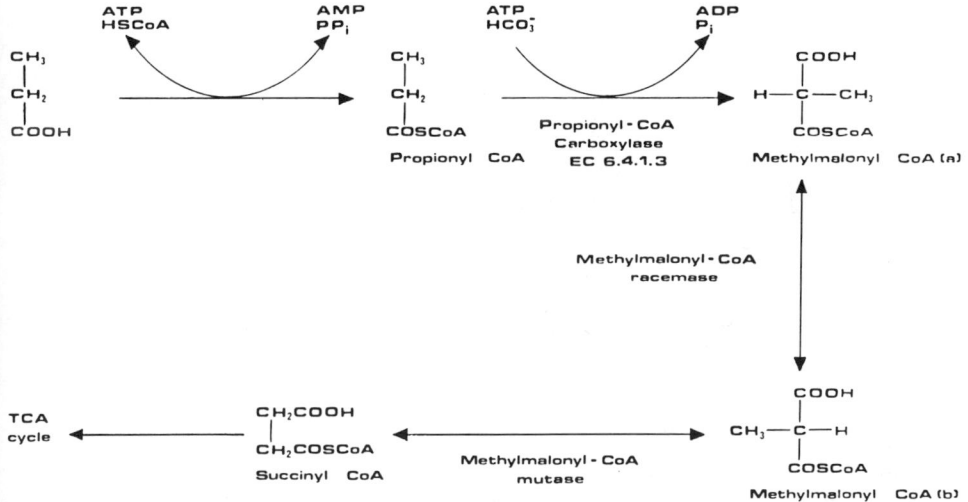

FIG. 5. Metabolism of propionate.

Since these early studies, a number of patients with PCC deficiency have been documented. It is now apparent that there is a wide variation in clinical presentations, and patients with developmental retardation without ketoacidosis, late-presenting variants, and indeed, patients with unremarkable plasma concentrations of propionate have been described (D2). This clinical heterogeneity is accompanied by an equally diverse pattern of abnormal urinary metabolites (D3) which are produced as a result of the intramitochondrial accumulation of propionyl CoA. Of these, 3-hydroxypropionate, methyl citrate, and propionylglycine are particularly prominent (A1, A2, R2). This clinical and metabolic diversity has been reemphasized by complementation studies, although even within the same complementation group there remains a large degree of clinical variation (G4).

In the majority of patients with isolated PCC deficiency, treatment has emphasized protein restriction (0.5–1.5 g/kg/day), with empirical treatment of the acidosis with alkali and peritoneal dialysis when biochemical control is lost. However, in two patients, treatment with large amounts of biotin (5–10 mg b.d.) has been shown to result in biochemical improvement (B5, H6). In the first documented use of pharmacological doses of biotin, only slight clinical improvement was noted (B5). However, a marked rise in plasma propionate following isoleucine loading (100 mg/kg) was greatly diminished after biotin treatment (Fig. 6). In the second case, a child with the rumination syndrome was found to have a moderately elevated plasma propionate (40 μM) which fell to 5 μM with biotin therapy. Leukocyte PCC activity was

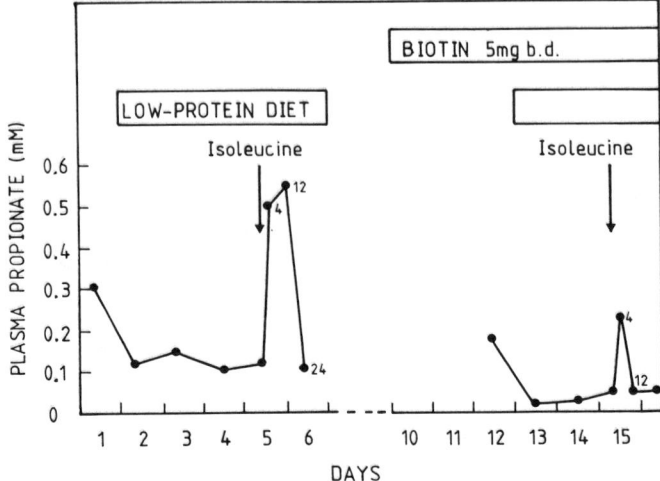

Fig. 6. Biotin-responsive isolated propionicacidemia. The response to an isoleucine load with and without biotin treatment. (Redrawn from B5 with permission.)

16% of control values prior to biotin and returned to normal with biotin treatment (H6). A biotin response was also noted in cultured fibroblasts. A combined carboxylase defect (see Section 2.6) was excluded because MCC activity was normal in fibroblasts.

2.3. 3-Methylcrotonyl-CoA Carboxylase Deficiency

This inborn error of leucine degradation (Fig. 7) was first known as 3-methylcrotonylglycine- and 3-hydroxyisovalericaciduria, and the first case, a child, was reported by Eldjarn and co-workers in 1970 (E1). A second child who excreted these metabolites as well as tiglylglycine was subsequently shown to have a combined defect of PCC and MCC (B6, G5), and illustrates the confusion which can arise when inborn errors are diagnosed purely on the basis of the abnormal urinary metabolites found. The first direct evidence for the enzyme defect was provided by Gompertz et al. (G6). Since these early cases appeared, about 20 patients have been described, having either isolated MCC deficiency or that combined with deficiencies of other carboxylases (G7, L1). The majority of cases of isolated MCC deficiency have presented in infancy, but presented with a clinical heterogeneity comparable to PCC deficiency (L1). All but three were clinically and biochemically responsive to biotin, and in contrast to isolated PCC deficiency, biotin responsiveness was the rule rather than the exception. In some cases of apparent biotin-responsive, isolated MCC deficiency, subsequent investigation

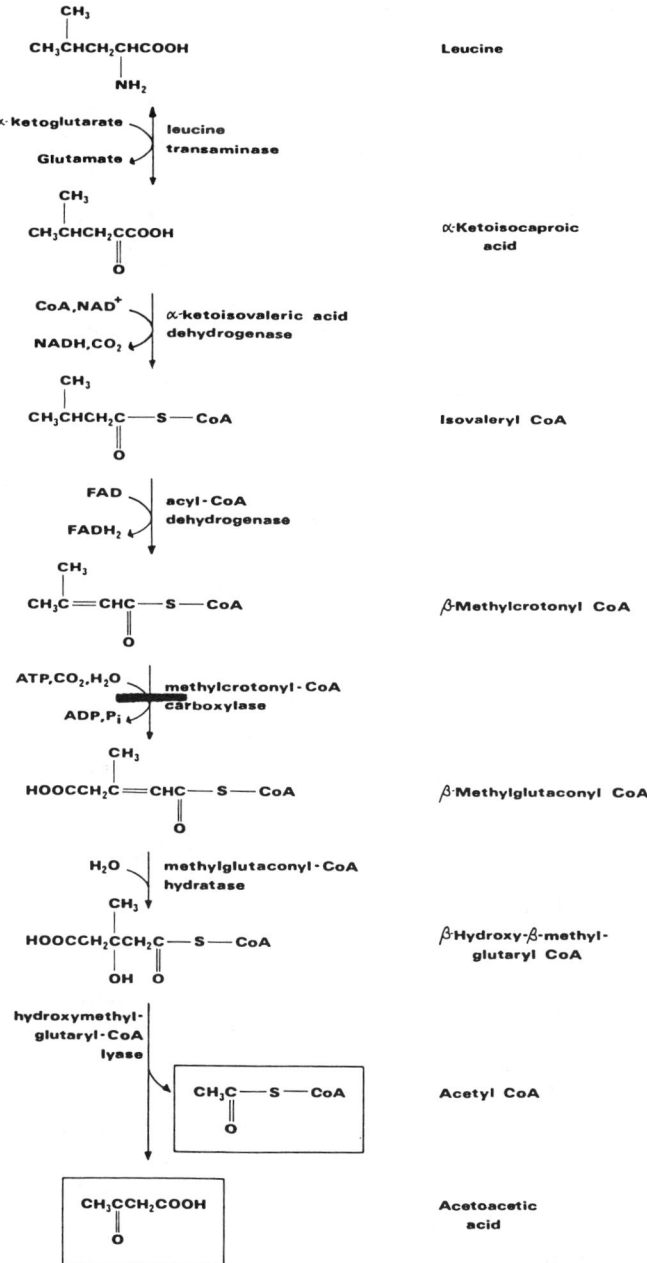

FIG. 7. The metabolism of leucine showing the site of the defect in 3-methylcrotyonyl-CoA carboxylase deficiency.

has revealed a combined defect (B5). It is therefore of importance to specifically exclude PCC deficiency when making a diagnosis of isolated MCC deficiency. Thus, whenever the metabolites characteristic of either PCC or MCC deficiency are found by routine gas-chromatographic screening, all four biotin-dependent enzymes should be measured. Either leukocytes or fibroblasts can be used, although both tissues have drawbacks. The culture of fibroblasts allows the manipulation of the biotin concentration and further investigation of the enzymology. However, this tissue has a disadvantage in that in cases of inherited defects of biotin absorption, normal enzyme activity is observed. In addition, in two cases the enzyme defect was not expressed in cultured cells (E1, L1). In these instances, leukocytes are the more appropriate tissue and can also be used to monitor biotin treatment. The methodology for the measurement of carboxylases in both of these tissues has been described elsewhere (B7). All of the carboxylase assays are based upon the substrate-dependent fixation of $^{14}CO_2$, although in the case of PC and AC, further complications are introduced by the requirement of these enzymes for activators—acetyl CoA and citrate, respectively. Indirect methods such as the incorporation of labeled precursors into trichloroacetic acid-insoluble material or the oxidation of labeled precursors to $^{14}CO_2$ are at best only confirmatory.

2.4. Pyruvate Carboxylase Deficiency

Isolated hepatic pyruvate carboxylase deficiency was first reported by Hommes et al. (H7) in a patient who presented with Leigh's encephalomyelopathy, hyperlacticacidemia, hyperpyruvicacidemia, and a moderate degree of hypoglycemia. Some subsequently described cases follow this pattern, with variations in the age of onset and in the degree of neurological involvement. Early studies relied upon liver biopsy for enzyme diagnosis, an approach which precluded family studies and antenatal diagnosis. In more recent studies, PC deficiency has been demonstrated in fibroblasts and leukocytes (A3, D4) and suggests an autosomal mode of inheritance, in common with PCC and MCC deficiencies. To date, no biotin-responsive variants of isolated PC deficiency have been reported, although biotin therapy has been attempted in a number of instances (D4, H8). Hyperlacticacidemia is a relatively common finding, and since there are a number of other possible causes, such as pyruvate dehydrogenase deficiency, phosphoenolpyruvate carboxykinase deficiency, glycogen storage disease type 1, cytochrome defects, and fructose biphosphatase deficiency, it is sometimes difficult to decide in which case PC should be measured. Indeed, it can be difficult to establish congenital lacticacidemia, as very high blood lactate levels have been observed in healthy but screaming infants (K1). Some have advocated the use of alanine loading tests to check the patency of gluconeogenesis (F1), but this approach has been criticized for the high incidence of false negatives and positives due to secondary endocrine effects (L2).

2.5. Acetyl-CoA Carboxylase Deficiency

One case of isolated AC deficiency has been reported (B8). A newborn girl with a hypotonic myopathy and severe brain damage was found to excrete 2-ethyl-3-ketohexanoic acid, 2-ethyl-3-hydroxyhexanoic acid, and 2-ethylhexanedioic acid. Blom *et al.* postulated that these metabolites arose from the condensation of two molecules of butyryl CoA, and that the butyryl CoA was derived from acetyl CoA by reverse β oxidation. The acetyl CoA accumulated because of the reduced activity of AC, which was demonstrated in a liver biopsy and in cultured skin fibroblasts. Biotin therapy was not attempted.

2.6. Combined Carboxylase Deficiency

As described in Section 2.2, the first demonstrated response to biotin was in a child with isolated PCC deficiency, where the response to an isoleucine load was attenuated by the administration of biotin (Fig. 6) (B5). The first patient in whom biotin had an obvious and dramatic effect was described by Gompertz and co-workers (G5) and was a child with 3-methylcrotonylglycinuria and tiglylglycinuria. The administration of biotin had an immediate

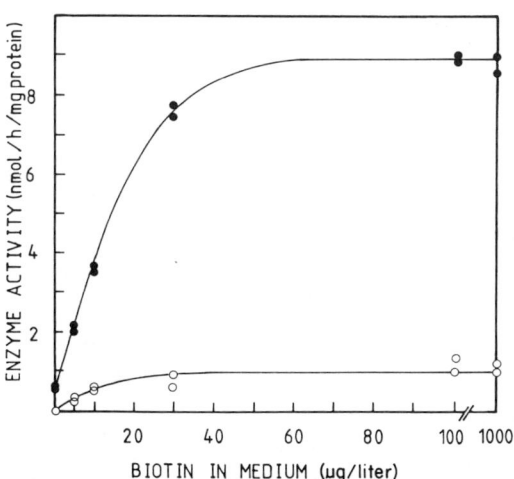

Fig. 8. Biotin-responsive combined carboxylase deficiency, the effects of biotin therapy on protein tolerance, urinary metabolites, and acid–base balance. (Redrawn from G5 with permission.)

effect on the metabolic acidosis and the abnormal urinary metabolites disappeared (Fig. 8) (G5). Investigations by others demonstrated that this child excreted metabolites characteristic of PCC deficiency such as propionylglycine, methyl citrate, and 3-hydroxy propionate (A1, A2, R2). Direct enzyme measurement confirmed this and led to the realization that this was the first case of a generalized defect of biotin-dependent enzymes (B6, W1).

The *in vivo* response to biotin was also demonstrated in cultured fibroblasts (Fig. 9) (B6). If fibroblasts are cultured in Eagle's minimum essential medium (MEM), the only source of biotin is the fetal calf serum which is usually present at a concentration of 10% (v/v). Since fetal calf serum has a biotin concentration of about 100 nM, the final medium biotin concentration is about 10 nM, which is 5–10 times higher than that in human plasma (B7). The specific activities of PCC and MCC in the patient's fibroblasts were low when cultured under these conditions. If, however, the culture medium was supplemented with additional biotin, the specific activities rose to about half

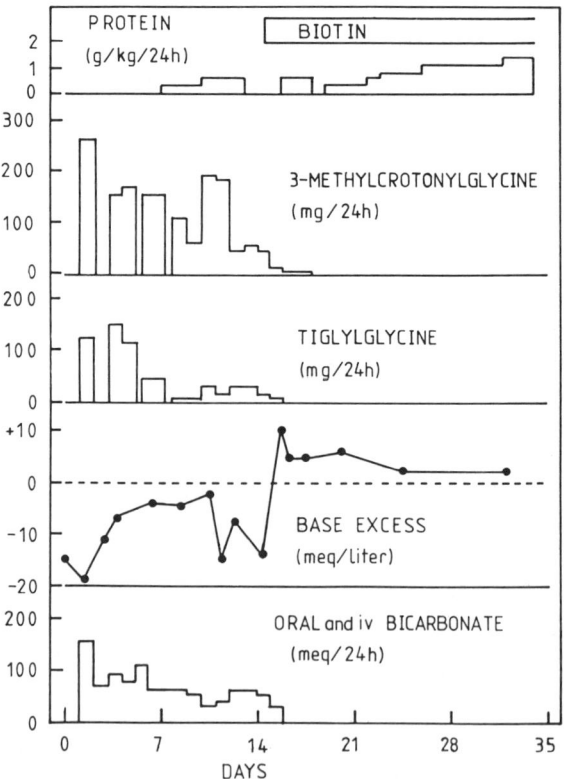

FIG. 9. Biotin responsiveness in cultured fibroblasts from a patient with combined carboxylase deficiency. (Redrawn from B6 with permission.)

of the normal values. This response to biotin could also be demonstrated by growing the cells in normal MEM and then exposing them for short periods of time to high concentrations of biotin. This response was not blocked by protein synthesis inhibitors (B9). These findings were interpreted as meaning that the apocarboxylases were being synthesized normally, a view which was supported by immunological studies in cells from another patient with the same defect (M4). It was suggested that the primary defect was either of cellular uptake, of intracellular transport, or of the holocarboxylase synthetase (B6). Evidence that it was the latter mechanism has been provided by Burri *et al.* (B10), who utilized the low specificity of the holocarboxylase synthetase with respect to the apocarboxylase (M5) to demonstrate an increased K_m with respect to biotin. In this study, fibroblasts from the patient originally described by Gompertz *et al.* (G5) were cultured in MEM. A homogenate of these cells was incubated with semipurified apoPCC isolated from biotin-deficient rat liver, ATP, Mg^{2+}, and varying concentrations of biotin, and the resultant holoPCC was assayed. They found a K_m with respect to biotin 60 times the value of that of the control fibroblasts. Although these workers have undoubtedly identified the primary defect in combined carboxylase deficiency, the experimental approach can be criticized on a number of grounds. First, it is probable that there are two holocarboxylase synthetases, one cytosolic and the other mitochondrial. Second, they report a pH optimum of 8.5, in contrast to the literature value of 6.5 for the rat liver synthetase. This latter point is of some importance, as biotinylation of apo-PCC will continue during the assay of holoPCC, which is done at the same pH and makes the kinetic data difficult to interpret. Using a different experimental approach, Ghneim and Bartlett have come to the same conclusion (G8) as Burri *et al.* It is possible to culture fibroblasts in media depleted of biotin (0.02 nM biotin) by passing the fetal calf serum through a column of immobilized avidin (G9). The specific activities of all four biotin-dependent enzymes in normal cells grown under these conditions are greatly reduced and return to normal by the addition of biotin. It is possible to measure the K_m with respect to biotin for the formation of holoPCC by using the endogenous, accumulated apoPCC in homogenates of mitochondria prepared from cells cultured in a biotin-depleted medium. When the K_m with respect to biotin was determined in mitochondria derived from biotin-deficient fibroblasts from a child with a combined defect of all three mitochondrial carboxylases (B11), a value 470 times that of normal was found (Fig. 10) (G8). Using this system, it was found that holoPCC synthetase was inactive at pH 8, and that the first stage of the assay had to be done at pH 6.5.

Dietary deficiency of biotin causes skin rash, hair loss, and, with prolonged deficiency, neurological symptons, and can be induced by the ingestion of raw egg white, which contains avidin (S3, S4). Biotin deficiency can also be a complication of parenteral nutriton (M6). Sweetman (P3) has made

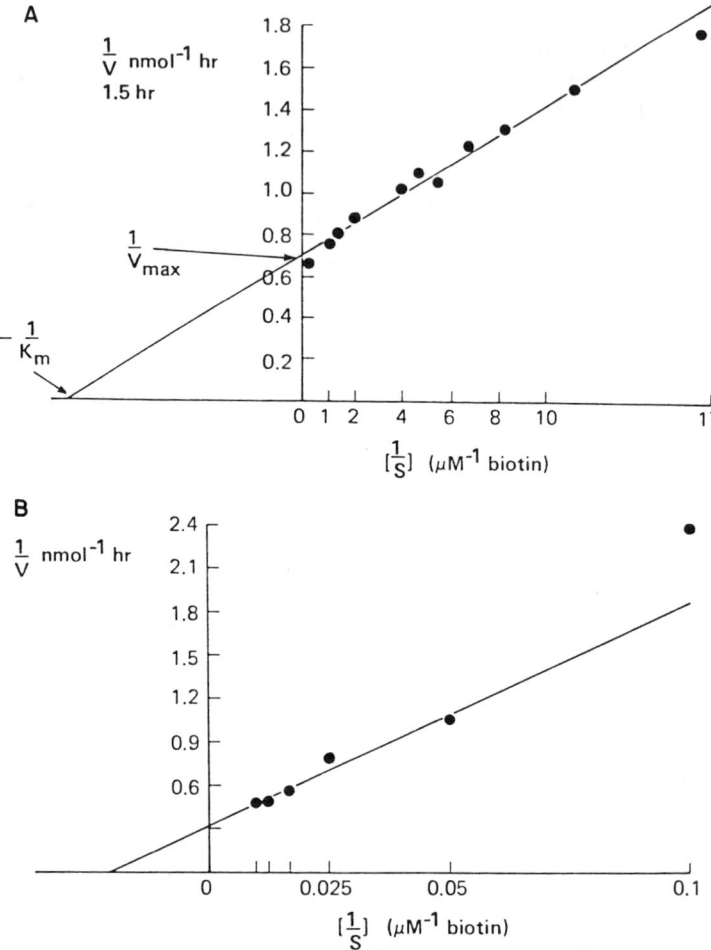

FIG. 10. The mechanism of biotin responsiveness in combined carboxylase deficiency. Lineweaver–Burke plots of holopropionyl-CoA carboxylase synthetase from biotin-deficient fibroblast mitochondria. (A) Control ($K_m = 0.11$ μM). (B) Patient ($K_m = 54.0$ μM). (Redrawn from G8 with permission.)

a distinction between neonatal and infantile forms of congenital combined carboxylase deficiency. Recently, a defect in the intestinal absorption of biotin has been reported (M7). In all of these instances, the specific activities of the carboxylases in leukocytes are reduced and return to normal following the administration of biotin. However, only in the congenital defects of intracellular biotinylation can a defect be demonstrated in cultured skin fibroblasts. Recent studies (W11) have shown that the late-onset form of

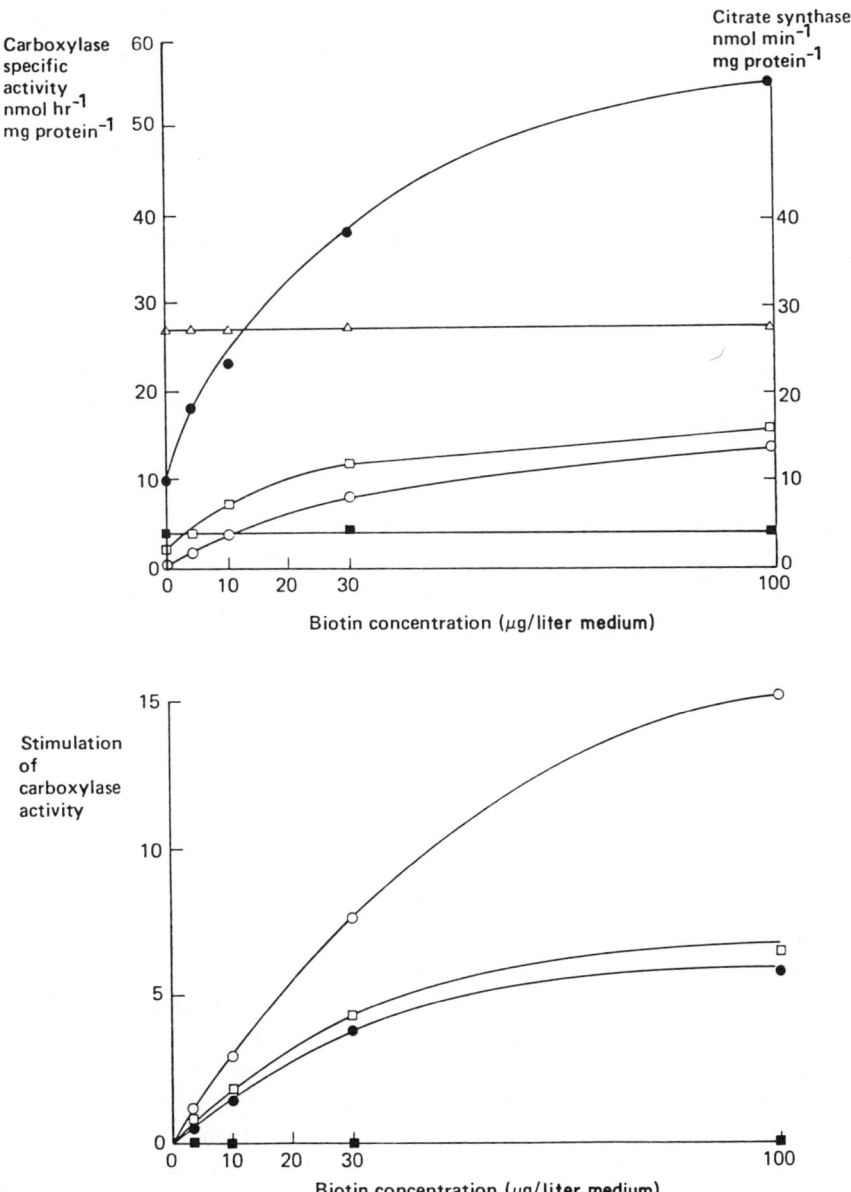

FIG. 11. Biotin responsiveness in fibroblasts from a patient with a combined defect of the three mitochondrial carboxylases. (Redrawn from B11 with permission.)

combined carboxylase deficiency is caused by biotinidase deficiency, that is, an inability to reutilize endogenous biotin derived from the turnover of biotin-dependent enzymes, and results in a functional biotin-deficiency state. Existing assays for biotin do not distinguish biotin and biocytin, resulting in the anomalous plasma biotin concentrations found in supposed defects of biotin transport. The degree of impairment is dependent upon the concentration of biotin in the culture medium (Fig. 11). Since the biotin content of fetal calf serum is high and variable (B7), it is advisable to use a biotin-depleted medium supplemented with known amounts of the vitamin.

Thus, it is apparent that a combined defect of the biotin-dependent carboxylases can arise as a result of either a defect of the enzyme which attaches biotin to the apocarboxylases, or as a result of a reduced availability of biotin. Combined carboxylase deficiency has been associated with defects of the immune system. Cowan et al. (C2) reported defects in T cell and B cell immunity in three patients. Furthermore, a deficiency of prostaglandin E_2 in monocytes from a patient with a combined deficiency of PCC, PC, and MCC has been demonstrated (M8), and suggests that AC might also be affected. Direct evidence for the involvement of AC has been provided in two patients (F2, M9). It has been suggested that a more appropriate name for congenital combined or multiple carboxylase deficiency is holocarboxylase synthetase deficiency (B10). This is correct only if the enzyme is actually measured, but not if, for example, the biochemical evidence is provided by analysis of urinary metabolites or measurement of leukocyte holocarboxylase activities.

3. Cobalamin (Vitamin B_{12})

3.1. FUNCTION AND METABOLISM

Adenosylcobalamin (adoCbl) is required for the activity of methylmalonyl-CoA mutase [MMM; methylmalonyl-CoA CoA carbonylmutase (EC 5.4.99.2)], whereas methylcobalamin (MeCbl) is required for the activity of homocysteine–methyltetrahydrofolate methyltransferase [HMMT; 5-methyltetrahydropteroyl-L-glutamate:L-homocysteine S-methyltransferase (EC 2.1.1.13)]. The positions of these two enzymes in intermediary metabolism are shown in Figs. 5 and 12, respectively, and have been known for some years (for reviews, see B12, R1, Z1). Recently, a third mammalian cobalamin-requiring enzyme, leucine-2,3-aminomutase, has been identified and catalyzes the conversion of β-leucine to α-leucine (P4, P5). This finding raises many interesting questions regarding, for example, the nutritional essentiality of leucine. Inherited defects of cobalamin-dependent enzymes resulting from impaired absorption, transport, and transformation of the vitamin, as well as apoenzyme mutations, have been described, and thus,

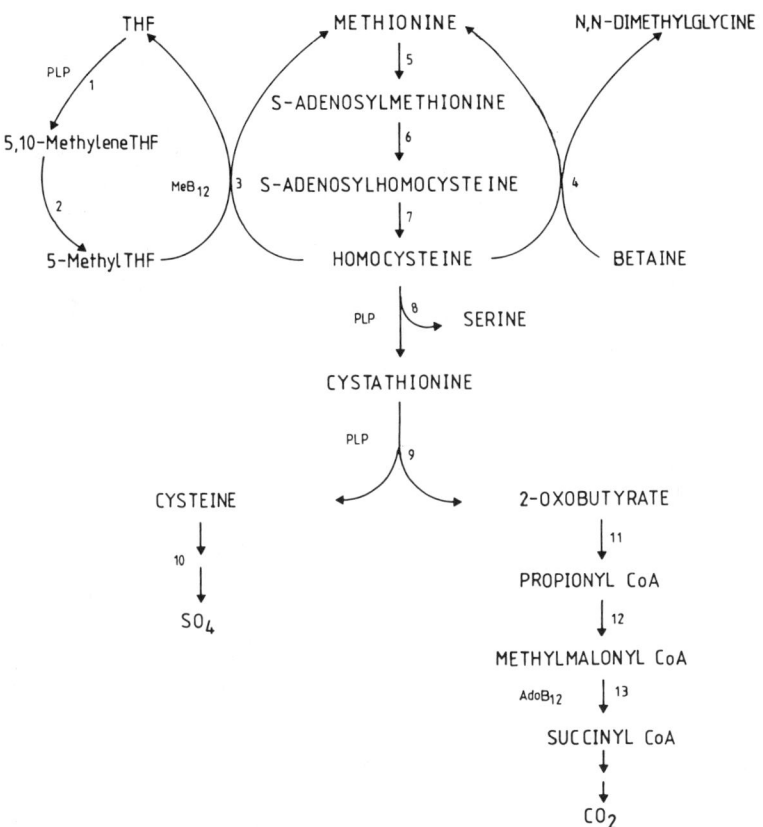

FIG. 12. The transsulfuration pathway—the metabolism of sulfur amino acids. THF, 5,6,7,8-Tetrahydrofolate; PLP, pyridoxal phosphate; AdoB$_{12}$, adenosylcobalamin; MeB$_{12}$, methylcobalamin; 1, serine hydroxymethyltransferase; 2, methyleneTHF reductase; 3, homocysteine methylase; 4, betaine–homocysteine methyltransferase; 5, methionine adenosyltransferase; 6, methyltransferase; 7, adenosylhomocysteinase; 8, cystathionine synthase; 9, cystathionine lyase; 10, cysteine oxidase; 11, 2-oxo-acid dehydrogenase; 12, propionyl-CoA carboxylase; 13, methylmalonyl-CoA mutase.

most of the theoretically possible defects outlined in Section 1 have been documented. Cyanocobalamin, although the most commonly used therapeutic form of the vitamin, has no function in mammals and arises from the methods used in the preparation of cobalamin (Z1).

The intestinal absorption of cobalamin is thought to occur as follows: Cobalamin is bound by R protein in gastric juice (M10, M11) and then by intrinsic factor (IF, a glycoprotein) in the jejunum following the partial proteolysis of R protein by pancreatic proteases (A4, A5, P6). The IF–Cbl complex is bound by a specific ileal receptor (D5, H9, H10, K2), which has been purified (S5, S6) and has a high affinity ($K_a = 1-4$ nM) for IF–Cbl (dog).

The fate of the IF–Cbl bound to the brush borders of ileal enterocytes is not clear, but available evidence suggests that endocytosis is followed by the formation of secondary lysosomes in which the IF–Cbl complex is split. The cobalamin is then released into the cytosol and enters the portal circulation (J1). This intralysosomal phase could account for the 2-hour delay which occurs between the ileal binding of IF–Cbl and the appearance of Cbl in the portal circulation (B13, S7).

Cobalamin is transported in the plasma by specific carrier proteins called the transcobalamins (TC), of which the best characterized is transcobalamin II (TC2). TC2 is involved in tissue uptake of Cbl (A6, A7, M1, Y1), and although quantitatively minor (M12), it is probably functionally the most important plasma Cbl binder. The TC2–Cbl complex is taken up by cultured fibroblasts by a process of absorbtive endocytosis (Y2), and studies have suggested the existence of a specific cellular receptor for TC2. However, recent studies have shown that cultured fibroblasts from normal and TC2-deficient subjects are able to absorb free Cbl in the absence of TC2 by a process which is saturable, low capacity, Ca^{2+} independent (in contrast to TC2-mediated endocytosis), cycloheximide sensitive, and sulfhydryl and energy dependent (B14). This mechanism is similar to the free Cbl-specific uptake found in bacteria (S8, T1) and to free Cbl binding by isolated rat liver membranes and HeLa cells (H11, F3). The biological significance of this mechanism and its overall contribution to tissue Cbl uptake remain in question. The transplacental transport of Cbl appears to be largely mediated by TC2 (F4). Two intracellular Cbl binders have been described in rat liver and rabbit liver and are identical to methylmalonyl-CoA mutase and homocysteine methylase (K3, M13), a finding which has been repeated in cultured human fibroblasts (M14). The transport of Cbl into hepatocytes is mediated by a granulocyte R-type binder, and a pinocytotic mechanism appears to operate (A8, B15). Following its internalization, the binder–Cbl complex is digested in secondary lysosomes and the released Cbl is used for the synthesis of its coenzyme forms. MeCbl is generated and utilized in the cytosol, whereas adoCbl is generated and utilized in the mitochondria. Presumably, there exists a transport process for the mitochondrial uptake of Cbl, although no studies of this have been reported. The formation of adoCbl includes the stepwise reduction of the cobalt atom from Co^{3+} to Co^{+} followed by adenosylation catalyzed by ATP:cob (I) alamin Co-β-adenosyltransferase (EC 2.5.1.17) (F5, F6, K4, M14–M16). The details of the biosynthesis of both adoCbl and MeCbl in mammalian cells remain unclear.

The daily requirement for cobalamin is about 2.5 μg in infants and 5 μg in adults. Although human feces contain cobalamin, it is not available for absorption. However, it has been recently suggested that in some individuals, microflora in the small intestine can produce Cbl which may contribute to Cbl nutriture (A9).

No attempt will be made here to review the methods for the estimation of the cobalamins. However, as with most vitamin estimations, bioassays have tended to be surplanted by competitive-binding methods. This has, in fact, led to difficulties where R proteins are used as cobalamin binders due to interference from biologically inactive cobalamin (C3, K5).

3.2. Deficiency of R-type Cobalamin Binders

The R-type binders of Cbl are a heterogeneous group of glycoproteins which are distinct from intrinsic factor and TC2. They have been isolated from human serum (transcobalamin I), granulocytes, bile, amniotic fluid, gastric juice, and milk (A6, A7, G10, S6). A family originally described by Carmel and Herbert (C4) in which two brothers had an almost complete absence of R-type binders has recently been reinvestigated (H12). The condition appears to be benign, and although one of the brothers had a neurological disease of unknown etiology, the authors concluded that it was unrelated to R-protein deficiency (H12). All other indexes of Cbl metabolism were unremarkable and this family remains a unique curiosity.

3.3. Defects of Intestinal Absorption of Cobalamins

Deficiency of cobalamins leads to the well-known picture of megaloblastic bone marrow cells, macrocytic anemia, and neurological symptoms. Tissue and blood levels of cobalamins are low and in some cases there is an increased excretion of homocysteine and methylmalonic acid, which is a consequence of the attenuated synthesis of the appropriate coenzyme forms of cobalamin, MeCbl and adoCbl, respectively. These abnormalities disappear with parenteral administration of physiological amounts of Cbl (1–5 µg/day). There appear to be three causes of inherited defective intestinal absorption of Cbl; these have been collectively termed juvenile pernicious anemia and have been reported as follows: (1) a total absence of IF (M17, S8), (2) IF is formed and can bind Cbl but has a greatly reduced ability to bind with ileal receptors (K6, K7), and (3) an ileal defect which results in impaired absorption of exogenous IF–Cbl complex (G11).

3.4. Transcobalamin II Deficiency

A number of patients with inherited deficiency of TC2 have been reported (B16, B17, F7, F8, G12, H11, H13–H15, S9). The disease is characterized by severe anemia, thrombocytopenia, leukopenia, growth retardation, and immunodeficiency. Complete recovery follows parenteral administration of large amounts of Cbl (500 µg) and presumably, tissue uptake occurs by the

TC2-independent route described earlier (Section 3.1). The condition is inherited as an autosomal recessive trait (H11). Attention has been drawn to genetic polymorphism in recent reports (F8–F10), and analysis of TC2 variants in heterozygotes allows identification of carriers in some individuals (F8).

3.5. The Methylmalonicacidurias

The first isolation of methylmalonic acid (MMA) from mammalian sources was reported by Boyland and Levi in 1936 (B18). Oberholzer et al. (O1) and Stokke et al. (S10), however, were the first to report acutely ill children with ketosis, acidosis, and developmental retardation who excreted large amounts of this acid, but who did not have the neurological or hematological symptoms associated with Cbl deficiency, and who did not respond to Cbl therapy. However, reports of patients who presented in the same way but who did respond to the administration of large amounts of Cbl soon followed (L3, L4, R3, R4). An example is shown in Fig. 13. An additional group of patients who presented without ketoacidosis but who excreted homocysteine, cystathionine, and who were hypomethioninemic were distinguished (G13, M18). In all of these patients, there was increased excretion of MMA, and indirect evidence such as the oxidation of [^{14}C]propionate and [^{14}C]MMA by intact leukocytes and fibroblasts implicated a defect at methylmalonyl-CoA mutase in the pathway of propionate oxidation (Fig. 5) (M19, R4). Space precludes a complete review of the unraveling of this group of disorders, and

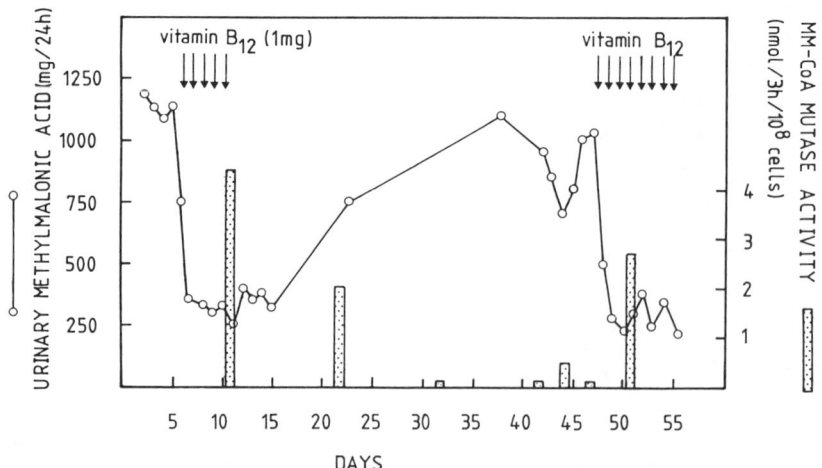

Fig. 13. Cobalamin-responsive methylmalonicaciduria. (Redrawn from R4 with permission.)

TABLE 2
COMPLEMENTATION ANALYSIS OF THE METHYLMALONICACIDURIAS

Complementation group	Metabolites excreted	Primary defect
CblA	MMA[a]	AdoCbl synthesis
CblB	MMA	AdoCbl synthesis
CblC	MMA and HC[b]	Ado and MeCbl synthesis
CblD	MMA and HC	Ado and MeCbl synthesis
Mut_0	MMA	Absent mutase
Mut_1	MMA	Mutase with high K_m

[a] MMA, Methylmalonic acid.
[b] HC, Homocystine.

the interested reader should consult a recent review of the subject (R1). However, it became apparent from the abnormal metabolites excreted, clinical presentation, and, especially, somatic cell hybridization (complementation) studies, that several different mutations would give rise to deficient activity of MMM. For example, in those patients with a combined defect of MMM and homocysteine methylase, it was demonstrated that a mutation resulted in a defect in intracellular Cbl metabolism which affected the production of both adoCbl and MeCbl (complementation groups CblC and CblD (M20). Another group of patients have been shown to produce an apoMMM with a decreased affinity for adoCbl (complementation group mut_1). ApoMMM from control cells binds adoCbl with a K_m = 60–70 nM, whereas in the patients' cells, K_ms of 17 and 280 μM were observed (W2).

There is no doubt that the elucidation of the methylmalonicacidurias has been greatly advanced by the technique of somatic cell hybridization. I have attempted to summarize these studies and give some indication of the known complementation groups and mutations in Table 2.

4. Folate

4.1. METABOLISM AND FUNCTION

Over the past 20 years, an immense amount of work on the metabolism of folic acid and its derivatives has been done. Several excellent reviews reflecting this wealth of information have appeared (B19, E2, R5) and the reader is referred to these accounts for details of the primary literature. However, some general observations and more recent studies will be given here.

The folates consist of a group of compounds which possess pteroic acid linked to a variable number of glutamate residues. Folic acid refers to the monoglutamate derivative. The pteridine moiety can exist in the fully oxidized, 7,8-dihydro (DHF) or 5,6,7,9-tetrahydro (THF) form. In addition, a number of one-carbon derivatives have been characterized. These structures are summarized in Fig. 14. There are some 15 mammalian folate-dependent enzymes which link amino acid, purine, and pyrimidine metabolism and are concerned with one-carbon transformations. These are summarized in Fig. 15. With the exception of dihydrofolate reductase, which catalyzes the conversion of folic acid to 7,8-dihydrofolate and 7,8-dihydrofolate to 5,6,7,8-tetrahydrofolate, all of the folate-requiring enzymes utilize the THF derivative. It is apparent that the folate-dependent metabolism interacts with a number of other vitamin-derived cofactors. Perhaps the most well known is the relationship between 5-methylTHF and MeCbl, and this has been termed the "methylfolate trap hypothesis." The arguments in this controversy have been recently restated (C7, S11).

The folates present in the diet are a complex mixture of pteroylglutamates,

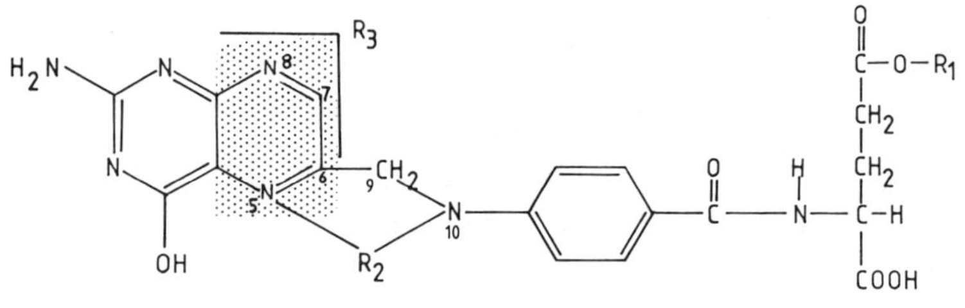

R_1 —OH or polyglutamate

R_2 Substituent	N_5	N_{10}	Name
	CH_3	H	5-Methyl
	CHO	H	5-Formyl
	CHNH	H	5-Formimino
	=CH—		5,10-Methenyl
	—CH_2—		5,10-Methylene
		CHO	10-Formyl

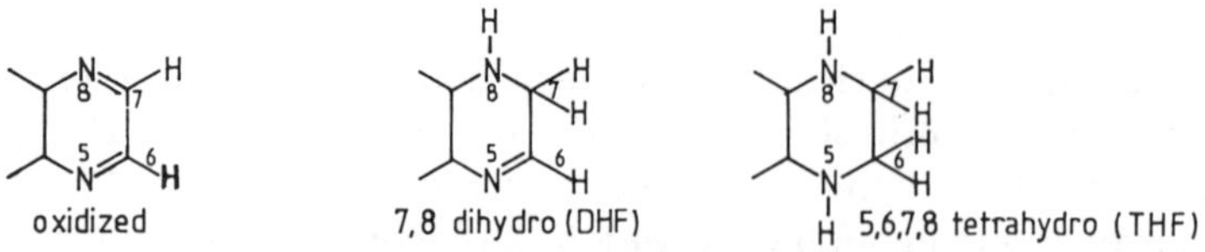

FIG. 14. Derivatives of folic acid.

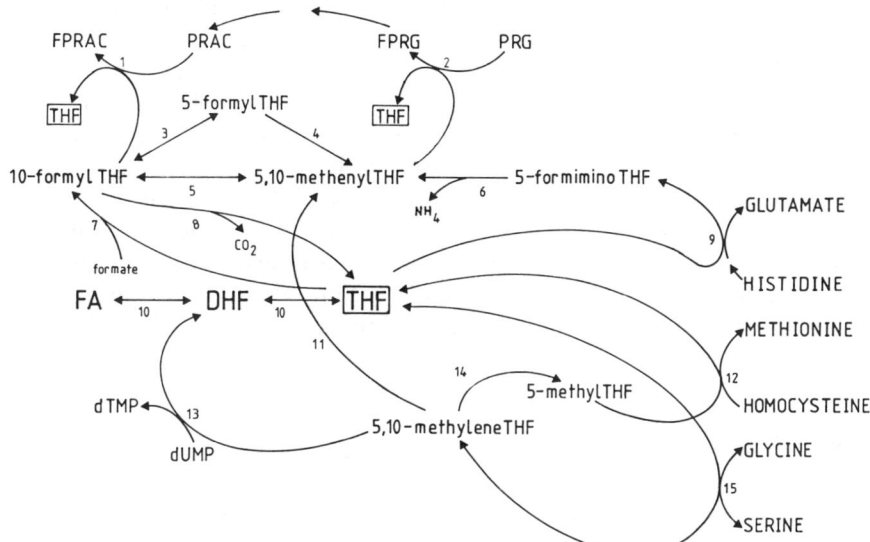

FIG. 15. Metabolism of folic acid and its derivatives—the relationship to amino acid, purine, and pyrimidine metabolism. FA, Folic acid; DHF, 7,8-dihydrofolate; THF, 5,6,7,8-tetrahydrofolate; PRG, 5′-phosphoribosylglycinamide; FPRG, formylPRG; PRAC, 5′-phosphoribosyl-5-aminoimidazole-4-carboxamide; FPRAC, formylPRAC; dTMP, thymidine 5′-phosphate; dUMP, deoxyuridine 5′-phosphate; 1, PRAC transformylase; 2, PRG transformylase; 3, formylTHF isomerase; 4, cyclodehydrase; 5, methenylTHF cyclohydrolase; 6, formiminoTHF cyclodeaminase; 7, 10-formylTHF synthetase; 8, formylTHF dehydrogenase; 9, glutamate formiminotransferase; 10, dihydrofolate reductase; 11, methyleneTHF dehydrogenase; 12, homocysteine methylase; 13, thymidylate synthetase; 14, methyleneTHF reductase; 15, serine hydroxymethyltransferase.

90% of which are polyglutamates, mainly the 5-methyltetrahydro derivative, while 10% are monoglutamates, mainly the 10-formyltetrahydro derivative. Only the monoglutamates are absorbed, and thus hydrolysis of the polyglutamates must occur before they can be internalized by the enterocytes. Several recent reviews have covered this topic (C5, H16, H17). The minimum daily requirement for folate is about 50 µg/day and the recommended dietary allowance is 300 µg/day (H18). Recent studies using isolated brush-border membrane vesicles have suggested that there is a specific, saturable (K_m = 1.5 µM), and pH-dependent transport mechanism in rat jejunum (S12). It is probable that following reconjugation within the enterocyte, folates are transported in blood by a number of protein binders of varying affinities (C6). Folate-binding proteins were first demonstrated in milk (G14), and it has been suggested that these binders have an antibacterial function; however, recent evidence suggests that a milk binder facilitates intestinal absorption (C8). Folate uptake by isolated chick intestinal epithelial cells is saturable ($K_m = 0.3$ μM) and appears to be coupled to Na$^+$ transport (E3). An enterohepatic cycle of folate has been demonstrated in rats and suggests that the liver may regulate the availability of folate to extrahepatic tissues (S13). Folate-binding proteins have been demonstrated in human placenta (A10), erythrocytes (A10), umbilical cord serum (H19), a variety of cultured cell types (M21, M22, W3), rabbit choroid plexus (S14), and rat liver mitochondria (W4, W5) and cytosol (S15). In the case of the rat liver mitochondria, the binders were identified as dimethylglycine dehydrogenase and sarcosine dehydrogenase. With the exception of the binders with enzyme activity, the biological significance of the folate-binding proteins has been the subject of much speculation but little-established fact.

4.2. Inherited Folate Malabsorption

Six patients with congenital folate malabsorption have been reported (L5, L6, P7, S16, S17). They present with severe megaloblastic anemia in infancy and with varying degrees of mental impairment. Although the hematological abnormality responds to parenteral folic acid and serum folic acid levels normalize, the CSF concentrations remain low, and in four of the six patients the mental retardation occurred in spite of folate therapy. These findings suggest that there is a defect both in the intestine and at the blood–brain barrier. In two patients, there was some response to large oral doses of folic acid (L5, L6) but in the others, frequent intramuscular administration was needed.

In addition to the above cases of defective folate absorption, a case was reported of a patient who presented with megaloblastic anemia and had an apparent congenital absence of a low-affinity folate binder (B20). Bentsen *et al.* speculate that a nutritional deficiency of folate may have been exacerbated by the absence of the binder.

4.3. Homocystinuria Due to 5,10-MethyleneTHF Reductase Deficiency

Deficiency of 5,10-methyleneTHF reductase was first described in two sisters by Mudd *et al.* (M23) and in a third patient by Shih *et al.* (S18). The homocystinuria observed in these patients arose from an impaired ability to remethylate homocysteine to methionine because of defective production of the cofactor 5-methylTHF. Ten other patients have been documented as having this defect (R6). In all of the patients, the predominant clinical findings were neurological. In infants, the defect is lethal, although there has been one patient in whom a combination of folinic acid, methionine, B_{12}, and B_6 led to marked clinical improvement (H20). In patients with a variant of the disease with later onset, folic acid administration corrects the homocystinuria but some neurological impairment remains. In general, the severity of the disease seems related to the age at presentation. However, in a comprehensive study of fibroblast cultures from nine patients, there was a good correlation between residual enzyme activity, the proportion of cellular folate present as 5-methylTHF, and the clinical severity (R6). The mechanism of the folate response seen in three patients is not understood, although with respect to the report of the single infant who was treated successfully with a combination of folinic acid, B_6, B_{12}, and methionine, Harpey et al. suggest that treatment should be aimed at increasing the alternative route of homocysteine dispersal via cystathionine synthase (B_6) and at correcting the hypomethioninemia by supplying exogenous methionine and B_{12}.

4.4. Dihydrofolate Reductase Deficiency

Only three patients with probable DHFR deficiency have been described (T2, W6). Their symptoms were varied, with anemia, megaloblastosis, and in one patient (T2) some mental impairment. All three responded to parenteral 5-formylTHF. However, the evidence for a primary defect of DHFR has been questioned (see ref. E2 pp. 324–327 for a fuller discussion). The mechanism of the response to 5-formylTHF is probably effected by circumventing the defective enzyme by supplying fully reduced folate.

4.5. Glutamate Formiminotransferase Deficiency

The degradation of histidine via urocanic acid and imidazalonepropionic acid results in the formation of formiminoglutamate (FIGLU), which in turn is converted to glutamate and formiminoTHF by a transferase. Cyclodeaminase then converts formiminoTHF to 5,10-methyleneTHF. The transferase and cyclodeaminase reactions are probably catalyzed by a single protein (T3). Two groups of patients with defective activity of this complex have been described. In the first group, patients excreted large amounts of FIGLU (0.5 g/day). Two patients described by Niederwieser et al. were unresponsive to folate (N1), but two similar patients described by Perry et al., when given folate, responded with a marked drop in FIGLU excretion (P8). In the second group of patients, FIGLU excretion was elevated only in response to a histidine load and high serum folate levels were observed (A11, A12). These findings have been interpreted as follows: In the first group, a single defect of formiminotransferase results in the excretion of large amounts of FIGLU. In the second group, a partial block of the transferase but a total block of the cyclodeaminase results in slightly elevated FIGLU excretion and the accumulation of 5-formiminoTHF (R5). However, direct enzyme confirmation for this model is lacking.

5. Pyridoxine (Vitamin B_6)

5.1. Function and Metabolism

Pyridoxine (Fig. 16) as its phosphorylated derivative pyridoxal 5'-phosphate (PLP) is required by a large number of enzymes, all of which are concerned with amino acid metabolism. The coenzyme participates with an amino acid in the formation of a Schiff base (aldimine). The way in which the complex is broken down governs the final product, e.g., amino acid (racemization), amine (decarboxylation), or 2-oxo-acid (decarboxylation). The various classes of PLP-dependent enzymes have been recently reviewed (B21) and accounts are also included in most biochemistry texts. In the present description I shall consider in detail only those enzymes, inborn errors of which have been shown to be responsive to pyridoxine therapy.

György was the first to report a water-soluble factor that would cure acrodynia (vitamin deficiency dermatitis) in the rat (G15). The first biochemical function of pyridoxine was described by Gunsalus (G16), who showed that PLP was a cofactor for amino acid decarboxylation. These and other early studies have been extensively reviewed (G17, H21).

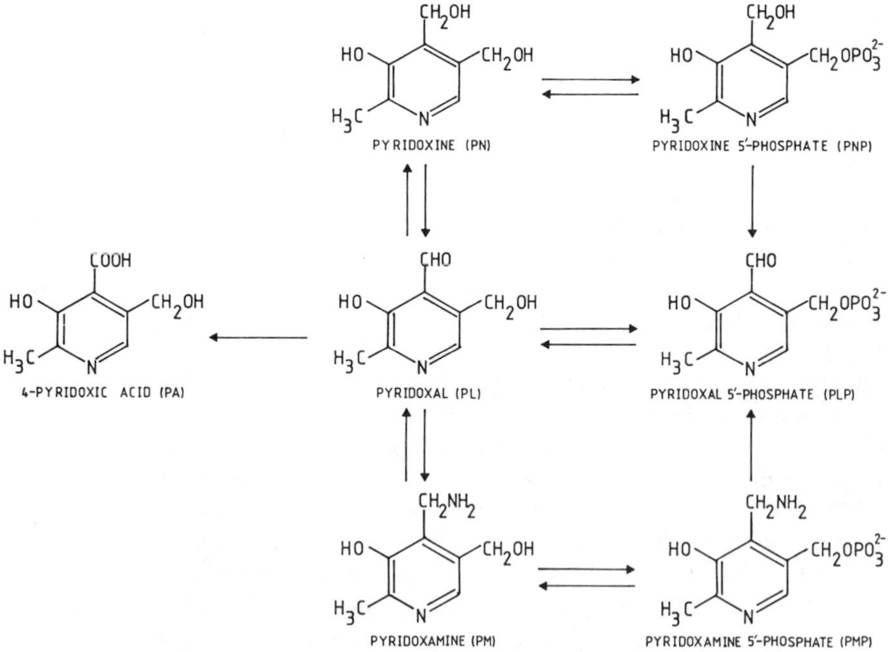

FIG. 16. Pyridoxine and its derivatives.

Dietary deficiency of pyridoxine is rare because of the large body stores and the contribution made by intestinal bacteria. However, infants breast fed by malnourished mothers or bottle fed on milk preparations in which the vitamin has been destroyed by heat treatment can become deficient (C9, M24). The clinical picture seen in pyridoxine deficiency in man and experimental animals has been reviewed (B21) and will not be further discussed here. The dietary requirement for the vitamin is difficult to determine because of the input from the intestinal flora but is thought to be 0.5 mg/day for infants and 2 mg/day for adults (S19, U1). Vitamin B_6 status can be determined by direct measurement of the vitamin and its congeners in blood by a variety of methods. These include microbiological methods, of which that based on *Lactobacillus casei* appears to be the most specific (R7), and fluorometry (C10, R8). Indirect methods such as the loading of a PLP-dependent pathway and the measurement of the urinary metabolites arising from its attenuation have been suggested (W7) and criticized (R9). Alternatively, the urinary output of 4-pyridoxic acid has been used as an index of pyridoxine status (S20).

There is general agreement that pyridoxine is taken up by the rat jejunum by a process of passive diffusion (B22, M25, S21, T4). This is followed by phosphorylation and dephosphorylation within the enterocyte (M26, M27), and although the biological significance of these events is unclear, the initial diffusional entry of pyridoxine appears to be independent of its subsequent metabolism. The extrapolation of these results from experimental animals to man is unwarranted in view of the known species differences in intestinal absorption of biotin (S1). The metabolism of pyridoxine and its derivatives has been studied in perfused liver, isolated hepatocytes, liver mitochondria, erythrocytes, and brain (D6, L7, L8, M28, S22, S23). In each case, the initial passage across a membrane is followed by trapping by phosphorylation and protein binding. About 80% of pyridoxine administered via a cannulated jejunal loop appears in the liver after 1 hour (S21). A schematic diagram of the relationship between gut, blood, liver, and muscle [Mehansho *et al.* (M28)] is shown in Fig. 17. Pyridoxine and its congeners are bound to albumin (D6) but the majority of these compounds in the blood are intracellular in erythrocytes and leukocytes.

5.2. Cystathionine Synthase Deficiency

Cystathionine synthase (CS) deficiency is one of several defects which can give rise to homocystinuria. Other inborn errors of the transsulfuration pathway such as N^5-methylTHF-homocysteine methyltransferase deficiency, N^5,N^{10}-methyleneTHF reductase deficiency, and failure to form MeCbl also result in homocystinuria (Fig. 12). The present discussion is concerned with

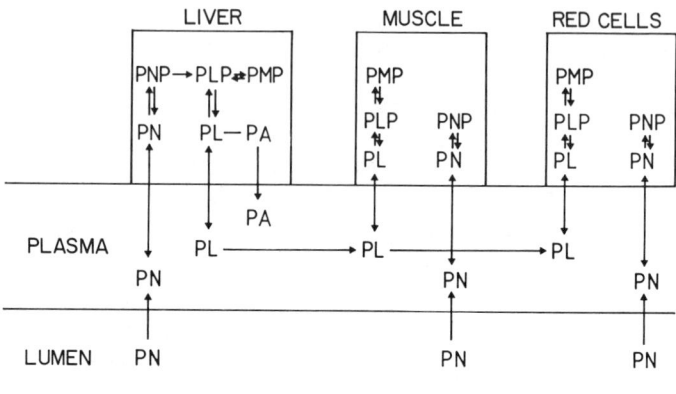

FIG. 17. Interorgan relationships of pyridoxine and its derivatives. Abbreviations as in Fig. 16. (Redrawn from M28 with permission.)

cystathionine synthase deficiency and its pyridoxine-responsive variants. Since its first description in 1962 by Field et al. (F11) and the localization of the enzyme defect in 1964 by Mudd et al. (M29), over 300 cases have appeared in the literature. The clinical picture and accompanying biochemical abnormalities have recently been exhaustively reviewed (M30). Typically, CS deficiency presents with mental retardation, ectopia lentis, skeletal abnormalities, and arterial and venous thromboemboli, which can be life threatening. Both homocystine and methionine accumulate in plasma (up to 2 mM methionine and 0.2 mM cystathionine). The renal reabsorption of homocystine is poor and large amounts are excreted in the urine (B23, W8). The elevated plasma levels of methionine are probably due to increased methylation of homocysteine because the route to cysteine is blocked. A variety of other sulfur-containing compounds have been characterized in patients with this disorder (M30). The enzyme defect has been demonstrated in liver, brain, cultured skin fibroblasts, and phytohemagglutinin-stimulated lymphocytes (M30).

Since CS requires PLP, Barber and Spaeth gave their patients with homocystinuria large amounts of pyridoxine and observed a reduction in the urinary excretion of homocystine and methionine (B24). Subsequently, many patients have been shown to be pyridoxine responsive (about 50% of cases) and a typical response is shown in Fig. 18.

The mechanism of the pyridoxine response remained in question for some years. Early studies using cultured fibroblasts were ambiguous. *In vivo* responsiveness appeared to correlate with the presence of residual activity of CS (U2), but there have been exceptions—some nonresponsive patients also

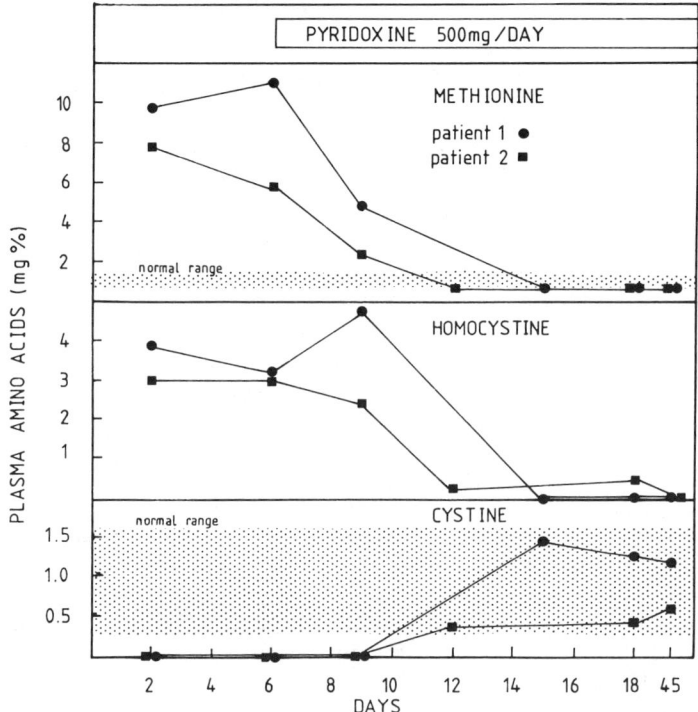

FIG. 18. The response of two patients with cystathionine synthase deficiency to therapy with pyridoxine. (Redrawn from S36 with permission.)

had residual activity (F12, U2). Increased thermolability of CS has been reported (F13, K8). However, the stimulation *in vitro* by PLP of the synthase is sometimes as great in patients' cells as in control cells (B25). Mudd (M30, M31) has argued that if reserve capacity is present in normals, then the small increases observed in mutant cell lines might be sufficient for a clinical response. However, it now seems probable that confusion has arisen from the use of culture media rich in pyridoxal, a situation reminiscent of the biotin-responsive disorders. This problem was solved by growing cells in a pyridoxal-depleted medium (M32), and demonstrated that the CS had a K_m with respect to PLP 2–4 times higher than controls (reduced affinity), whereas the nonresponsive cell lines with residual activity had K_ms 16–63 times higher (Fig. 19). Mipson *et al.* suggested that *in vivo* responsiveness is observed if (1) some residual activity is present, (2) the K_m is only moderately increased, and (3) the cells' ability to accumulate PLP is intact. On the other hand, nonresponsiveness is observed either because there is no re-

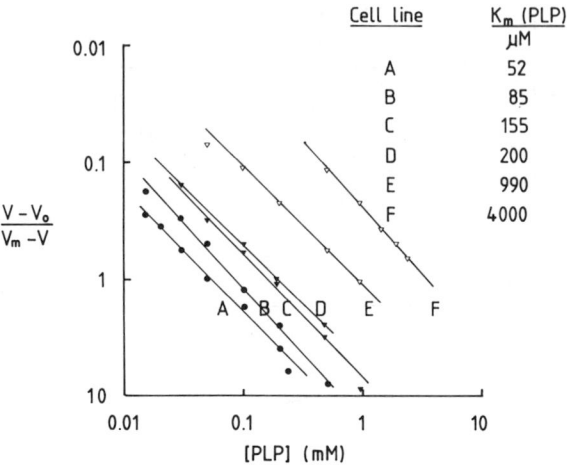

FIG. 19. Estimation of the binding constant (K_m) of cystathionine synthase for pyridoxal phosphate in fibroblasts from controls (A and B), *in vivo* pyridoxine-responsive CS-deficient patients (C and D), and nonresponsive patients (E and F). Cells cultured in pyridoxine-deficient medium. (Redrawn from M32 with permission.)

sidual activity as a result of a structural gene defect (G18), or because the K_m for PLP is so elevated that the intracellular PLP concentration never reaches high enough values.

5.3. Cystathionase Deficiency

Cystathionine formed by the condensation of serine and homocysteine is cleaved by cystathionase, another PLP-dependent enzyme, to yield cysteine and 2-oxobutyrate (Fig. 12). The first patient with an inherited defect of this enzyme was described by Harris *et al.* (H22). A 60-year-old mentally retarded woman was found to excrete >2 mmol/day cystathionine. The enzyme defect was proved some years later (F14) in liver extracts from different patients, but it is also expressed in lymphocyte cultures (P9). The results from cultured fibroblasts are conflicting (B26, M33, P10), a result, perhaps, of variable concentrations of pyridoxal in the culture medium. The clinical picture in this disorder is far from uniform, and mental retardation or neurological involvement is by no means a constant feature (M30). The majority of patients are pyridoxine responsive, a finding that was first noted by Frimpter *et al.* (F15). The same group has demonstrated *in vitro* activation of liver cystathionase by the addition of PLP to the assay mixture (F16). As in the case of pyridoxine-responsive cystathionine synthase deficiency, the simplest explanation is that the mutant apoenzyme has a reduced affinity for PLP. However, unlike CS deficiency, in the presence of high PLP, near normal holoenzyme activities are observed (F16).

5.4. KYNURENINASE DEFICIENCY

Xanthurenic aciduria arises from defective activity of kynureninase (Fig. 20) (H23, K9, O2, T5, T6) and kynurenine and hydroxykynurenine are also excreted. The patients described to date are all pyridoxine responsive. The defect was demonstrated in liver extracts, with complete restoration of enzyme activity by PLP in two patients (T6). This presumably reflects decreased affinity in the apoenzyme for its cofactor. In most of the reported cases, dietary deficiency or malabsorption of pyridoxine was excluded because the urinary concentrations of 4-pyridoxic acid were normal. This point is of some importance since the excretion of xanthurenic acid after a tryptophan load is one of the first abnormalities seen in pyridoxine deficiency (C11, F16).

5.5. GLYOXYLATE:2-OXOGLUTARATE CARBOLIGASE DEFICIENCY

Deficiency of glyoxylate:2-oxoglutarate carboligase (primary hyperoxaluria Type 1) has been shown to be pyridoxine responsive in a number of patients (G19, G20, S24), insofar as the urinary output of oxalate was decreased. Of

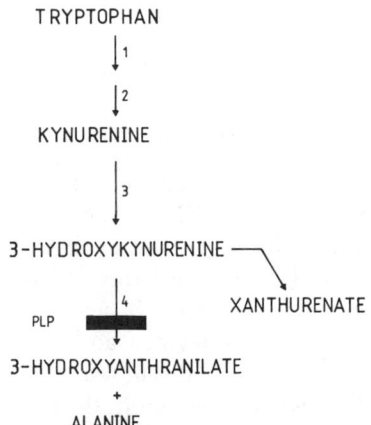

FIG. 20. The metabolism of tryptophan. The horizontal bar indicates the enzyme defect in xanthurenic aciduria. PLP, Pyridoxal phosphate; 1, tryptophan 2,3-dioxygenase; 2, formamidase; 3, kynurenine 3-monooxygenase; 4, kynureninase.

relevance to this finding is that pyridoxine deficiency in man and in experimental animals leads to increased urinary oxalate (F16, G21) and that pyridoxine administration to normal subjects lowers urinary oxalate (G22). Although glyoxylate:2-oxoglutarate carboligase is a PLP-dependent enzyme, it is not clear whether the effect of large doses of pyridoxine are due to a direct effect on the defective enzyme or to augmentation of an alternative pathway.

5.6. Ornithine–2-Oxo-acid Aminotransferase

Gyrate atrophy of the choroid and retina in association with hyperornithinemia was first described by Simell and Takki (S25). The disease is characterized by a constellation of ocular defects which are manifested in the first decade of life and can lead to blindness in later life. In addition, abnormal EEGs, glucose tolerance tests, and lowered intelligence have been reported. A total of 34 cases have been documented (K10). Defective activity of ornithine–2-oxo-acid aminotransferase was first shown in liver by Takki (T7) and has since been demonstrated in transformed lymphocytes and cultured fibroblasts (K11, V1). Four patients have responded to pyridoxine therapy (B27, K10) with a lowering of the plasma concentration of ornithine. There is also evidence to suggest some improvement in the visual defect in two patients (K10). A pyridoxine response has been demonstrated in cultured cells from five patients (B27, K10, T7), although in one instance the patient was not clinically responsive.

6. Riboflavin

6.1. Function and Metabolism

Riboflavin is present within the cell in two major forms, riboflavin 5'-phosphate (flavin mononucleotide, FMN) and complexed to adenosine 5'-monophosphate (flavin adenine dinucleotide, FAD). Both of these species are present in the protein-bound form, the flavoproteins, and function as hydrogen carriers in redox reactions. No attempt will be made here to document the flavin-dependent enzymes and only those of relevance to inborn errors of metabolism will be discussed.

The mechanism of riboflavin absorption in the gut has been reviewed (J2), and it appears to follow saturation kinetics and is localized in the proximal small intestine. Although there is good evidence that riboflavin absorption in the rat involves a phosphorylation–dephosphorylation mechanism, there are considerable species differences, and in man, the precise mechanism is not known (C12, Y3). The majority of riboflavin found in blood is associated with albumin (C13, J2). However, it has been recently shown that other plasma proteins, mainly immunoglobulins, have the capacity to bind the vitamin (M34). In addition, a binder has been identified in fetal and pregnant women's plasma and appears to be rather more specific for riboflavin (K_d = 0.4 μM). A similar protein has been identified in pregnant rat serum (M35) and Muniyappa and Adiga suggest that it is related to the transplacental transport of the vitamin. The daily requirement for riboflavin is 1.5–2.0 mg. The effects of riboflavin deficiency are well known and have been extensively reviewed (G23). It has been suggested that the activity of the erythrocyte flavin-dependent enzyme glutathione reductase is a good index of riboflavin status in man and in experimental animals (B28, G24). There are many methods for the assay of riboflavin and its metabolites, and these have been reviewed in detail elsewhere (M36).

6.2. Congenital Methemoglobinemia

Congenital methemoglobinemia is caused by a deficiency of erythrocyte NADH:methemoglobin reductase (NADH diaphorase, soluble NADH:cytochrome b_5 reductase) (H24, L9, S26). Although the methemoglobinemia can be controlled by the administration of either methylene blue or ascorbate (S27), prolonged treatment with high doses of ascorbate (0.5–1.0 g/day) can result in hyperoxaluria and urolithiasis (H25, T8). Erythrocytes contain an NADH-flavin reductase which can also contribute to the reduction of methemoglobin (Y4) and this effect has been demonstrated *in vitro* (M37). Riboflavin therapy was first reported by Kaplan and Chirouze in two patients (K12), and more recently, a third patient has been described by Hirano *et al.* (H26). In all three cases, methemoglobin was reduced to about 5% by the administration of riboflavin (50–120 mg/day).

The precise mechanism of this effect is not certain, but it has been suggested that the flavin is reduced by NADH-flavin reductase, which in turn reduces methemoglobin. Thus the mechanism of the riboflavin response is rather different from the other vitamin-responsive inborn errors in that the therapeutic effect is not as a result of the restoration of the defective enzyme activity, but rather by stimulation of an alternative pathway.

6.3. Pyruvate Kinase Deficiency

Congenital erythrocyte pyruvate kinase [PK; ATP:pyruvate 2-O-phosphotransferase (EC 2.7.1.40)] deficiency is a well known cause of hemolytic anemia (V2). It has been shown that the kinetic properties of PK are altered by incubation with reduced glutathione (V3). Van Berkel *et al.* suggest that

PK is maintained in its normal "reduced" form within the erythrocyte by reduced glutathione. Glutathione is maintained in the reduced state by glutathione reductase, a flavin-dependent enzyme. Thus, in the event of diminished activity of glutathione reductase, PK might alter its properties. Such a case has been described (S28) and the patient responded to riboflavin treatment (36 mg/day for 6 months). The glutathione reductase activity returned to normal (compared to 50% of normal prior to treatment) and the secondary effect on PK disappeared. This was accompanied by a clinical improvement insofar as the repeated transfusions required prior to riboflavin treatment were no longer necessary and the hematological parameters normalized.

6.4. Glutaryl-CoA Dehydrogenase Deficiency

Goodman *et al.* were the first to describe a disorder of lysine, hydroxylysine, and tryptophan metabolism which results in the excretion of large amounts of glutaric acid, 3-hydroxyglutaric acid, and glutaconic acid (G25). The patients, two siblings, were characterized clinically by progressive neurological degeneration and chronic metabolic acidemia (G26). The oxidation of [1,5-^{14}C]glutaryl CoA by leukocytes and fibroblasts was totally absent, and accordingly, a diagnosis of glutaryl-CoA dehydrogenase deficiency was made (G27). Subsequently, three patients who excreted the same abnormal metabolites were described (C14, G28) and appeared to have some residual enzyme activity. Glutaryl-CoA dehydrogenase catalyzes the conversion of glutaryl CoA to crotonyl CoA and CO_2 (B29), with the intermediate formation of glutaconyl CoA. Attempts to resolve this reaction into dehydrogenase- and decarboxylase-mediated steps have not been successful in preparations from rat liver mitochondria, and it has been assumed that the same protein catalyzes both steps. The detection of glutaconic and 3-hydroxyglutaric acids has led to the suggestion that in man there may be two separate enzymes, and that it is the decarboxylase which is defective (S29). However, more recent enzymic studies have shown that it is the first, FAD-dependent step which is defective (C14). Human glutaryl-CoA dehydrogenase has a K_m of 3.6 μM with respect to FAD, and the enzyme is assayed in the presence of saturating amounts of the cofactor. The addition of 0.1 mM of the cofactor to lysates of leukocytes from patients with glutaricacidemia resulted in a 1.6-fold stimulation of activity, whereas a 20-fold stimulation was seen in controls. This result was interpreted as evidence for a defect of FAD binding by the apodehydrogenase. It has been suggested (S30) that the neurological symptoms, which are reminiscent of those seen in the juvenile form of Huntington's chorea, may be caused by an inhibition of glutamate decarboxylase, an enzyme involved in the synthesis of 4-aminobutyric acid, by glutaric acid, glutaconic acid, and 3-hydroxyglutarate (K_i = 1.3, 0.75, and 2.5 mM, respectively). Furthermore, treatment with the γ-aminobutyric acid analog 4-amino-3-(4-chlorophenyl)butyric acid (Lioresal) resulted in neurological improvement. Treatment with low-protein, -tryptophan, and -lysine diets decreased the excretion of urinary metabolites (B30). Riboflavin treatment has been attempted in three patients and resulted in lowered excretion of metabolites with some suppression of symptoms. In a recent case report, dietary restriction of lysine and tryptophan had little clinical impact, although again, the urinary metabolites were decreased to 40% of the prediet level (L10).

6.5. Defects of β Oxidation

The pathway of mitochondrial β oxidation is well known and there have been a number of reports of patients in whom this pathway is disordered (G32, N2, T9). These patients excrete a constellation of abnormal metabolites which include the C_6–C_{10} dicarboxylic acids, ethylmalonic acid, and hexanoylglycine. Indirect evidence supports the hypothesis of a defect of the short-chain acyl-CoA dehydrogenase(s) (N2), flavoproteins which catalyze the formation of 2,3-enoyl CoA esters from the corresponding acyl-CoA esters (G29). The unoxidized butyryl CoA can be carboxylated to yield ethyl malonyl CoA (K13), the hexanoylglycine arises from the action of glycine N-acylase (B31) and hexanedioic acid, and the other dicarboxylic acids arise from the ω oxidation of the corresponding monocarboxylic acids (W9). Recently, a patient has been described who excreted the above metabolites and who presented with a Reye's-like syndrome. Cultured fibroblasts showed an impaired ability to oxidize [U-^{14}C]palmitate. Treatment with riboflavin (100 mg t.d.s.) resulted in a dramatic improvement. The abnormal metabolites were decreased, the hepatomegaly resolved, and the elevated plasma transaminases returned to near normal levels. In the 6 months following the start of riboflavin treatment there has been no further Reye's-like crises (G30).

7. Thiamin

7.1. Function and Metabolism

Thiamin (Fig. 21) as its pyrophosphate (TPP) is an essential cofactor in a number of oxidative decarboxylation reactions. Pyruvate dehydrogenase complex, oxoglutarate dehydrogenase, and branched-chain ketoacid dehydrogenase require TPP. In addition, TPP is required for the activity of transketolase, which catalyzes two reactions in the hexose monophosphate

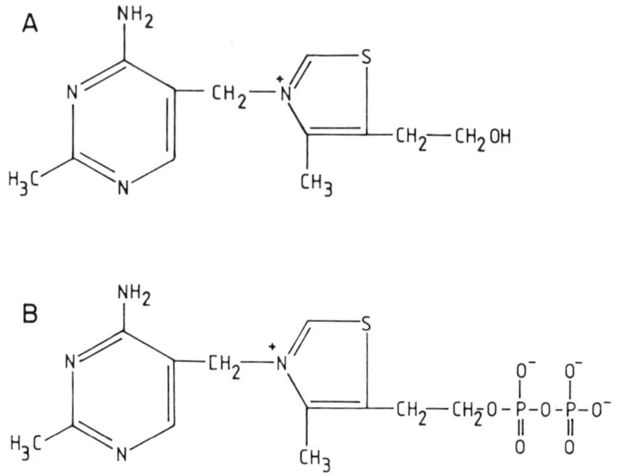

FIG. 21. Structure of thiamin (A) and thiamin pyrophosphate [TPP (B)].

shunt. In addition to TPP, the oxidative decarboxylation reactions also require FAD, NAD, coenzyme A, and lipoic acid (Fig. 22). The most intensively studied enzyme, pyruvate dehydrogenase complex (PDH), consists of three nonidentical subunits with a final molecular weight of about 4×10^6. In the case of PDH and the branched-chain ketoacid dehydrogenase complexes, phosphorylation and dephosphorylation by a protein kinase and phosphatase allow inactivation and activation, respectively (L11).

Thiamin, a water-soluble vitamin, is present in many foods and dietary deficiency (beriberi) is well known. It is absorbed from the intestinal tract as the free vitamin, and some studies have suggested an active transport pro-

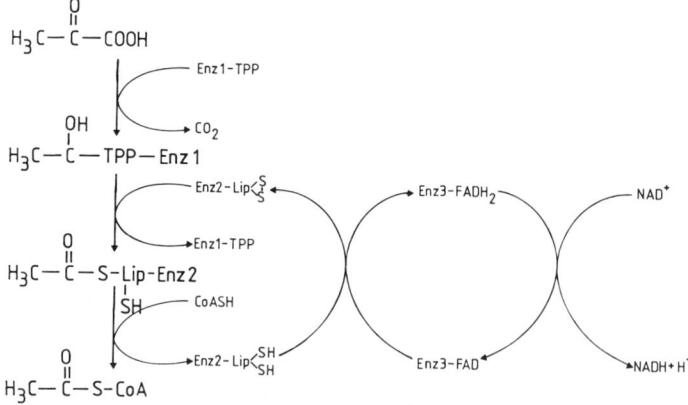

FIG. 22. Pyruvate dehydrogenase complex.

cess at low concentrations (2 μM), while at higher concentrations passive diffusion takes place (H27, K14, R10, R11, V4). These studies used the everted sac or tissue accumulation method to study transport, and it is impossible by these methods to separate the transport of thiamin from its subsequent metabolism. More recently, it has been shown that thiamin transport into membrane vesicles prepared from guinea pig jejunal brush borders occurs by simple diffusion over the concentration range 0.06–10 μM (H28). The relationship between the transport and consequent phosphorylation of thiamin remains controversial. For example, Chen (C15) asserts that following its uptake by isolated hepatocytes, thiamin remains unphosphorylated for up to 1 hour, whereas Lumeng et al. report that it is immediately converted to its phosphate ester (L12).

Thiamin status can be readily evaluated by the direct assay of the vitamin in blood—microbial bioassays (B32), by measurement of blood pyruvate following a glucose load (H29), by estimation of urinary thiamin (S31), or by assay of erythrocyte transketolase (A13) and its stimulation by the addition of TPP (B33, L13). There seems to be general agreement that a daily intake of about 0.5 mg/1000 kcal is sufficient in adults (S32).

7.2. Maple Syrup Urine Disease

Maple syrup urine disease or branched-chain ketoaciduria (BCKA) was the first inborn error of the degradation of the branched-chain amino acids leucine, isoleucine, and valine to be recorded and was first described by Menkes et al. (M38). In its severe neonatal form, the disease presents in the first week of life with vomiting, lethargy, a high concentration of the three branched-chain amino acids and their 2-oxoacyl analogs in plasma, and a distinctive maple syrup-like body odor. Following Menkes' study, it rapidly became apparent that the primary defect was of the branched-chain ketoacid dehydrogenase complex (for a review of the elucidation of the enzyme defect and other clinical details, see Ref. D7). Since this inborn error, unlike other defects of these pathways, is characterized by elevated plasma concentrations of the parent amino acids, it is detected by routine amino acid screening and many cases have been reported in the literature. It became apparent that, as has now become a truism with inborn errors of metabolism, variants of BCKA exist. Thus late-presenting, intermittent (D8), and "mild" (S33) phenotypes of the condition have been described. In addition, several cases of BCKA have been reported which are responsive to thiamin therapy (D9, E4, K15, P11, S34).

Some patients with the thiamin-responsive variant of BCKA also require a protein-restricted diet to avoid life-threatening metabolic disturbances (D9, P11). However, in one case, thiamin alone appeared to correct the defect

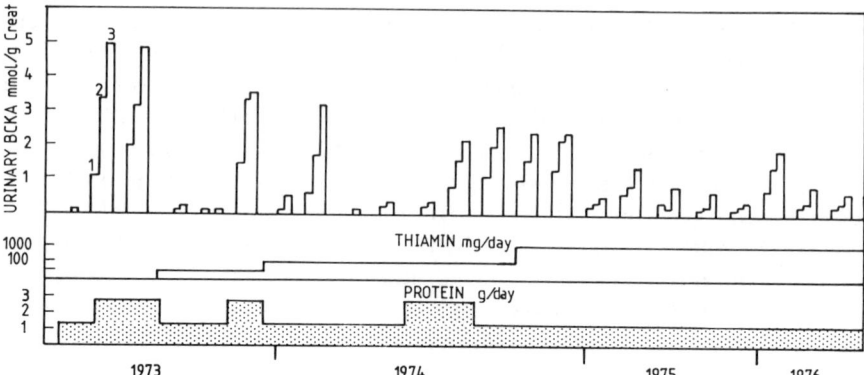

FIG. 23. Thiamin-responsive branched-chain ketoaciduria. (1) α-Ketoisovalerate; (2) α-ketoisocaproate; (3) α-keto-β-methylvalerate. (Redrawn from D9 with permission.)

(Fig. 23) (S34), although in this instance the defect was a mild variant (40% normal rate of oxidation of 0.1 mM leucine by fibroblasts). The precise molecular mechanism of the thiamin response is far from clear. A defect in thiamin pyrophosphorylation can be excluded since other TPP-dependent enzymes are normal. A lessening of the potency of thiamin has been reported and necessitated an increased dose (P11). Studies on *in vitro* activation in fibroblasts by thiamin have proved unhelpful (P11), although there is some evidence for the activation of hepatic branched-chain ketoacid dehydrogenase in normals by large doses of thiamin (D10). Thus it is possible that only in the milder forms of the disease is thiamin beneficial and that it acts in the same way as in the normal physiological response to large amounts of thiamin.

7.3. Congenital Lacticacidosis

The majority of the organicacidemias are characterized by the excretion of an abnormal metabolite (e.g., propionic acid), which gives a fairly unambiguous indication of the defective enzyme (i.e., propionyl-CoA carboxylase). However, in the group of inborn errors known collectively as the congenital lacticacidoses this is not the case. The excretion of lactate or elevated blood lactate concentration is associated with a heterogeneous group of disorders including glucose 6-phosphatase deficiency (type 1 glycogen storage disease), pyruvate carboxylase deficiency, PDH deficiency, phosphoenolpyruvate carboxykinase deficiency, fructose bisphosphatase deficiency, inborn errors of terminal electron transport, and a rather ill-defined group of patients with ataxias and chronic neurological diseases. This area has recently been reviewed and space precludes a full discussion of these defects. However, thiamin treatment has been tried in PDH deficiency, pyruvate carboxylase deficiency, and in a number of patients with lacticacidosis in whom the primary enzyme defect was not defined.

The administration of 1.8 g/day of thiamin to a child with moderate lacticacidosis resulted in biochemical normalization and an improvement in his development quotient (W10). Withdrawal of thiamin led to a relapse which was reversed by reintroduction of thiamin therapy. Thiamin therapy has been used in patients with proved pyruvate carboxylase deficiency with some success. When a child with lacticacidosis due to PC deficiency was given 20 mg thiamin iv, the anion gap, urinary lactate, and acid–base disturbance were corrected. These findings were interpreted as an activation of PDH which allowed increased disposal of pyruvate. Acute metabolic crises had been avoided for 2 years following the start of thiamin therapy, although there was still pronounced mental and psychomotor retardation (B34). Thiamin has been tried in a number of other cases with varying success. Some have shown that in conjunction with aspartate (to replace oxaloacetate), thiamin treatment results in both clinical and biochemical improvement (B35, D11, S35), biochemical improvement only, i.e., reduction in blood lactate concentrations (G31), while others report no effect with thiamin (D4, M39).

7.4. Thiamin-Responsive Megaloblastic Anemia

Two cases of megaloblastic anemia which responded to thiamin have been reported (R12, V5). The underlying defect in both is unclear, but all of the usual causes of megaloblastic anemia were excluded. In the first case, blood thiamin, leukocyte PDH, 2-oxoglutarate dehydrogenase, and erythrocyte transketolase were normal. Furthermore, pyridoxine, riboflavin, ascorbic acid, pantothenic acid, niacin, and biotin had no effect, whereas thiamin (50 mg/day) resulted in an immediate reticulocytosis and rise in hemoglobin. Diabetes and deafness were also present. In the second case, the presentation was very similar, with megaloblastic anemia, sensorineural deafness, and diabetes. The hereditary nature of the disease is open to question although in the second case, there was consanguinity—the parents being first cousins.

8. Conclusions

This account of the vitamin-responsive inborn errors of metabolism has entailed a degree of selection of the literature. In those areas where reviews are available they were cited along with limited references to the most recent developments. A total of about 25 vitamin-responsive inherited defects have been described, and it is apparent that almost any aspect of absorption, transport, intracellular processing, or apoenzyme binding which is dependent upon genetically determined proteins can be affected. In many instances, vitamin responsiveness arises because a protein has a reduced affinity for the vitamin or vitamin-derived cofactor in question, and elevated concentrations are required to generate sufficient holoenzyme. In other cases, the response to vitamin therapy appears to result from the augmentation of an alternative pathway so that abnormal metabolites do not accumulate to toxic levels.

Although the water-soluble vitamins are generally regarded as cofactor precursors, this is an oversimplification. The folates, for example, should more properly be termed cosubstrates since they are consumed in the course of reaction.

A large variety of biochemical and chemical methods have been used in the investigation of these defects and some mention has been made of them. Broadly speaking, they fall into three areas. First, the analytical methods used to detect inherited metabolic defects: These are nearly all chromatographic methods aimed at detecting the presence of abnormal metabolites in a variety of biological materials, although urine and plasma are most commonly used. For example, thin-layer and liquid chromatography with ninhydrin detection of amino acids in plasma and urine, and more recently, gas chromatography and gas chromatography–mass spectrometry for organic acids have been used to detect disorders of amino acid, lipid, and carbohydrate metabolism. Ideally, the detection of an abnormal metabolite should be pathognomonic, but in practice this is only sometimes the case. Thus homocystinuria can arise from several quite distinct enzyme defects. Second, the assay of enzymes in a range of tissues has proved to be an indispensable part of the investigation of an inborn error. Allied with tissue culture and somatic cell hybridization, an extraordinary range of defects has been revealed, as is typified in the inborn errors of cobalamin metabolism. Third, the investigation of a vitamin-responsive disorder is impossible without techniques for the estimation of the vitamin in question and I have detailed the methods available.

Finally, it is apparent that although there have been remarkable advances in the investigation of vitamin-responsive inherited defects of metabolism over the past few years, there remain many unanswered questions and, in fact, areas of total ignorance. Perhaps one of the most challenging problems concerns the regulation of the whole-body economy of vitamins and the balance between the turnover of vitamin-dependent enzymes and vitamin requirement.

References

A1. Ando, T., Rasmussen, K., Nyhan, W. L., and Hull, D., 3-Hydroxypropionate: Significance of β-oxidation of propionate in patients with propionic acidaemia and methylmalonic acidaemia. *Proc. Natl. Acad. Sci. U.S.A.* **69**, 2807–2811 (1972).

A2. Ando, T., Rasmussen, K., Wright, J. M., and Nyhan, W. L., Isolation and purification of methylcitrate, a major metabolic product of propionate in patients with propionicacidaemia. *J. Biol. Chem.* **247**, 2200–2204 (1972).

A3. Atkin, B. M., Buist, N. R. M., Utter, M. F., Leiter, A. B., and Banber, B. Q., Pyruvate carboxylase deficiency and lactic acidois in a retarded child without Leigh's disease. *Pediatr. Res.* **13**, 109–116 (1979).

A4. Allen, R. H., Seetharam, B., Podell, E. R., and Alpers, D. H., Effect of proteolytic enzymes on the binding of cobalamin to R protein and intrinsic factor. *J. Clin. Invest.* **61**, 47–54 (1978).

A5. Allen, R. H., Seetharam, B., Allen, N. C., Podell, E. R., and Alpers, D. H., Correction of cobalamin malabsorption in pancreatic insufficiency with a cobalamin analogue that binds with high affinity to R protein but not to intrinsic factor. *J. Clin. Invest.* **61**, 1628–1634 (1978).

A6. Allen, R. H., *Prog. Hematol.* **9**, 57–84 (1975).

A7. Allen, R. H., The plasma transport of vitamin B12. *Br. J. Haematol.* **33**, 161–171 (1976).

A8. Ashwell, G., and Morell, A. G., The role of surface carbohydrate in the hepatic recognition and transport of circulating glycoprotein. *Adv. Enzymol.* **41**, 99–128 (1974).

A9. Albert, M. J., Mathan, V. I., and Baker, S. J., Vitamin B12 synthesis by human small intestinal bacteria. *Nature (London)* **283**, 781–782 (1980).

A10. Antony, A. C., Utley, C., Van Horne, K. C., and Kolhouse, J. F., Isolation and characterisation of a folate receptor from human placenta. *J. Biol. Chem.* **256**, 9684–9692 (1981).

A11. Arakawa, T., Tamura, T., Ohara, K., Narisawa, K., Tanno, K., Honda, Y., and Higashi, O., Familial occurrence of formiminotransferase deficiency syndrome. *Tohoku J. Exp. Med.* **96**, 211–217 (1968).

A12. Arakawa, T., Congenital and acquired disturbances of histidine metabolism. *Clin. Endocrinol. Metab.* **3**, 17–35 (1974).

A13. Akbarian, M., and Dreyfus, P. M., Blood transketolase activity in beriberi heart disease. *JAMA, J. Am. Med. Assoc.* **203**, 23–26 (1968).

B1. Berger, E., Long, E., and Semenza, G., The sodium activation of biotin absorption in hamster small intestine in vitro. *Biochim. Biophys. Acta* **255**, 873–887 (1972).

B2. Bhagavan, H. N., and Coursin, D. B., Biotin content of blood in normal infants and adults. *Am. J. Clin. Nutr.* **20**, 903–906 (1967).

B3. Baker, H., Frank, O., Matovitch, V., Pasher, I., Aaronson, S., Hunter, S., and Sabotka, H., A new assay method for biotin in blood, serum, urine and tissue. *Anal. Biochem.* **3**, 31–39 (1962).

B4. Bonjour, J. P., Biotin in man's nutrition and therapy—a review. *Int. J. Vitam. Nutr. Res.* **47**, 107–17 (1977).

B5. Barnes, N. D., Hull, D., Balgobin, L. and Gompertz, D., Biotin-responsive propionicacidaemia. *Lancet* **2**, 244–245 (1970).

B6. Bartlett, K., and Gompertz, D., Combined-carboxylase defect: Biotin-responsiveness in cultured fibroblasts. *Lancet* **2**, 804 (1976).

B7. Bartlett, K., Ng, H., Dale, G., Green, A., and Leonard, J. V., Studies on cultured fibroblasts from patients with defects of biotin-dependent carboxylation. *J. Inherited Metab. Dis.* **4**, 183–189 (1981).

B8. Blom, W., De Muink Keizer, S. M. P. F., and Scholte, H. R., Acetyl CoA carboxylase deficiency: An inborn error of de novo fatty acid synthesis. *N. Engl. J. Med.* **305**, 465–466 (1981).

B9. Bartlett, K., and Gompertz, D., Biotin activation of carboxylase activity in cultured fibroblasts from a child with a combined carboxylase defect. *Clin. Chim. Acta* **84**, 399–401 (1978).

B10. Burri, B. J., Sweetman, L., and Nyhan, W. L., Mutant holocarboxylase synthetase: Evidence for the enzyme defect in early infantile biotin-responsive multiple carboxylase defect. *J. Clin. Invest.* **68**, 1491–1495 (1981).

B11. Bartlett, K., Ng, H., and Leonard, J. V., A combined defect of three mitochondrial carboxylases presenting as biotin responsive 3-methylcrotonylglycinuria and 3-hydroxyisovalericaciduria. *Clin. Chim. Acta* **100**, 183–186 (1980).

B12. Babior, B. M., ed., "Cobalamin: Biochemistry and Pathophysiology." Wiley, New York, 1975.

B13. Booth, C. C., Chanarin, I., Anderson, B. B., and Mollins, D. L., The site of absorbtion and tissue distribution of orally administered 56Co-labelled vitamin B12 in the rat. *Br. J. Haematol.* **3**, 253–261 (1957).

B14. Berliner, N., and Rosenberg, L. E., Uptake and metabolism of free cyanocobalamin by cultured human fibroblasts from controls and a patient with transcobalamin II deficiency. *Metab., Clin. Exp.* **30**, 230–236 (1981).

B15. Burger, R. L., Schneider, R. J., Mehlmann, C. S., and Allen, R. H., Human plasma R-type vitamin B12-binding protein. *J. Biol. Chem.* **250**, 7707–7713 (1975).

B16. Burman, J. F., Mollin, D. L., Sladden, R. A., Sourial, N., and Greany, M., Inherited deficiency of transcobalamin II causing megaloblastic anaemia. *Br. J. Haematol.* **35**, 676–677 (1977).

B17. Burman, J. F., Mollin, D. L., Sourial, N. A., and Sladden, R. A., Inherited lack of transcobalamin II in serum and megaloblastic anaemia: A further patient. *Br. J. Haematol.* **43**, 27–48 (1979).

B18. Boyland, E., and Levi, A. A., The isolation of methylmalonic acid from rat urine. *Biochem. J.* **30**, 2007–2008 (1936).

B19. Blakely, R. L., The biochemistry of folic acid and related pteridine, North-Holland Research Monographs. *Front. Biol.* **13**, (1969).

B20. Bentsen, K. D., Hansen, S. I., Holm, J., and Lyngbye, J., Abnormalities in folate binding pattern of serum from a patient with megaloblastic anaemia and folate deficiency. *Clin. Chim. Acta* **109**, 225–228 (1981).

B21. Barber, B. M., and Bender, D. A., Vitamin B6. *In* "Vitamins in Medicine" (B. M. Barber and D. A. Bender, eds.), pp. 248–280. Heinemann, London, 1980.

B22. Booth, C. C., and Brain, M. C., The absorption of tritium-labelled pyridoxine hydrochloride in the rat. *J. Physiol. (London)* **64**, 282–294 (1962).

B23. Brenton, D. P., Cusworth, D. C., and Gaull, G. E., Homocystinuria: Metabolic studies in 3 patients. *J. Pediatr.* **67**, 58–68 (1965).

B24. Barber, G. W., and Spaeth, G. L., Pyridoxine therapy in homocystinuria. *Lancet* **1**, 337 (1967).

B25. Bittles, A. H., and Carson, N. A. J., Homocystinuria: The effect of pyridoxine supplementation on cultured skin fibroblasts. *J. Inherited Metab. Dis.* **4**, 7–9 (1981).

B26. Bittles, A. H., and Carson, N. A. J., Cystathionase deficiency in fibroblast cultures from a patient with primary cystathioninuria. *J. Med. Genet.* **11**, 121–122 (1974).

B27. Berson, E. L., Schmidt, S. Y., and Shih, V. E., Ocular and biochemical abnormalities in gyrate atrophy of the choroid and retina. *Ophthalmology* **85**, 1018–1027 (1978).

B28. Beutler, E., Effect of flavin compounds on glutathione reductase activity: In vivo and in vitro studies. *J. Clin. Invest.* **48**, 1957–1966.

B29. Besrat, A., Polan, C. E., and Henderson, L. M., Mammalian metabolism of glutaric acid. *J. Biol. Chem.* **244,** 1461–1467 (1969).
B30. Brandt, N. J., Gregersen, N., Christensen, E., Gron, I. H., and Rasmussen, K., Treatment of glutaryl-CoA dehydrogenase deficiency (glutaricaciduria). *J. Pediatr.* **94,** 669–673 (1979).
B31. Bartlett, K., and Gompertz, D., The specificity of glycine-N-acylase and acylglycine excretion in the organicacidaemias. *Biochem. Med.* **10,** 15–23 (1974).
B32. Baker, H., Frank, O., and Fenelly, J. J., A method for assessing thiamine status in man and animals. *Am. J. Clin. Nutr.* **14,** 197–203 (1964).
B33. Brin, M., Transketolase (sedoheptulose-7-phosphate, D-glyceraldehyde-3-phosphate dihydroxyacetone-transferase, EC 2.2.1.1) and the TPP effect in assessing thiamine adequacy. *In* "Methods in Enzymology" (D. B. McCormick and L. D. Wright, eds.), Vol. 18, pp. 125–133. Academic Press, New York, 1970.
B34. Brunetti, M. G., Delvin, E., Hazel, B., and Scriver, C. R., Thiamine responsive lactic acidosis in a patient with deficient low-K_m pyruvate carboxylase activity in liver. *Pediatrics* **50,** 702–711 (1972).
B35. Baal, M. G., Gabreels, F. J. M., Renier, W. D., Hommes, F. A., Gijsbers, T. H. J., Lamers, K. J. B., and Kok, J. C. N., A patient with pyruvate carboxylase deficiency in the liver: Treatment with aspartic acid and thiamine. *Dev. Med. Child Neurol.* **23,** 521–530 (1982).
C1. Childs, B., Nyhan, W. L., Barden, M., Bard, L., and Cooke, R. E., Idiopathic hyperglycinaemia and hyperglycinuria: A new disorder of amino acid metabolism. I. *Pediatrics* **27,** 522–538 (1961).
C2. Cowan, M. J., Wara, D. W., Packman, S., Ammann, A. J., Yoshino, M., Sweetman, L., and Nyhan, W. L., Multiple biotin-dependent carboxylase deficiencies associated with defects in T-cell and B-cell immunity. *Lancet* **2,** 115–118 (1979).
C3. Cooper, B. A., and Whitehead, V. M., Evidence that some patients with pernicious anemia are not recognized by radiodilution assay for cobalamin in serum. *N. Engl. J. Med.* **299,** 816–818 (1978).
C4. Carmel, R., and Herbert, V., Deficiency of vitamin B12 binding alpha globulin in two brothers. *Blood* **33,** 1–12 (1969).
C5. Cooper, B. A., Physiology of absorption of monoglutamyl folates from the gastrointestinal tract. *In* "Folic Acid: Biochemistry and Physiology in Relation to the Human Nutrition Requirement," p. 188. Natl. Acad. Sci., Washington, D.C., 1975.
C6. Colman, N., and Herbert, V., Folate-binding proteins. *Annu. Rev. Med.* **31,** 433–439 (1980).
C7. Chanarin, I., The methylfolate trap and the supply of S-adenosylmethionine. *Lancet* **2,** 755 (1981).
C8. Colman, N., Hettiarachy, N., and Herbert, V., Detection of a milk factor that facilitates folate uptake by intestinal cells. *Science* **211,** 1427–1428 (1981).
C9. Coursin, D. B., Convulsive seizures in infants with pyridoxine deficient diets. *JAMA, J. Am. Med. Assoc.* **154,** 406–408 (1954).
C10. Chauhan, M. S., and Dakshinamurti, K., Fluorometric assay of pyridoxal. *In* "Methods in Enzymology" (D. B. McCormick and L. D. Wright, eds.), Vol. 62, pp. 405–407. Academic Press, New York, 1979.
C11. Coursin, D. B., Recommendations for the standardization of the tryptophan load test. *Am. J. Clin. Nutr.* **14,** 50–61 (1964).
C12. Chen, C., and Yamauchi, K., Absorption of riboflavin in the isolated intestine of rats. *J. Vitaminol.* **6,** 247–250 (1960).
C13. Cavalli-Sforza, L. L., Daiger, S. P., and Rummel, D. P., Detection of genetic variation

with radioactive ligands. I. Electrophoretic screening of plasma proteins with a selected panel of compounds. *Am. J. Hum. Genet.* **29**, 581–592 (1977).

C14. Christensen, E., and Brandt, N. J., Studies on glutaryl CoA dehydrogenase in leucocytes, fibroblasts and amniotic fluid cells. The normal enzymes and the mutant form in patients with glutaricaciduria. *Clin. Chim. Acta* **88**, 267–276 (1978).

C15. Chen, C.-P., *J. Nutr. Sci. Vitaminol.* **24**, 351–362 (1978).

D1. Dakshinamurti, K., and Chalifour, L. E., Biotin requirement of HeLa cells. *J. Cell. Physiol.* **107**, 427–438 (1981).

D2. "Dietary Standard for Canada." Department of National Health and Welfare, Ottawa, 1975.

D3. Duran, M., Gompertz, D., Bruinvis, L., Ketting, D., and Wadman, S. K., The variability of metabolite excretion in propionicacidaemia. *Clin. Chim. Acta* **82**, 93–99 (1978).

D4. De Vivo, D. C., Haymond, M. W., Leckie, M. P., Bussman, Y. L., McDougal, D. B., and Pagliara, A. S., The clinical and biochemical implications of pyruvate carboxylase deficiency. *J. Clin. Endocrinol. Metab.* **45**, 1281–1296 (1977).

D5. Donaldson, R. M., Mackenzie, I. L., and Trier, J. S., Intrinsic factor mediated attachment of vitamin B12 to brush borders and microvillous membranes of hamster intestine. *J. Clin. Invest.* **46**, 1215–1228 (1967).

D6. Dempsey, W. B., and Christensen, H. N., The specific binding of pyridoxal-5'-phosphate to bovine plasma albumin. *J. Biol. Chem.* **237**, 1113–1120 (1962).

D7. Dancis, J., and Levitz, M., Abnormalities of branched-chain amino acid metabolism. *In* "The Metabolic Basis of Inherited Disease" (J. B. Stanbury, J. B. Wyngaarden, and D. S. Fredrickson, eds.), 4th Ed., pp. 397–410. McGraw-Hill, New York, 1978.

D8. Dancis, J., Hutzler, J., and Rokkones, T., Intermittent branched-chain ketonuria. *N. Engl. J. Med.* **276**, 84–89 (1967).

D9. Duran, M., Tielens, A. G. M., Wadman, S. K., Stigter, J. C. M., and Kleijer, W. J., Effects of thiamine in a patient with a variant form of branched-chain ketoaciduria. *Acta Paediatr. Scand.* **67**, 367–372 (1978).

D10. Danner, D. J., Davidson, E. D., and Elsas, L. J., Thiamine increases the specific activity of human liver branched chain α-ketoacid dehydrogenase. *Nature (London)* **254**, 529–530 (1975).

D11. De Groot, C. J., and Hommes, F. A., Further speculation on the pathogenesis of Leigh's encephalomyelopathy. *J. Paediatr.* **82**, 541–542 (1973).

E1. Eldjarn, L., Jellum, E., Stokke, O., Pande, H., and Waaler, P. E., β-Hydroxyisovalericaciduria and β-methylcrotonylglycinuria: A new inborn error of metabolism. *Lancet* **2**, 521–522 (1970).

E2. Erbe, R. W., Genetic aspects of folate metabolism. *Adv. Hum. Genet.* **9**, 293–254 (1979).

E3. Eilam, Y., Ariel, M., Jablonska, M., and Grossowicz, N., On the mechanism of folate transport in isolated intestinal epithelial cells. *Am. J. Physiol.* G170–175.

E4. Elsas, L., Blocker, T., Wheeler, F., Pask, B., Perl, D., and Trusler, S., Variant "classical" maple syrup urine disease: Cofactor response. *Clin. Res.* **20**, 43–48 (1972).

F1. Fernandes, J., and Blom, W., The intravenous L-alanine tolerance test as a means for investigating gluconeogenesis. *Metab., Clin. Exp.* **23**, 1149–1156 (1974).

F2. Feldman, G. L., and Woolf, B., Deficient acetyl CoA carboxylase activity in multiple carboxylase deficiency. *Clin. Chim. Acta* **111**, 147–151 (1981).

F3. Fiedler-Nagy, C., Rosley, G. R., and Coffey, J. W., Binding of vitamin B12-rat transcobalamin II and free vitamin B12 to plasma membranes isolated from rat liver. *Br. J. Haematol.* **31**, 311–321 (1975).

F4. Fernandez-Costa, F., and Metz, J., Transplacental transport in the rabbit of vitamin B12 bound to human transcobalamins I, II and III. *Br. J. Haematol.* **43**, 625–630 (1979).

F5. Fenton, W. A., Ambani, L. M., and Rosenberg, L. E., Uptake of hydroxycobalamin by rat liver mitochondria. *J. Biol. Chem.* 251, 6616–6623 (1976).

F6. Fenton, W. A., and Rosenberg, L. E., Mitochondrial metabolism of hydroxocobalamin: Synthesis of adenosylcobalamin by intact rat liver mitochondria. *Arch. Biochem. Biophys.* 189, 441–447 (1978).

F7. Frater-Schroder, M., Linnell, J. C., Huser, H.-J., Galle, J., Wildfeuer, A., Seger, R., and Hitzig, W. H., Abnormal biochemistry and morphology of erythrocytes in a case of transcobalamin II deficiency under treatment. *J. Inherited Metab. Dis.* 4, 143–144 (1981).

F8. Frater-Schroder, M., Hitzig, W. H., and Sacher, M., Inheritance of transcobalamin II (TCII) in two families with TCII deficiency and related immunodeficiency. *J. Inherited Metab. Dis.* 4, 165–166 (1981).

F9. Frater-Schroder, M., Hitzig, W. H., and Butler, R., Studies on transcobalamin (TC): Detection of TCII isoproteins in human serum. *Blood* 53, 193–203 (1979).

F10. Frater-Schroeder, M., Luthy, R., Havrani, F. I., and Hitzig, W. H., Transcobalamin II Polymorphismus: Biochemische und klinische aspekt seltener Varianten. *Schweiz. Med. Wochenschr.* 109, 1373–1375 (1979).

F11. Field, C. M. B., Carson, N. A. J., Cusworth, D. C., Dent, C. E., and Neill, D. W., Homocystinuria; A new disorder of metabolism. *Abstr. Int. Congr. Paediatricians, 10th,* p. 274 (1962).

F12. Fowler, B., Kraus, J., Packman, S., and Rosenberg, L. E., Homocystinuria: Evidence for three distinct classes of cystathionine β-synthase mutants in cultured fibroblasts. *J. Clin. Invest.* 61, 645–653 (1978).

F13. Fleischer, L. D., Longhi, R. C., Tallan, H. H., and Gaull, G. E., Cystathionine β-synthase deficiency: Differences in thermostability between normal and abnormal enzyme from cultured human cells. *Pediatr. Res.* 12, 293–296 (1978).

F14. Frimpter, G. W., Cystathioninuria: Nature of the defect. *Science* 149, 1095–1096 (1965).

F15. Frimpter, G. W., Haymovitz, A., and Horwith, M., Cystathioninuria. *N. Engl. J. Med.* 268, 333 (1963).

F16. Faber, S. R., Feitler, W. W., Bleiler, R. E., Ohlson, M. A., and Hodges, R. E., The effects of an induced pyridoxine and pantothenic acid deficiency on excretions of oxalic and xanthurenic acid in the urine. *Am. J. Clin. Nutr.* 12, 406–411 (1963).

G1. Garrod, A. E., Inborn errors of metabolism. *Lancet* 2(1), 73, 142, 214 (1908).

G2. Ghneim, H. K., and Bartlett, K., unpublished results.

G3. Gompertz, D., Storrs, C. N., Bau, D. C. K., Peters, T. J., and Hughes, E. A., Localisation of enzymic defect in propionicacidaemia. *Lancet* 1, 1140–1143 (1970).

G4. Gravel, R. A., Lam, K.-F., Scully, K. J., and Hsia, Y. E., Genetic complementation of propionyl CoA carboxylase deficiency in cultured human fibroblasts. *Am. J. Hum. Genet.* 29, 378–388 (1977).

G5. Gompertz, D., Draffan, G. H., Watts, J. L., and Hull, D., Biotin responsive β-methylcrotonyl-glycinuria. *Lancet* 2, 22–24 (1971).

G6. Gompertz, D., Goodey, P. A., and Bartlett, K., Evidence for the enzymic defect in β-methylcrotonylglycinuria. *FEBS Lett.* 32, 13–14 (1973).

G7. Gompertz, D., Bartlett, K., Blair, D., and Stern, C. M. M., Child with a defect in leucine metabolism associated with β-hydroxyisovaleric aciduria and β-methylcrotonylglycinuria. *Arch. Dis. Child.* 48, 975–977 (1973).

G8. Ghneim, H. K., and Bartlett, K., The mechanism of biotin responsiveness in biotin-responsive combined carboxylase deficiency. *Lancet* 1, 1187–1188 (1982).

G9. Ghneim, H. K., Noy, G. A., and Bartlett, K., Biotin-dependent carboxylases and cultured human fibroblasts. *Biochem. Soc. Trans.* 9, 405–406 (1981).

G10. Grasbeck, R., Absorption and transport of vitamin B12. *Br. J. Haematol.* **31**, 103–110 (1975).
G11. Grasbeck, R., Gordin, R., and Kantero, I., Selective vitamin B12 malabsorption and proteinuria in young people: A syndrome. *Acta Med. Scand.* **167**, 289–296 (1960).
G12. Gimpert, E., Jakob, M., and Hitzig, W. H., Vitamin B12 transport in blood. I. Congenital deficiency of TCII. *Blood* **45**, 71–82 (1975).
G13. Goodman, S. I., Moe, P. G., Hammond, K. B., Mudd, S. H., and Uhlendorf, B. W., Homocystinuria with methylmalonic aciduria: Two cases in a sibship. *Biochem. Med.* **4**, 500–515 (1970).
G14. Ghitis, J., The folate binding in milk. *Am. J. Clin. Nutr.* **20**, 1–14 (1967).
G15. György, P., Vitamin B_2 and pellagra-like dermatitis in rats. *Nature (London)* **133**, 498–499 (1934).
G16. Gunsalus, I. C., Bellamy, W. D., and Umbreit, W. W., A phosphorylated derivative of pyridoxal as the coenzyme of tyrosine decarboxylase. *J. Biol. Chem.* **155**, 685–686 (1944).
G17. György, P., The history of vitamin B_6. *Vitam. Horm. (N.Y.)* **22**, 361–365 (1964).
G18. Griffiths, R., and Tudball, N., The molecular defect in a case of (cystathionine β-synthase)-deficient homocystinuria. *Eur. J. Biochem.* **74**, 269–273 (1977).
G19. Gibbs, D., and Watts, R. W. E., The action of pyridoxine in primary hyperoxaluria. *Clin. Sci.* **38**, 277–286 (1970).
G20. Giertz, G., "Hyperoxaluria." Urologists Correspondence Club, Karolinska Sjukhuset, Stockholm, Sweden, 1970.
G21. Gershoff, S. N., Farragalla, F. F., Nelson, D. A., and Andrus, S. E., Vitamin B_6 deficiency and oxalate nephrocalcinosis in the cat. *Am. J. Med.* **27**, 72–80 (1959).
G22. Gershoff, S. N., Vitamin B_6 and oxalate metabolism. *Vitam. Horm. (N.Y.)* **22**, 581–593 (1964).
G23. Goldsmith, G. A., Riboflavin deficiency. *In* "Riboflavin" (R. S. Rivlin, ed.), pp. 221–244. Plenum, New York, 1975.
G24. Glatzle, D., Weber, F., and Wiss, O., Enzymatic test for the detection of a riboflavin deficiency. *Experientia* **24**, 1122 (1968).
G25. Goodman, S. I., Markey, S. P., Moe, P. G., Miles, D. S., and Teng, C. C., Glutaricaciduria; A "new" disorder of amino acid metabolism. *Biochem. Med.* **12**, 12–21 (1975).
G26. Goodman, S. I., Norenberg, M. D., Strikes, R. H., Breslich, D. J., and Moe, P. G., Glutaric aciduria: Biochemical and morphologic considerations. *J. Pediatr.* **90**, 746–750 (1977).
G27. Goodman, S. I., and Kohlhoff, J. G., Glutaric aciduria. Inherited deficiency of glutaryl-CoA dehydrogenase activity. *Biochem. Med.* **13**, 138–140 (1975).
G28. Gregersen, N., Brandt, N. J., Christensen, E., Gron, I., Rasmussen, K., and Brandt, S., Glutaricaciduria: Clinical and laboratory findings in two brothers. *J. Pediatr.* **90**, 740–745 (1977).
G29. Green, D. E., Mii, S., Mahler, M. R., and Bock, R. M., Studies on the fatty acid oxidizing system of animal tissues. III. Butyryl coenzyme A dehydrogenase. *J. Biol. Chem.* **206**, 1–12 (1954).
G30. Gregersen, N., Wintzensen, H., Kolvraa, S., Christensen, E., Christensen, M. F., Brandt, N. J., and Rasmussen, K., Riboflavin and treatment of a patient with β-oxidation defect. *In* "Advances in the Treatment of Inborn Errors of Metabolism" (M.d'A. Crawford, D. A. Gibbs, and R. W. E. Watts, eds.), p. 311. Wiley, New York, 1982.
G31. Grobe, H., von Bassevitz, D. B., Dominick, H.-C., and Pfeiffer, R. A., Subacute necrotizing encephalomyelopathy. *Acta Paediatr. Scand.* **64**, 755–762 (1975).

G32. Gregerson, N., Lauritzen, R., and Rasmussen, K., Suberylglycine excretion in the urine from a patient with dicarboxylicaciduria. *Clin. Chim. Acta* **70**, 417–425 (1976).
H1. Hood, R., A radiochemical assay for biotin in biological materials. *J. Sci. Food Agric.* **26**, 1847–1852 (1975).
H2. Hosborough, T., and Gompertz, D., A protein-binding assay for measurement of biotin in physiological fluids. *Clin. Chim. Acta* **82**, 215–223 (1978).
H3. Hsia, Y. E., Scully, K. J., and Rosenberg, L. E., Defective propionate carboxylation in ketotic hyperglycinaemia. *Lancet* **1**, 757–758 (1969).
H4. Hsia, Y. E., Scully, K. J., and Rosenberg, L. E., Inherited propionyl CoA carboxylase deficiency in "ketotic hyperglycinaemia." *J. Clin. Invest.* **50**, 127–130 (1971).
H5. Hommes, F. A., Kuipers, J. R. G., Elema, J. D., Jansen, J. F., and Jonxis, J. J. P., Propionicacidaemia, a new inborn error of metabolism. *Pediatr. Res.* **2**, 519–524 (1968).
H6. Hillman, R. E., Keating, J. P., and Williams, J. C., Biotin-responsive propionicacidaemia presenting as the rumination syndrome. *J. Pediatr.* **92**, 439–441 (1978).
H7. Hommes, F. A., Polman, H. A., and Reerink, J. D., Leigh's encephalomyelopathy: An inborn error of gluconeogenesis. *Arch. Dis. Child.* **43**, 423–426 (1968).
H8. Haworth, J. C., Robinson, B. H., and Perry, T. L., Lacticacidosis due to pyruvate carboxylase deficiency. *J. Inherited Metab. Dis.* **4**, 57–58 (1981).
H9. Hooper, D. D., Alpers, D. H., Burger, R. L., Mehlmann, G., and Allen, R. H., Characterisation of ileal vitamin B_{12} binding using homogeneous human and hog intrinsic factor. *J. Clin. Invest.* **52**, 3074–3083 (1973).
H10. Hagedorn, C. H., and Alpers, D. H., Distribution of intrinsic factor—vitamin B_{12} receptors in human intestine. *Gastroenterology* **73**, 1019–1022 (1977).
H11. Hall, C. A., Hitzig, W. H., Green, P. D., and Begley, J. A., Transport of therapeutic cyanocobalamin in the congenital deficiency of transcobalamin II (TCII). *Blood* **53**, 251–263 (1979).
H12. Hall, C. A., and Begley, J. A., Congenital deficiency of human R-type binding of cobalamin. *Am. J. Hum. Genet.* **26**, 619–626 (1977).
H13. Hakami, N., Neiman, P. E., Canellos, G. P., and Lazerson, J., Neonatal megaloblastic anemia due to inherited transcobalamin II deficiency in two siblings. *N. Engl. J. Med.* **285**, 1163–1170 (1971).
H14. Hitzig, W. H., Dohmann, U., Pluss, H. J., and Vischer, D., Hereditary transcobalamin II deficiency: Clinical findings in a new family. *J. Pediatr.* **85**, 622–628 (1972).
H15. Hitzig, W. H., and Kenny, A. B., The role of vitamin B_{12} and its transport globulins in the production of antibodies. *Clin. Exp. Immunol.* **20**, 105–111 (1975).
H16. Halstead, C. H., The intestinal absorption of folates. *Am. J. Clin. Nutr.* **32**, 846–855 (1979).
H17. Halstead, C. H., Intestinal absorption and malabsorption of folates. *Annu. Rev. Med.* **31**, 79–87 (1980).
H18. Herbet, V., *In* "Folic Acid" (H. P. Broquist, C. E. Butterworth, and C. Wagner, eds.), pp. 277–295. Natl. Acad. Sci., Washington, D.C., 1977.
H19. Holm, J., Hansen, S. I., and Lyngbye, J., A high affinity folate binding protein in umbilical cord serum. *Scand. J. Clin. Lab. Invest.* **40**, 523–527 (1980).
H20. Harpey, J.-P., Rosenblatt, D. S., Cooper, B. A., LeMoel, G., Roy, C., and Lafourcade, J., Homocystinuria caused by 5,10-methylene-tetrahydrofolate deficiency: A case in an infant responding to methionine, folinic acid, pyridoxine, and vitamin B_{12} therapy. *J. Pediatr.* **98**, 275–278.
H21. Harris, S. A., Industrial preparation. *In* "The Vitamins" (W. H. Sebrell, Jr. and R. S. Harris, eds.), 2nd Ed., Vol. 2, pp. 3–18. Academic Press, New York, 1968.

H22. Harris, H., Penrose, L. S., and Thomas, D. H. H., Cystathioninuria. *Ann. Hum. Genet.* **23**, 442–451 (1959).
H23. Heeley, A. F., McCubbing, D. G., and Shepherd, J., Effect of pyridoxine on the metabolism of tryptophan and branched chain amino acids in two mentally retarded sibs. *Arch. Dis. Child.* **41**, 652–657 (1966).
H24. Hultquist, D. E., and Passon, P. G., Catalysis of methaemoglobin reductions by erythrocyte cytochrome b_5 and cytochrome b_5-reductase. *Nature (London)* **229**, 252–254 (1971).
H25. Hodgkinson, A., "Oxalic Acid in Biology and Medicine." London, 1977.
H26. Hirano, M., Matsuki, T., Tanischima, K., Takeshita, M., Shimizu, S., Nagamura, Y., and Yoneyama, Y., Congenital methaemoglobinaemia due to NADH methaemoglobin reductase deficiency: Successful treatment with oral riboflavin. *Br. J. Haematol.* **47**, 353–359 (1981).
H27. Hoyumpa, A. M., Middleton, H. M., Wilson, F. A., and Schenker, S., Thiamine transport across the rat intestine. I. Normal characteristics. *Gastroenterology* **68**, 1218–1223 (1975).
H28. Hayashi, K., Yoshida, S., and Kawasaki, T., Thiamine transport in the brush border membrane vesicles of the guinea-pig jejunum. *Biochim. Biophys. Acta* **641**, 106–113 (1981).
H29. Horwitt, M. K., and Kreisler, O., The determination of early thiamine-deficient status by estimation of blood lactic and pyruvic acids after glucose administration. *J. Nutr.* **37**, 411–427 (1949).
J1. Jenkins, W. J., Empson, R., Jewell, D. P., and Taylor, K. B., Subcellular localisation of vitamin B_{12} during absorption in the guinea-pig ileum. *Gut* **22**, 617–622 (1981).
J2. Jusko, W. J., and Levy, G., Absorption, protein binding and elimination of riboflavin. In "Riboflavin" (R. S. Rivlin, ed.), pp. 99–152. Plenum, New York, 1975.
K1. Kollee, L. A. A., Willems, J. L., De Kort, A. F. M., Monnens, L. A. H., and Trijbels, J. M. F., Blood sampling technique for lactate and pyruvate estimation in children. *Ann. Clin. Biochem.* **14**, 285–287 (1977).
K2. Katz, M., and Cooper, B. A., Solubilized receptor for intrinsic factor—vitamin B_{12} complex from guinea pig intestinal mucosa. *J. Clin. Invest.* **54**, 733–739 (1974).
K3. Kolhouse, J. F., and Allen, R. H., Recognition of two intracellular cobalamin binding proteins and their identification as methylmalonyl CoA mutase and methionine synthetase. *Proc. Natl. Acad. Sci. U.S.A.* **74**, 921–925 (1977).
K4. Kerwar, S. S., Spears, C., McAuslan, B., and Weissbach, H., Studies on vitamin B_{12} metabolism in HeLa cells. *Arch. Biochem. Biophys.* **142**, 231–237 (1971).
K5. Kolhouse, J. F., Kando, H., Allen, N. C., Podell, E., and Allen, R. H., Cobalamin analogues are present in human plasma and can mask cobalamin deficiency because current radioisotope dilution assays are not specific for true cobalamin. *N. Engl. J. Med.* **299**, 785–792 (1978).
K6. Katz, M., Lee, S. F., and Cooper, B. A., Vitamin B_{12} malabsorption due to a biologically inert intrinsic factor. *N. Engl. J. Med.* **287**, 425–429 (1972).
K7. Katz, M., Mehlman, C. S., and Allen, R., Isolation and characterisation of an abnormal intrinsic factor. *J. Clin. Invest.* **53**, 1274–1283 (1974).
K8. Kim, Y. J., and Rosenberg, L. E., On the mechanism of pyridoxine-responsive homocystinuria. II. Properties of normal and mutant cystathionine β-synthase from cultured fibroblasts. *Proc. Natl. Acad. Sci. U.S.A.* **71**, 4821–4825 (1974).
K9. Knapp, A., A new hereditary disturbance of tryptophan metabolism dependent on vitamin B_6. *Clin. Chim. Acta* **5**, 6–13 (1960).
K10. Kennaway, N. G., Weleber, R. G., and Buist, N. R. M., Gyrate atrophy of the choroid

and retina with hyperornithinemia: Biochemical and histologic studies and response to Vitamin B_6. *Am. J. Hum. Genet.* **32**, 529–541 (1980).

K11. Kennaway, N. G., Weleber, R. G., and Buist, N. R. M., Gyrate atrophy of the choroid and retina: Deficient activity of ornithine ketoacid aminotransferase in cultured skin fibroblasts. *N. Engl. J. Med.* **297**, 1180 (1977).

K12. Kaplan, J. C., and Chirouze, M., Therapy of recessive congenital methaemoglobinaemia by oral riboflavine. *Lancet* **2**, 1043–1044 (1978).

K13. Kaziro, Y., Ochoa, S., Warner, R. C., and Chen, J.-Y., Metabolism of propionic acid in animal tissues. VII. Crystalline propionyl CoA carboxylase. *J. Biol. Chem.* **236**, 1917–1923 (1961).

K14. Komai, T., Kawai, K., and Shindo, H., Active transport of thiamine from rat small intestine. *J. Nutr. Sci. Vitaminol.* **20**, 163–177 (1974).

K15. Kodama, S., Seki, A., Hanabuda, M., Morisita, Y., Sakurai, T., and Matsuo, T., Mild variant of maple syrup urine disease. *Eur. J. Pediatr.* **124**, 31–36 (1976).

L1. Leonard, J. V., Seakins, J. W. T., Bartlett, K., Hyde, J., Wilson, J., and Clayton, B., Inherited disorders of 3-methylcrotonyl CoA carboxylation. *Arch. Dis. Child.* **56**, 53–59 (1981).

L2. Leonard, J. V., Problems in the congenital lacticacidoses. *In* "Metabolic Acidosis" (R. Porter and G. Lawrenson, eds.), pp. 340–356. Pitman, London, 1982.

L3. Lindblad, B., Olin, P., Svanberg, B., and Zetterström, R., Methylmalonicacidaemia. *Acta Paediatr. Scand.* **57**, 417–424 (1968).

L4. Lindblad, B., Lindstrand, K., Svanberg, B., and Zetterström, R., The effect of cobamide coenzyme in methylmalonic acidemia. *Acta Paediatr. Scand.* **58**, 178–180 (1969).

L5. Luhby, A. L., and Cooperman, J. M., Congenital megaloblastic anaemia and progressive central nervous system degeneration: Further clinical and physiological characterisation and therapy of syndrome due to inborn error of folate transport. *Am. Pediatr. Soc., Annu. Meet.* (1968).

L6. Lanzkowsky, P., Congenital malabsorption of folate. *Am. J. Med.* **48**, 580–583 (1970).

L7. Lui, A., Lumeng, L., and Li, T.-K., Metabolism of vitamin B_6 in rat liver mitochondria. *J. Biol. Chem.* **256**, 6041–6046 (1981).

L8. Lumeng, L., Lui, A., and Li, T.-K., Plasma content of B_6 vitamins and its relationship to hepatic B_6 metabolism. *J. Clin. Invest.* **66**, 688–695 (1980).

L9. Leroux, A., Torlinski, L., and Kaplan, J. C., Soluble and microsomal forms of NADH-cytochrome$_{b5}$ reductase from human placenta. Similarity with NADH methaemaglobin reductase from human erythrocytes. *Biochim. Biophys. Acta* **481**, 50–62 (1977).

L10. Leibel, R. L., Shih, V. E., Goodman, S. I., Bauman, M. L., McCabe, E. R. B., Zwerdling, R. G., Bergman, I., and Castello, C., Glutaric acidemia: A metabolic disorder causing progressive choreoathetosis. *Neurology* **30**, 1163–1168 (1980).

L11. Lehninger, A. L., "Biochemistry," pp. 450–453. Worth Publ., New York, 1975.

L12. Lumeng, L., Edmondson, J. W., Schenker, S., and Li, T.-K., Transport and metabolism of thiamine in isolated rat hepatocytes. *J. Biol. Chem.* **254**, 7265–7268 (1979).

L13. Lonsdale, D., and Shamberger, R. J., Red cell transketolase as an indicator of nutritional deficiency. *Am. J. Clin. Nutr.* **33**, 205–211 (1980).

M1. Moss, J., and Lane, M. D., The biotin-dependent enzymes. *Adv. Enzymol.* **35**, 321–442 (1971).

M2. Murthy, P. N. A., and Mistry, S. P., Synthesis of holoacetyl CoA carboxylase in vitro by cytosolic preparations from chicken liver. *Proc. Soc. Exp. Biol. Med.* **147**, 114–117 (1974).

M3. Mandella, R. D., Meslar, H. W., and White, H. E., The relationship between biotin binding proteins from chick plasma and egg yolk. *Biochem. J.* **175**, 629–633 (1978).

M4. McKeon, C., Eanes, R. Z., Fall, R. R., Tasset, D. M., and Woolf, B., Immunological studies of propionyl CoA carboxylase in livers and fibroblasts of patients with propionicacidemia. *Clin. Chim. Acta* **101**, 217–223 (1980).

M5. McAllister, H. C., and Coon, M. J., Further studies of liver propionyl CoA holocarboxylase synthetase and the specificity of holocarboxylase formation. *J. Biol. Chem.* **241**, 2855–2861 (1966).

M6. Mock, D. M., DeLorimer, A. A., Liebman, W. M., Sweetman, L., and Baber, H., Biotin deficiency: An unusual complication of parenteral alimentation. *N. Engl. J. Med.* **304**, 820–822 (1981).

M7. Munnich, A., Saudubray, J.-M., Carre, G., Coude, F. X., Ogier, H., Charpentier, C., and Frezal, J., Defective biotin absorption in multiple carboxylase deficiency. *Lancet* **2**, 263 (1981).

M8. Munnich, A., Fischer, A., Saudubray, J. M., Grinselli, C., Coude, F. X., Ogier, H., Charpentier, C., and Frezal, J., Biotin-responsive immunoregulatory dysfunction in multiple carboxylase deficiency. *J. Inherited Metab. Dis.* **4**, 113–114 (1981).

M9. Munnich, A., Saudubray, J. M., Coude, F. X., Charpentier, C., Saurat, J. M., and Frezal, J., Fatty-acid-responsive alopecia in multiple carboxylase deficiency. *Lancet* **1**, 1080–1081 (1980).

M10. Marcoullis, G., Parmentier, Y., Nicholas, J.-P., Jimenez, M., and Gerard, P., Cobalamin malabsorption due to non-degradation of R-proteins in the human intestine. *J. Clin. Invest.* **66**, 430–440 (1980).

M11. Mahoney, M. J., and Rosenberg, L. E., Inborn errors of cobalamin metabolism. *In* "Cobalamin: Biochemistry and Pathophysiology" (B. M. Babior, ed.), pp. 369–402. Wiley, New York, 1975.

M12. MacDonald, C. M. L. A., Farquharson, J., Bessent, R. G., and Adams, J. F., The forms of vitamin B_{12} on the transcobalamins. *Clin. Sci.* **52**, 215–218 (1977).

M13. Mellman, I. S., Youngdahl-Turner, P., Willard, H. F., and Rosenberg, L. E., Intracellular binding of radioactive hydroxocobalamin to cobalamin-dependent apoenzymes in rat liver. *Proc. Natl. Acad. Sci. U.S.A.* **74**, 916–920 (1977).

M14. Mellman, I., Willard, H. F., and Rosenberg, L. E., Cobalamin binding and cobalamin-dependent enzyme activity in normal and mutant human fibroblasts. *J. Clin. Invest.* **62**, 952–960 (1978).

M15. Mahoney, M. J., Hart, A. C., Steen, V. D., and Rosenberg, L. E., Methylmalonicacidaemia: Biochemical heterogeneity in defects of 5'-deoxyadenosylcobalamin synthesis. *Proc. Natl. Acad. Sci. U.S.A.* **72**, 2799–2803 (1975).

M16. Mellman, I., Willard, H. F., Youngdahl-Turner, P., and Rosenberg, L. E., Cobalamin coenzyme synthesis in normal and mutant human fibroblasts. *J. Biol. Chem.* **254**, 11847–11853 (1979).

M17. Mohammed, S. D., McKay, E., and Galloway, W. H., Juvenile familial megaloblastic anemia due to selective malabsorption of vitamin B_{12}. *Q. J. Med.* **35**, 433–453 (1966).

M18. Mudd, S. H., Levy, H. L., and Abeles, R. H., A derangement in B_{12} metabolism leading to homocystinuria, cystathioninemia and methylmalonicaciduria. *Biochem. Biophys. Res. Commun.* **35**, 121–126 (1969).

M19. Morrow, G., Mellman, W. J., Barnes, L. A., and Dimitrov, N. V., Propionate metabolism in cultured cells from a patient with methylmalonic acidemia. *Pediatr. Res.* **3**, 217–219 (1969).

M20. Mahoney, M. J., Rosenberg, L. E., Mudd, S. H., and Uhlendorf, B. W., Defective metabolism of vitamin B_{12} in fibroblasts from patients with methylmalonicaciduria. *Biochem. Biophys. Res. Commun.* **44**, 375–381 (1971).

M21. McCormick, J. I., Susten, S. S., Rader, J. I., and Freisheim, J. H., Studies of a meth-

otrexate binding protein fraction from L1210 lymphocyte plasma membranes. *Eur. J. Cancer* **15**, 1377–1386 (1979).
M22. McHugh, M., and Cheng, Y.-C., Demonstration of a high affinity folate binder in human cell membranes and its characterisation in cultured human KB cells. *J. Biol. Chem.* **254**, 11312–11318 (1979).
M23. Mudd, S. H., Uhlendorf, B. W., Freeman, J. M., Finklestein, J. D., and Shih, V. E., Homocystinuria associated with decreased methylenetetrahydrofolate reductase activity. *Biochem. Biophys. Res. Commun.* **46**, 905–912 (1972).
M24. Molony, C. J., and Parmalee, A. H., Convulsions in young infants as a result of pyridoxine deficiency. *JAMA, J. Am. Med. Assoc.* **154**, 405–406 (1954).
M25. Middleton, H. M., Uptake of pyridoxine hydrochloride by the rat jejunal mucosa in vitro. *J. Nutr.* **107**, 126–131 (1977).
M26. Middleton, H. M., Jejunal phosphorylation and dephosphorylation of absorbed pyridoxine·HCl in vitro. *Am. J. Physiol.* **235**, E272–E278 (1978).
M27. Middleton, H. M., In vivo absorption and phosphorylation of pyridoxine·HCl in rat jejunum. *Gastroenterology* **76**, 43–49 (1979).
M28. Mehansho, H., Buss, D. D., Hamm, M. W., and Henderson, L. M., Transport and metabolism of pyridoxine in rat liver. *Biochim. Biophys. Acta* **631**, 112–123 (1980).
M29. Mudd, S. H., Finklestein, J. D., Irreverre, F., and Laster, L., Homocystinuria: An enzymatic defect. *Science* **143**, 1443–1445 (1964).
M30. Mudd, S. H., and Levy, H. L., Disorders of transsulphuration. *In* "The Metabolic Basis of Inherited Disease" (J. B. Stanbury, J. B. Wyngaarden, and D. S. Frederickson, eds.), 4th Ed., pp. 458–503. McGraw-Hill, New York, 1978.
M31. Mudd, S. H., Diseases of sulphur metabolism: Implications for the methionine-homocysteine cycle and vitamin responsiveness. *Ciba Found. Symp.* [N.S.] **72** (1980).
M32. Mipson, M. H., Kraus, J., and Rosenberg, L. E., Affinity of cystathionine β-synthase for pyridoxal-5'-phosphate in cultured cells. *J. Clin. Invest.* **66**, 188–193 (1980).
M33. Mudd, S. H., Discussion. *In* "Inherited Disorders of Sulphur Metabolism" (N. A. J. Carson and D. N. Raine, eds.), p. 311. Churchill-Livingstone, Edinburgh and London, 1971.
M34. Merrill, A. H., Froehlich, J. A., and McCormick, D. B., Isolation and identification of alternative riboflavin-binding proteins from human plasma. *Biochem. Med.* **25**, 198–206 (1981).
M35. Muniyappa, K., and Adiga, P. R., Isolation and characterisation of riboflavin-binding protein from pregnant rat serum. *Biochem. J.* **187**, 537–540 (1980).
M36. McCormick, D. B., and Wright, L. D., eds., "Methods in Enzymology," Vol. 66, pp. 217–429. Academic Press, New York, 1980.
M37. Matsuki, T., Yubisui, T., Tomoda, A., Yoneyama, Y., Tabeshita, M., Hirano, M., Kobayashi, K., and Tani, Y., Acceleration of methaemoglobin reduction by riboflavin in human erythrocytes. *Br. J. Haematol.* **39**, 523–528 (1978).
M38. Menkes, J. H., Hurst, P. L., and Craig, J. M., New syndrome: Progressive familial infantile cerebral disfunction associated with an unusual urinary substance. *Pediatrics* **14**, 462–467 (1954).
M39. Maesaka, H., Komiya, K., Misugi, K., and Tada, K., Hyperalaninemia, hyperpyruvicemia and lactic acidosis due to pyruvate carboxylase deficiency of the liver: Treatment with thiamine and lipoic acid. *Eur. J. Pediatr.* **122**, 159–168 (1976).
N1. Niederwieser, A., Giliberti, P., Matasovic, A., Pluznik, S., Steinman, B., and Baerlocher, K., Folic acid non-dependent formiminoglutamicaciduria in two siblings. *Clin. Chim. Acta* **54**, 293–316 (1974).
N2. Naylor, E. W., Masovitch, L. L., Guthrie, R., Evans, J. E., and Tieckelmann, H.,

Intermittent non-ketotic dicarboxylic aciduria in two siblings with hypoglycaemia: An apparent defect in β-oxidation of fatty acids. *J. Inherited Metab. Dis.* **3**, 19–24 (1980).

O1. Oberholzer, V. G., Levin, B., Burgess, E. A., and Young, W. F., Methylmalonic aciduria: An inborn error of metabolism leading to chronic metabolic acidosis. *Arch. Dis. Child.* **42**, 492–504 (1967).

O2. O'Brien, D., and Jensen, C. B., Pyridoxine dependency in two mentally retarded subjects. *Clin. Sci.* **24**, 179–186.

P1. Patrelli, F., Moretti, P., and Paparelli, M., Intracellular distribution of (^{14}COOH)-biotin in rat liver. *Mol. Biol. Rep.* **4**, 247–252 (1979).

P2. Pispa, J., Biotinidase. *Ann. Med. Exp. Biol. Fenn.* **43**, Suppl. 5 (1965).

P3. Packman, S., Sweetman, L., Baber, S., and Well, S., The neonatal form of biotin-responsive multiple carboxylase deficiency. *J. Pediatr.* **99**, 418–420 (1981).

P4. Poston, J. M., Leucine-2,3-aminomutase, an enzyme of leucine catabolism. *J. Biol. Chem.* **251**, 1859–1863 (1976).

P5. Poston, J. M., Cobalamin-dependent formation of leucine and β-leucine by rat and human tissue. *J. Biol. Chem.* **255**, 10,067–10,072 (1980).

P6. Parmentier, Y., Marcoullis, G., and Nicholas, J. P., The intraluminal transport of vitamin B_{12} and the exocrine pancreatic insufficiency. *Proc. Soc. Exp. Biol. Med.* **160**, 396–400 (1979).

P7. Ponez, M., Colman, N., Herbert, V., Schwartz, E., and Cohen, A. R., Therapy of congenital folate malabsorption. *J. Pediatr.* **98**, 76–79 (1981).

P8. Perry, T. L., Applegarth, D. E., Evans, M. E., Hansen, S., and Jellum, E., Metabolic studies of a family with massive formiminoglutamic aciduria. *Pediatr. Res.* **9**, 117–122 (1975).

P9. Pascal, T. A., Gaull, G. E., Beratis, N. G., Gillam, B. M., Tallan, H. H., and Hirschhorn, K., Vitamin B_6-responsive and unresponsive cystathioninuria: Two variant molecular forms. *Science* **190**, 1209–1210 (1975).

P10. Pascal, T., Gaull, G. E., Beratis, N., Gillam, B., Tallan, H., Hirschhorn, K., and Parker, C., Cystathioninase in long term lymphoid cell lines: Evidence for altered enzyme protein in cystathioninuria. *Pediatr. Res.* **9**, 315–319 (1975).

P11. Pueschel, S. M., Bresnan, M. J., Shih, V. E., and Levy, H. L., Thiamine responsive intermittent branched-chain ketoaciduria. *J. Pediatr.* **94**, 628–631 (1979).

R1. Rosenberg, L. E., Disorders of propionate, methylmalonate and cobalamin metabolism. *In* "The Metabolic Basis of Inherited Disease" (J. B. Stanbury, J. B. Wyngaarden, and D. S. Frederickson, eds.), 4th Ed., pp. 411–429. McGraw-Hill, New York, 1978.

R2. Rasmussen, K., Ando, T., Nyhan, W. L., Hull, D., Cottom, D., Donnell, G., Wadlington, W., and Kilroy, A. W., Excretion of propionylglycine in propionic acidaemia. *Clin. Sci.* **42**, 665–571 (1972).

R3. Rosenberg, L. E., Lilljequist, A. C., and Hsia, Y. E., Methylmalonic aciduria: An inborn error leading to metabolic acidosis, long-chain ketonuria and intermittent hyperglycinemia. *N. Engl. J. Med.* **278**, 1319–1322 (1968).

R4. Rosenberg, L. E., Lilljequist, A., and Hsia, Y. E., Methylmalonic aciduria: Metabolic block localization and vitamin B_{12} dependency. *Science* **162**, 805–807 (1968).

R5. Rowe, P. B., Inherited disorders of folate metabolism. *In* "The Metabolic Basis of Inherited Disease" (J. B. Stanbury, J. B. Wyngaarden, and D. S. Frederickson, eds.), 4th Ed., pp. 430–457. McGraw-Hill, New York, 1978.

R6. Rosenblatt, D. S., Cooper, B. A., Lue-Shing, S., Wong, P. W. K., Berlow, S., Narisawa, K., and Baumgartner, R., Folate distribution in cultured human cells, studies on 5,10-CH_2-H_4 PTEGLU reductase deficiency. *J. Clin. Invest.* **63**, 1019–1025 (1979).

R7. Rabinowitz, J. C., Mondy, N. I., and Snell, E. E., The vitamin B_6 group: An improved procedure for determination of pyridoxal with *Lactobacillus casei*. *J. Biol. Chem.* **175**, 147–153 (1948).

R8. Reddy, S. K., Reynolds, M. S., and Price, J. M., The determination of 4-pyridoxic acid in human urine. *J. Biol. Chem.* **233**, 691–696 (1958).

R9. Rosenberg, L. E., Vitamin responsive inherited metabolic disorders. *Adv. Hum. Genet.* **6**, 1–73 (1976).

R10. Rose, R. C., Water-soluble vitamin absorption in intestine. *Annu. Rev. Physiol.* **42**, 157–171 (1980).

R11. Rindi, G., and Ventura, U., Thiamine intestinal transport. *Physiol. Rev.* **52**, 821–827 (1972).

R12. Rogers, L. E., Porter, F. S., and Sidbury, J. B., Jr., Thiamine-responsive megaloblastic anemia. *J. Pediatr.* **74**, 494–504 (1969).

S1. Spencer, R. P., and Brody, K. R., Biotin transport by the small intestine of rat, hamster and other species. *Am. J. Physiol.* **206**, 653–657 (1964).

S2. Suchy, S. F., and Woolf, B., Protein-bound biotin: A consideration in multiple carboxylase deficiency. *Lancet* **1**, 108 (1982).

S3. Sydenstricker, V. P., Observations on 'egg white injury' in man and its cure with biotin concentrate. *Science* **95**, 176–177 (1942).

S4. Sydenstricker, V. P., Singal, S. A., Briggs, A. P., De Vaughn, N. M., and Isbell, H., Observations on the 'egg white injury' in man. *JAMA, J. Am. Med. Assoc.* **118**, 1199–1200 (1942).

S5. Seetharam, B., Alpers, D. H., and Allen, R. H., Isolation and characterisation of the ileal receptor for intrinsic factor-cobalamin. *J. Biol. Chem.* **256**, 3785–3790 (1981).

S6. Stenman, U.-H., Vitamin B-12 binding proteins of R-type, cobalphilin: Characterisation and comparison of cobalphilin from different sources. *Scand. J. Haematol.* **14**, 91–107 (1975).

S7. Spurling, C. L., Sacks, M. S., and Jigi, R. M., Juvenile pernicious anemia. *N. Engl. J. Med.* **271**, 995–1003 (1964).

S8. Sasaki, T., Further studies on the binding of vitamin B_{12} to the cell wall of a B_{12}-requiring *Lactobacillus*. *J. Bacteriol.* **109**, 169–178 (1972).

S9. Scott, C. R., Hakami, N., Teng, C. C., and Sagerson, R. N., Hereditary transcobalamin II deficiency: The role of transcobalamin II in vitamin B_{12} mediated reactions. *J. Pediatr.* **81**, 1106–1110 (1972).

S10. Stokke, O., Eldjarn, L., Norum, K. R., Steen-Johnsen, J., and Halvorsen, S., Methylmalonic aciduria: A new inborn error of metabolism which may cause fatal acidosis in the neonatal period. *Scand. J. Clin. Lab. Invest.* **20**, 313–328 (1967).

S11. Scott, J. M., and Weir, D. G., The methylfolate trap. *Lancet* **2**, 337–340 (1981).

S12. Selhub, J., and Rosenberg, I. H., Folate transport in isolated brush border membrane vesicles from rat intestine. *J. Biol. Chem.* **256**, 4489–4493 (1981).

S13. Steinberg, S. E., Campbell, C. L., and Hillman, R. S., Kinetics of the normal folate enterohepatic cycle. *J. Clin. Invest.* **64**, 83–88 (1979).

S14. Spector, R., Identification of folate binding macromolecule in rabbit choroid plexus. *J. Biol. Chem.* **252**, 3364–3370 (1977).

S15. Suzuki, N., and Wagner, C., Purification and characterisation of a folate binding protein from rat liver cytosol. *Arch. Biochem. Biophys.* **199**, 236–248 (1980).

S16. Santiago-Borrero, P. J., Santini, R., Jr., Perez-Santiago, E., and Maldonado, N., Congenital isolated defect of folic acid absorption. *J. Pediatr.* **82**, 450–455 (1973).

S17. Su, P. C., Congenital folate deficiency. *N. Engl. J. Med.* **294**, 1128 (1976).

S18. Shih, V. E., Salam, M. Z., Mudd, S. H., Uhlendorf, B. W., and Adams, R. D., A new form of homocystinuria due to $N^{5,10}$-methylenetetrahydrofolate reductase deficiency. *Pediatr. Res.* **6,** 395 (abstr.) (1972).

S19. Scriver, C. R., and Hutchinson, J. H., The vitamin B_6 deficiency syndrome in human infancy, biochemical and clinical observation. *Pediatrics* **31,** 240–250 (1963).

S20. Scriver, C. R., and Cullen, A. M., Urinary vitamin B_6 and 4-pyridoxic acid in health and in vitamin B_6 dependency. *Pediatrics* **36,** 14–20 (1965).

S21. Serebro, H. A., Solomon, H. M., Johnson, J. M., and Hendrix, T. R., The intestinal absorption of vitamin B_6 compounds by the rat and hamster. *Bull. Johns Hopkins Hosp.* **119,** 166–171 (1966).

S22. Solomon, L. R., and Hillman, R. S., Pyridoxine kinase activity in human erythrocytes and leucocytes: Assay and properties. *Biochem. Med.* **16,** 223–233 (1976).

S23. Spector, R., and Greenwald, L. L., Transport and metabolism of vitamin B_6 in rabbit brain and choroid plexus. *J. Biol. Chem.* **253,** 2373–2379 (1978).

S24. Smith, L. H., and Williams, H. E., Treatment of hyperoxaluria *Mod. Treat.* **4,** 522–533 (1967).

S25. Simell, O., and Takki, K., Raised plasma-ornithine and gyrate atrophy of the choroid and retina. *Lancet* **1,** 1031–1033 (1973).

S26. Scott, E. M., and Griffith, I. V., The enzymic defect of hereditary methaemaglobinaemia: Diaphorase. *Biochim. Biophys. Acta* **34,** 584–586 (1959).

S27. Schwartz, J. M., and Jaffe, E. R., Hereditary methemoglobinemia with deficiency of NADH dehydrogenase. *In* "The Metabolic Basis of Inherited Disease" (J. B. Stanbury, J. B. Wyngaarden, and D. S. Frederickson, eds.), 4th Ed., pp. 1452–1464. McGraw-Hill, New York, 1978.

S28. Staal, G. E. J., Van Berbel, T. J. C., Nijessen, J. G., Koster, J. F., and Wensink-Vander Loo, A., Normalisation of red blood cell pyruvate kinase in pyruvate kinase deficiency by riboflavin treatment. *Clin. Chim. Acta* **60,** 323–327 (1975).

S29. Stokke, O., Goodman, S. I., Thompson, J. A., and Miles, B. S., Glutaric aciduria; presence of glutaconic and β-hydroxyglutaric acids in urine. *Biochem. Med.* **12,** 386–391 (1975).

S30. Stokke, O., Goodman, S. I., and Moe, P. G., Inhibition of brain glutamate decarboxylase by glutarate, glutaconate, and β-hydroxyglutarate: Explanation of the symptoms in glutaric aciduria. *Clin. Chim. Acta* **66,** 411–415 (1976).

S31. Sauberlich, H. E., Biochemical alterations in thiamine deficiency—their interpretation. *Am. J. Clin. Nutr.* **20,** 528–535 (1967).

S32. Sauberlich, H. E., Herman, Y. F., Stevens, C. O., and Herman, R. H., Thiamin requirement of the adult human. *Am. J. Clin. Nutr.* **32,** 2237–2248 (1979).

S33. Schulman, J. D., Lustberg, T. J., Kennedy, J. L., Museles, M., and Seegmiller, J. E., A new variant of maple syrup urine disease (branched chain ketoaciduria): Clinical and biochemical evaluation. *Am. J. Med.* **49,** 118–124 (1970).

S34. Scriver, C. R., Mackenzie, S., Clow, C. L., and Delvin, E., Thiamine-responsive maple syrup urine disease. *Lancet* **1,** 310–313 (1971).

S35. Saudubray, J.-M., Marsoac, C., Charpentier, C., Cathelineau, L., Besson Leaud, M., and Leroux, J. P., Neonatal congenital lacticacidosis with pyruvate carboxylase deficiency in two siblings. *Acta Paediatr. Scand.* **65,** 717–724 (1976).

S36. Seashore, M. R., Durant, J. L., and Rosenberg, L. E., Studies on the mechanism of pyridoxine-responsive homocystinuria. *Pediatr. Res.* **6,** 187–192 (1972).

T1. Taylor, R. T., Stephen, A. N., and Hanna, M. L., Uptake of cobalamin by *Escherichia coli* B: Some characteristics and evidence for a binding protein. *Arch. Biochem. Biophys.* **148,** 366–381 (1978).

T2. Tauro, G. P., Danks, D. M., Rowe, P. B., Van der Weyden, M. B., Schwartz, M. A.,

Collins, V. L., and Neal, B. W., Dihydrofolate reductase deficiency causing megaloblastic anemia in two families. *N. Engl. J. Med.* **294**, 299–307 (1972).

T3. Tabor, H., and Wyngarden, L., The enzymatic formation of formiminotetrahydrofolic acid, 5,10-methenyltetrahydrofolic acid and 10-formyl-tetrahydrofolic acid in the metabolism of formiminoglutamic acid. *J. Biol. Chem.* **234**, 1830–1846 (1959).

T4. Tsuji, T., Yamada, R., and Nase, Y., Intestinal absorption of vitamin B_6. I. Pyridoxol uptake by rat intestinal tissue. *J. Nutr. Sci. Vitaminol.* **19**, 401–417 (1973).

T5. Tada, K., Yokoyama, Y., Nakagawa, H., Yoshida, T., and Arakawa, T., Vitamin B_6 dependent xanthurenic aciduria. *Tohoku J. Exp. Med.* **93**, 115–124 (1967).

T6. Tada, K., Yokoyama, Y., Nakagawa, H., and Arakawa, T., Vitamin B_6 dependent xanthurenic aciduria (the second report). *Tohoku J. Exp. Med.* **95**, 107–114 (1968).

T7. Takki, K., Gyrate atrophy of the choroid and retina associated with hyperornithinemia. *Br. J. Ophthalmol.* **58**, 3–23 (1974).

T8. Tiselius, H. G., and Almgard, L. E., The diurnal urinary excretion of oxalate and the effect of pyridoxine and ascorbate. *Eur. Urol.* **3**, 41–46 (1977).

T9. Tanaka, K., Mantagos, S., Genel, M., Seashore, M. R., Billings, B. A., and Baretz, B. H., A new defect in fatty acid metabolism with hypoglycaemia and organic aciduria. *Lancet* **2**, 986–987 (1977).

U1. United States Food and Nutrition Board, "Recommended Daily Dietary Allowances of Vitamins." Natl. Acad. Sci., Washington, D.C., 1968.

U2. Uhlendorf, B. W., Conerly, E. B., and Mudd, S. H., Homocystinuria: Studies in tissue culture. *Pediatr. Res.* **7**, 645–658 (1973).

V1. Valle, D., Kaiser-Kupfer, M. I., and Del Valle, L. A., Gyrate atrophy of the choroid and retina: Deficiency of ornithine aminotransferase in transformed lymphocytes. *Proc. Natl. Acad. Sci. U.S.A.* **74**, 5159–5161 (1977).

V2. Valentine, W. N., and Tanaka, K. R., Pyruvate kinase and other enzyme deficiency hereditary hemolytic anemias. *In* "The Metabolic Basis of Inherited Disease" (J. B. Stanbury, J. B. Wyngaarden, and D. S. Frederickson, eds.), 4th Ed., pp. 1410–1429. McGraw-Hill, New York, 1978.

V3. Van Berkel, T. J. C., Staal, G. E. J., Koster, J. F., and Nijessen, J. G., On the molecular basis of pyruvate kinase deficiency. II. Role of thiol groups in pyruvate kinase from pyruvate kinase-deficient patients. *Biochim. Biophys. Acta* **334**, 361–367 (1974).

V4. Ventura, U., and Rindi, G., Transport of thiamine by the small intestine in vitro. *Experientia* **21**, 645–646 (1965).

V5. Viana, M. B., and Carvalho, R. I., Thiamine-responsive megaloblastic anemia, sensorineural deafness, and diabetes mellitus: A new syndrome. *J. Pediatr.* **93**, 235–238 (1978).

W1. Weyler, W., Sweetman, L., Maggio, D. C., and Nyhan, W. L., Deficiency of propionyl CoA carboxylase and 3-methylcrotonyl CoA carboxylase in a patient with 3-methylcrotonylglycinuria. *Clin. Chim. Acta* **76**, 321–328 (1977).

W2. Willard, H. F., and Rosenberg, L. E., Inherited deficiencies of human methylmalonyl CoA mutase activity: Reduced affinity of mutant apoenzyme for adenosylcobalamin. *Biochem. Biophys. Res. Commun.* **78**, 927–934 (1977).

W3. Waxman, S., Folate binding proteins. *Br. J. Haematol.* **29**, 23–29 (1975).

W4. Wittwer, A. J., and Wagner, C., Identification of the folate-binding proteins of rat liver mitochondria as dimethylglycine dehydrogenase and sarcosine dehydrogenase. I. *J. Biol. Chem.* **256**, 4102–4108 (1981).

W5. Wittwer, A. J., and Wagner, C., Identification of the folate binding proteins of rat liver mitochondria as dimethylglycine dehydrogenase and sarcosine dehydrogenase. II. *J. Biol. Chem.* **256**, 4109–4115 (1981).

W6. Walters, T. R., Congenital megaloblastic anemia responsive to N^5-formyltetrahydrofolate administration. *J. Pediatr.* **70**, 686–687 (abstr.) (1967).

W7. Weiser, H., Brubacher, G., and Wiss, O., Standardization of activity. *In* "The Vitamins" (W. H. Sebrell, Jr. and R. H. Harris, eds.), Vol. 2, pp. 29–31. Academic Press, New York, 1968.

W8. Werder, E. A., Curtius, H.Ch., Tancredi, F., Anders, P. W., and Prader, A., Homocystinuria. *Helv. Paediatr. Acta* **21**, 1–8 (1966).

W9. Wakabayashi, K., and Shimazono, N., Studies on ω-oxidation of fatty acids in vitro. I. Overall reaction and intermediates. *Biochim. Biophys. Acta* **70**, 132–142 (1963).

W10. Wick, H., Schweizer, K., and Baumgartner, R., Thiamine-dependency in a patient with congenital lactic acidaemia due to pyruvate dehydrogenase deficiency. *Agents Actions* **7**, 405–410 (1977).

W11. Wolf, B., Grier, R. E., Allen, R. J., Goodman, S. I., and Kien, C. L., Biotinidase deficiency: The enzymatic defect in late-onset multiple carboxylase deficiency. *Clin. Chem. Acta* **131**, 273–281.

Y1. Youngdahl-Turner, P., Mellman, I. S., Allen, R. H., and Rosenberg, L. E., Protein mediated vitamin uptake. *Exp. Cell Res.* **118**, 127–134 (1979).

Y2. Youngdahl-Turner, P., Rosenberg, L. E., and Allen, R. H., Binding and uptake of transcobalamin II by human fibroblasts. *J. Clin. Invest.* **61**, 133–141 (1978).

Y3. Yagi, K., and Okuda, J., Phosphorylation of riboflavin by transferase action. *Nature (London)* **181**, 1663–1664 (1958).

Y4. Yubisui, T., Matsuki, T., Tanishima, K., Takeshita, M., and Yoneyama, Y., NADPH-flavin reductase in human erythrocytes and the reduction of methemoglobin through flavin by the enzyme. *Biochem. Biophys. Res. Commun.* **76**, 174–182 (1977).

Z1. Zagalak, B., and Friedrich, W., eds., "Proceedings of the Third European Symposium on Vitamin B_{12} and Intrinsic Factor." de Gruyter, Berlin, 1979.

SPECTROPHOTOMETRY OF HEMOGLOBIN AND HEMOGLOBIN DERIVATIVES

E. J. van Kampen* and W. G. Zijlstra†

*Clinical Chemical Laboratory,
Diaconessenhuis,
and
†Department of Physiology,
University of Groningen,
Groningen, The Netherlands

1. Introduction	200
2. Determination of Total Hemoglobin	202
2.1. Spectral Properties of Hemiglobincyanide	202
2.2. Reagent Solutions	205
2.3. Handling of Blood Samples	210
2.4. Measurement of the Diluted Hemiglobincyanide Solution	211
2.5. Hemiglobincyanide Reference Solutions	213
2.6. Quality Control	215
2.7. Other Methods for the Determination of Total Hemoglobin	219
3. Absorption Spectra and Millimolar Absorptivities of Hemoglobin and Hemoglobin Derivatives	220
3.1. Sample Preparation and Measurement Procedure	220
3.2. Deoxygenated Hemoglobin (Hb)	223
3.3. Oxyhemoglobin (HbO_2)	225
3.4. Carboxyhemoglobin (HbCO)	225
3.5. Hemiglobin (Methemoglobin; Hi)	228
3.6. Hemiglobincyanide (HiCN)	228
3.7. Hemiglobinnitrite ($HiNO_2$) and Hemiglobinazide (HiN_3)	229
3.8. Sulfhemoglobin (SHb), Sulfhemiglobin (SHi), and Sulfhemiglobincyanide (SHiCN)	230
4. Determination of Hemoglobin Derivatives	232
4.1. Theoretical Considerations	232
4.2. Determination of the Oxygen Saturation (S_{O_2})	235
4.3. Determination of the Carbon Monoxide Fraction (F_{HbCO})	239
4.4. Determination of the Hemiglobin Fraction (F_{Hi})	241
4.5. Multicomponent Analysis of Hemoglobin Derivatives	242
5. Concluding Remarks	248
6. Appendix	251
References	253

Copyright © 1983 by Academic Press, Inc.
All rights of reproduction in any form reserved.
ISBN 0–12–010323–0

1. Introduction

Since our first review of the subject appeared in 1965 (58), (spectro)-photometric methods have been increasingly used for the determination of the total hemoglobin concentration in blood (c_{Hb}^*) and for measuring the fractions of the common hemoglobin derivatives (oxyhemoglobin, HbO_2; carboxyhemoglobin, HbCO; hemiglobin or methemoglobin, Hi; sulfhemoglobin, SHb) normally or abnormally—though not infrequently—present in human blood. As for the determination of total hemoglobin, the hemiglobincyanide (cyanmethemoglobin, HiCN) method eventually proved to be superior to the various other methods and came to be increasingly used in clinical chemical laboratories (Fig. 1). After it had been adopted as the preferred method by the Standardizing Committee of the European Society of Haematology (39, 40), it was finally accepted for worldwide use by the International Committee for Standardization in Haematology (ICSH) (15, 16). One of the major reasons for selecting the HiCN method was the suitability of HiCN for the preparation of stable and reliable reference solutions. Prescriptions for the production, distribution, and use of such a reference solution were issued by ICSH together with instructions for carrying out the standardized procedure (15, 16).

Numerous (spectro)photometric methods have been developed for the determination of oxyhemoglobin, either as oxyhemoglobin fraction (c_{HbO_2}/c_{Hb}^*) or as oxygen saturation [$S_{O_2} = c_{HbO_2}/(c_{HbO_2} + c_{Hb})$]. Among these methods are many conventional two-wavelength spectrophotometric methods, most of them involving the use of an isobestic point in the absorption

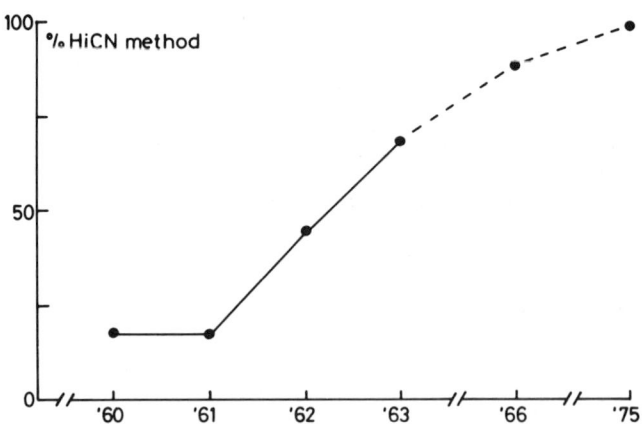

FIG. 1. Relative increase in the use of the HiCN method for the determination of the total hemoglobin concentration in blood, 1960–1975 (55).

spectra of Hb and HbO_2, i.e., a wavelength at which the absorptivity (cf. Section 4.1) of the two components has the same value (58). This yields a single, linear relationship between the oxygen saturation and the ratio of the light absorbances of the sample at the two wavelengths used (65). An important factor in selecting the most suitable pair of wavelengths is avoiding the influence of other absorbing substances in the sample such as bilirubin, lipids, and HbCO (34), or indocyanine green, the commonly used indicator in dye dilution studies (25). Various types of special photometers (oximeters) have been constructed for measuring oxygen saturation *in vitro* as well as *in vivo* (31). In some of these instruments, light reflection is used instead of light transmission (65, 67). The use of reflected light ultimately led to the development of fiber optic oximeters (18, 26, 33), which, though too complicated for routine clinical chemical use, are very suitable for some special applications (77).

Many two-wavelength methods are also used in the determination of the common dyshemoglobins in human blood (HbCO, Hi, SHb). In all of these methods, the samples undergo some kind of pretreatment so that, besides the dyshemoglobin species to be determined, only one other hemoglobin derivative is present. In most of these methods an isobestic point is used. Thus, HbCO can be measured in the system $HbCO/HbO_2$ at $\lambda = 562$ and 540 nm, and in the system HbCO/Hb at $\lambda = 538$ and 578 nm (58). In the former procedure, the blood is gently oxygenated to convert all non-HbCO to HbO_2, while in the latter procedure the sample is deoxygenated using sodium dithionite. A two-wavelength method ($\lambda = 558$ and 523 nm) for the determination of Hi in the system Hi/HbO_2 again involves oxygenation of the sample (58).

Until recently, measuring more than two components in a mixture was much more complicated. The methods involved rather laborious pretreatment of the samples and quite extensive calculations (1, 37, 38, 68). Through the availability of more reliable data as to the absorption spectra of all common hemoglobin derivatives (9, 45) and with the advent of simple and powerful calculators, multicomponent analysis has become much more practicable. It is carried out using conventional spectrophotometers (75) and special instruments (3, 74).

The availability of cheap calculating power led to the tendency of calculating quantities which can be measured directly. An example of this is the calculation of oxygen saturation from oxygen tension. When this is done in addition to the direct measurement, it affords an excellent opportunity for interparametric quality control (54). When it is done instead of the direct measurement, however, it can occasionally be the cause of grave errors (32).

This article deals primarily with spectrophotometric methods that can be carried out with general purpose spectrophotometers. The many special photometers (hemoglobinometers, oximeters, etc.) presently available for the determination of hemoglobin and hemoglobin derivatives will be mentioned only incidentally.

2. Determination of Total Hemoglobin

This section deals with the determination of the total hemoglobin concentration (c_{Hb}^*) in human blood by means of the standardized HiCN method (15, 16, 57, 71–73). The principles and practice of the method are presented in detail. Also, advice is given as to how the many pitfalls contained in even a strictly standardized procedure can be avoided. The section concludes with a brief description of quality control in hemoglobinometry and a few remarks on methods for the determination of c_{Hb}^* other than the standardized HiCN method.

2.1. Spectral Properties of Hemiglobincyanide

Besides its stability and the relative ease with which other hemoglobin derivatives are converted to it, a reason for selecting HiCN as the preferred compound for hemoglobinometry was its favorable absorption spectrum, with its flat absorption maximum around $\lambda = 540$ nm (Fig. 2). The fundamental quantity on which the HiCN method is based is obviously the (milli-

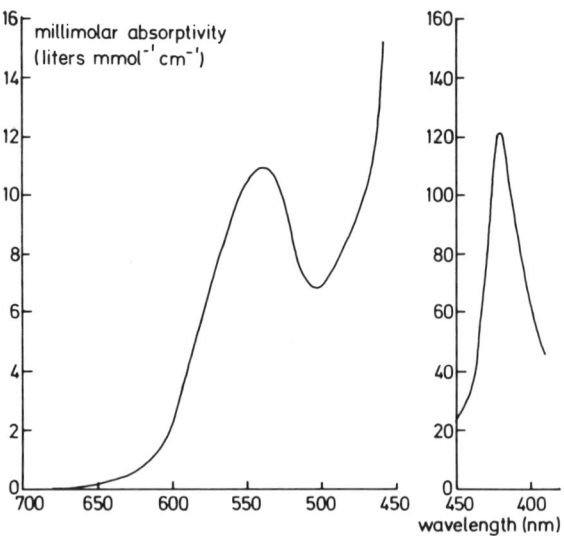

Fig. 2. Absorption spectrum of HiCN. (Based on data from refs. 45 and 58.)

molar) absorptivity at the wavelength of maximum absorption: $\epsilon_{\text{HiCN}}^{540}$ [cf. Eq. (9) in Section 4.1]. This quantity has been determined by several investigators using different methods. In most investigations, the iron content of the hemoglobin molecule was taken as the basis of the determination of $\epsilon_{\text{HiCN}}^{540}$ (22, 24, 29, 35, 36, 41, 59, 63, 71). The results thus obtained have been confirmed by the determination of $\epsilon_{\text{HiCN}}^{540}$ on the basis of both nitrogen (43) and carbon (17). Recently, Hoek et al. (14) determined $\epsilon_{\text{HiCN}}^{540}$ by direct titration of hemiglobin with cyanide. In Table 1, most of the determinations of $\epsilon_{\text{HiCN}}^{540}$ are summarized. Only the results of Minkowski and Swierczewski (24) and Wootton and Blevin (63) have been omitted. The former have not been included because the standard error of the mean value is not given in the paper, the latter because these results [$\epsilon_{\text{HiCN}}^{540}$ = 10.68 ± 0.04 (SEM) (n = 14)] are significantly different from all other results taken together. A total mean value for $\epsilon_{\text{HiCN}}^{540}$ of 10.99 with a total standard error of the mean of 0.01 was calculated for a total of 521 determinations using the equations given in refs. 52 and 73. These data constitute the solid experimental basis on which rests the internationally accepted value for $\epsilon_{\text{HiCN}}^{540}$ = 11.0.

An essential assumption underlying the determination of the total hemoglobin concentration by absorption photometry—and also the analysis of mixtures of two or more hemoglobin derivatives—is the applicability of Lambert–Beer's law [Eq. (9) in Section 4.1]. The validity of Lambert–Beer's law for HiCN solutions has been repeatedly demonstrated (58, 70). A more severe test has been carried out by Drabkin (11) for HbO_2 solutions. No significant difference in the absorptivity of HbO_2 at λ = 578, 562, and 542 nm was found when measuring over a concentration range of 0.077 to 38.2 mM, with a lightpath length varying from 1.0 to 0.007 cm. Nevertheless, a recent paper (4) has cast some doubt on the general applicability of Lambert–Beer's law to solutions of human hemoglobin. The investigators claim that the absorption spectra of human (oxy)hemoglobin are strongly dependent on the total hemoglobin concentration. Only data pertinent to HbO_2 are given, but it is implied that a similar dependence of the absorption spectrum on the total hemoglobin concentration might be found for HiCN.

Although the experimental evidence in support of this opinion is scanty, the importance of the matter for the whole of hemoglobin spectrophotometry is so great that an experimental reinvestigation seemed unavoidable. Therefore, Zijlstra et al. (66) have measured the absorptivity of HbO_2 and HiCN for a wide range of total hemoglobin concentrations. The total hemoglobin concentration in blood was varied from about 0.004 to 10 mM in 10 approximately equally distributed steps; the lightpath length varied from 0.007 to 1.000 cm. No concentration-dependent differences in the absorptivity of HbO_2 and HiCN were observed over the wavelength range of 450 to 750 nm. Figure 3 shows that even between the samples with the

TABLE 1
ϵ_{HiCN}^{540} AS REPORTED BY DIFFERENT AUTHORS

Authors	Material	ϵ_{HiCN}^{540} [a]	SEM[b]	n[c]	Method
Meyer-Wilmes and Remmer (22)	Horse Hb	11.0	0.04	12	Fe; o-phenanthroline
	Horse Hb	11.0	0.04	12	Fe; TiCl$_3$
Remmer (35)	Human whole blood	11.09	0.03	11	Fe; TiCl$_3$
	Human whole blood	11.19	0.065	4	Fe; complexon
Zijlstra and van Kampen (71)	Human Hb, toluene hemolysis	10.99	0.01	123	Fe; α,α'-dipyridyl
		10.94	0.03	35	Fe; α,α'-dipyridyl
		11.05	0.02	101	Fe; TiCl$_3$
Van Oudheusden et al. (59)	Human whole blood	10.99	0.05	10	Fe; α,α'-dipyridyl
	Human whole blood	11.06	0.08	8	Fe; α,α'-dipyridyl
Salvati et al. (36)	Human Hb purified on CMC column	10.95	0.03	46	Fe; α,α'-dipyridyl
Morningstar et al. (29)	Human whole blood	11.02	0.03	10	Fe; X-ray emission spectrography
	Human washed cells	10.97	0.07	6	
Stigbrand (41)	Human Hb purified on CMC or Sephadex column or by dialysis against Na$_2$-EDTA	11.00	0.02	55	Fe; sulfosalicylic acid
Tentori et al. (43)	Human Hb purified on CMC column	10.90	0.05	55	N analysis
Itano (17)	Human Hb A purified by chromatography	10.88	0.04	16	C analysis
Hoek et al. (14)	Human Hb, toluene hemolysis	11.01	0.03	17	Titration of Hi with CN$^-$

[a] ϵ_{HiCN}^{540}, Millimolar absorptivity (liters mmol^{-1} cm^{-1}) of hemiglobincyanide at $\lambda = 540$ nm.
[b] SEM, Standard error of the mean.
[c] n, Number of determinations.

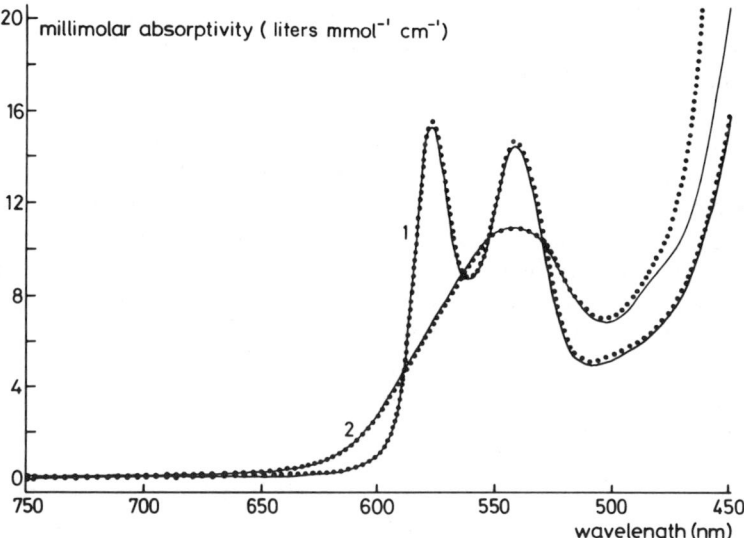

FIG. 3. Millimolar absorptivity of HbO_2 (1) and HiCN (2) as a function of wavelength. Solid lines: $l = 0.0071$ cm, $c_{HbO_2} = 10.609$ mM, $c_{HiCN} = 10.086$ mM. Dotted lines: $l = 1.000$ cm, $c_{HbO_2} = 0.0042$ mM, $c_{HiCN} = 0.0040$ mM. Measurements with a reversed-optics spectrophotometer (HP8450A) for every 2 nm in the wavelength range of 450 to 750 nm. The deviation of the low concentration HiCN spectrum at $\lambda < 500$ nm is due to absorption by the hexacyanoferrate(III) present in the diluent (78). [From Zijlstra et al. (66).]

highest and lowest total hemoglobin concentration, no differences in the absorption spectra occur. These results confirm the generally accepted view that, at least in the visible range, Lambert–Beer's law is valid for solutions of hemoglobin and its common derivatives.

2.2. REAGENT SOLUTIONS

For routine hemoglobinometry, the native blood must be diluted some 200–250 times to lower the hemoglobin concentration to such a level that the absorbance enters a range suitable for measurement with a lightpath length of 1 cm in a general purpose (spectro)photometer. Ideally, the diluent should have such properties that hemolysis occurs immediately, that all hemoglobin derivatives present are rapidly converted into HiCN, and that no turbidity ensues from the presence of plasma proteins and erythrocyte stromata (15). No reagent solution devised up to now strictly fulfills all requirements. This is because these requirements are to some degree mutually incompatible.

Oxidation of the hemoglobin iron, i.e., formation of Hi, is best accomplished with the aid of potassium hexacyanoferrate(III) [potassium ferricyanide, $K_3Fe(CN)_6$]. The CN^- ions for the second step of the reaction, the conversion of Hi to HiCN, can be provided by the addition of KCN. Therefore, all reagent solutions contain $K_3Fe(CN)_6$ and KCN, usually at 200 and 50 mg/liter, respectively. Because HCN is a very weak acid ($pK_a = 9.1$), solutions of $K_3Fe(CN)_6$ and KCN are quite alkaline (pH ≈ 9.6). At this pH, the formation of Hi from Hb or HbO_2 is very slow. The reagent solutions therefore usually contain a third compound to bring the pH so far down that the time necessary for the conversion to Hi becomes reasonably short. Drabkin's solution, the reagent solution in general use before 1961 (71), contains 1.0 g $NaHCO_3$ and has a pH of 8.6. With this reagent solution, the conversion time from Hb or HbO_2 to HiCN is 20 minutes and is mainly determined by the formation of Hi, the addition of CN^- to Hi being very rapid. To attain a further shortening of the conversion time, Chilcote and O'Dea (5) introduced a diluent with pH 7.4 which contains KH_2PO_4 instead of $NaHCO_3$. With this reagent, complete conversion to HiCN takes only a few minutes.

Whereas the requirement of a short conversion time can be achieved to a reasonable degree by lowering the pH of the diluent, the very same change in the composition of the solution favors the appearance of turbidity. This turbidity, which is mainly caused by precipitated plasma protein (primarily γ-globulins, some having an isoelectric point near the pH of the reagent solution) and, to a lesser degree, by erythrocyte stromata, can be effectively prevented by the addition of a small amount of a nonionic detergent such as Sterox SE[1] In the concentration needed, this detergent has no influence on the absorption spectrum of HiCN and it increases the overall reaction velocity by its strong hemolytic action (57).

On the basis of the above considerations, the reagent solution recommended by ICSH (16) has the following composition: 200 mg $K_3Fe(CN)_6$, 50 mg KCN, 140 mg KH_2PO_4, 0.5 ml Sterox SE, and deionized or distilled water to 1000 ml. Sterox SE is an alkylphenol(thiol) polyethylene oxide detergent. It may be replaced by similar detergents such as Nonidet P40,[2] Quolac Nic-218,[3] or Nonic 218.[4] This reagent solution reasonably fulfills the requirements mentioned above. The conversion time is 3 minutes (cf. Fig. 4 in ref. 58). If stored at room temperature in a brown borosilicate glass bottle, the solution keeps for several months. However, there are some imperfections which, in special circumstances, may be the cause of small and sometimes even large errors. Most of these errors can be prevented by quite

---

[1] Hartmann-Leddon Company, Philadelphia, Pennsylvania.
[2] Shell International Chemical Company, The Hague, The Netherlands.
[3] Unibasic Inc., Arnold, Maryland.
[4] Pennsalt Chemicals Corporation, Philadelphia, Pennsylvania.

simple measures. All errors can be detected if the operator has a good understanding of the procedure. In the following, the possible errors connected with the properties of the reagent solutions are discussed.

The only hemoglobin derivative which is not converted to HiCN is SHb; SHiCN is formed instead. The absorption spectrum of this compound is different from that of HiCN (cf. Figs. 2 and 17). Since ϵ_{SHiCN}^{540} is 8.0, the effect of SHb on the measured value of c_{Hb}^* is slight. It can be calculated that the presence of 1% SHb in the blood causes c_{Hb}^* to be underestimated by only 0.27%. With an SHb fraction of 5% and $c_{Hb}^* = 15$ g/dl, the measured value of c_{Hb}^* is thus 14.8 g/dl. In practice, the effect may accordingly be considered negligible, as SHb fractions exceeding 5% are seldom encountered.

HbCO is completely converted into HiCN, but the conversion time is considerably longer than for other hemoglobin derivatives (5, 42). As ϵ_{HbCO}^{540} is about 14.3, nonconversion of HbCO to HiCN results in too high values for c_{Hb}^*. The upper limit of this error can be calculated on the basis of the assumption that none of the HbCO present is converted to HiCN during the 3 minutes usually allowed for the reaction (53). For a sample with $c_{Hb}^* = 15$ g/dl, the maximum error in c_{Hb}^* caused by 10 and 20% HbCO is 3 and 6%, respectively. In practice, the error will be smaller. It can be prevented by increasing the time between the mixing of blood and reagent and the measurement from 3 to 30 minutes. The conversion time can be shortened by increasing the $K_3Fe(CN)_6$ concentration in the reagent solution. However, the number of samples with a HbCO fraction high enough to cause a considerable error in c_{Hb}^* is too small to make worthwhile the use of a special reagent such as the one of Taylor and Miller (42), which contains 1000 mg $K_3Fe(CN)_6$.

In some pathological conditions such as Kahler's disease (multiple myeloma), in which a pathological protein species (paraprotein) is present in the plasma, or in severe infectious diseases in which γ-globulins constitute some 30% of total plasma protein, visible turbidity will develop in the diluted HiCN solution, causing too high values for c_{Hb}^*. In these cases, the addition of one drop of 25% ammonia solution—after the conversion to HiCN has been completed—clears the blood–reagent mixture. In severe lipidemia of the Frederickson V type, chylomicrons may cause turbidity in the diluted HiCN solution. The lipid particles can be removed by diethyl ether extraction and centrifugation. This procedure does not influence the HiCN concentration. That, in the vast majority of cases, no appreciable turbidity develops after mixing blood and reagent solution was shown by the following experiment (46). In 150 consecutive hemoglobin determinations, one drop of ammonia solution was added to the diluted HiCN solution after the absorbance measurement and, thereafter, the absorbance was measured again. In only two cases was there any appreciable drop in the measured value of c_{Hb}^*.

Thus, there is no urgent need for the modified reagent solutions in which no turbidity, or less turbidity, develops when severe plasma abnormalities are present. All proposed modifications narrow down to increasing the ionic strength, which is accomplished either by the addition of NaCl (21, 61) or by increasing the concentration of phosphate buffer (20). The use of a very high NaCl concentration (50 g/liter) in the reagent has been advocated to obtain clear HiCN solutions even in the case of severe leukocytosis (21). However, it seems more convenient to clear the blood–reagent mixture by centrifugation before the absorbance is measured.

The concentration of CN^- in a freshly prepared reagent solution is 0.769 mM, which ensures complete conversion of Hi to HiCN (14). However, it has been found repeatedly that the CN^- concentration of a reagent solution decreases as the solution ages (53). CN^- may escape from the solution as HCN, especially when plastic containers are used. When loss of CN^- occurs, the other chemicals in the reagent remaining intact, all hemoglobin in the sample will be converted to Hi, but the subsequent formation of HiCN will be incomplete. As $\bar{\epsilon}_{Hi}^{540} = 6.8$ (cf. Fig. 14), c_{Hb}^* values up to 40% too low may be found when no HiCN is formed from Hi. The presence of a sufficient amount of CN^- in the reagent solution can be easily checked by repeating the determination of c_{Hb}^* after adding some extra KCN to the diluent. Other methods for this purpose are the use of an ion-selective electrode for CN^- (69) and the Liebig–Denigès argentometric titration of CN^-.

The toxicity of the KCN contained in the reagent solution has been a matter of concern. This has even led to the proposal to substitute a method involving hemiglobinazide (HiN$_3$) for the hemiglobincyanide method (60). Experience has shown, however, that the risks involved in handling KCN are not too serious. The reagent solution itself is harmless because of the low concentration of KCN. Handling of solid KCN in preparing the reagent solution requires the usual laboratory precautions for toxic substances, which can be easily taken. Moreover, many laboratories use packaged reagents which obviate the handling of solid KCN. As to the proposed HiN$_3$ method, the problems in substituting HiN$_3$ for HiCN have been much underestimated (49). Little would be gained by the adoption of this method and many new problems would arise.

The concentration of K$_3$Fe(CN)$_6$ in the reagent solution is 0.607 mM, which is amply sufficient for the oxidation of all hemoglobin iron contained in the diluted blood sample. The amount of $[Fe(CN)_6]^{3-}$ in the reagent solution is stable as long as the solution is not frozen. On freezing and thawing, $[Fe(CN)_6]^{3-}$ is converted to $[Fe(CN)_6]^{4-}$, and no formation of Hi occurs on mixing blood and reagent solution. Consequently, the hemoglobin in the solution will remain mainly in the form of HbO$_2$. As $\bar{\epsilon}_{HbO_2}^{540} = 14.3$ (cf. Fig. 12 and Table 5), the measured value for c_{Hb}^* will be about 30% too high. The occurrence of chemical conversions in $K_3Fe(CN)_6$/KCN reagents on freezing and thawing has been known since the investigations of Michelsen et al. (23) and Weatherburn and Logan (62). They reported independently that $K_3Fe(CN)_6$/KCN reagents, when frozen, loose their color and that after thawing, c^*_{Hb} values obtained with these reagents differ from the expected values. Figure 4 shows absorption spectra of a reagent solution before and after freezing and thawing. The absorption spectrum before freezing is identical with that of hexacyanoferrate(III) and the spectrum after thawing with that of hexacyanoferrate(II) [ferrocyanide; $[Fe(CN)_6]^{4-}$].

Zweens et al. (78) have demonstrated that the overall reaction occurring in the HiCN reagent on freezing is as follows:

$$2\,[Fe(CN)_6]^{3-} + 2\,CN^- \rightarrow 2\,[Fe(CN)_6]^{4-} + (CN)_2 \tag{1}$$

Bubbles of $(CN)_2$ can be seen to form on the phase boundary between water and ice. At much higher concentrations of the reactants than present in the HiCN reagent, reaction (1) also proceeds in simple solution at room temperature. However, during freezing, the reaction proceeds at low overall concentrations of the reactants because these become highly concentrated at the moving phase boundary. When ethanol, methanol, ethylene glycol (20 ml/liter), or glycerol (5 ml/liter) is added to the HiCN reagent solution,

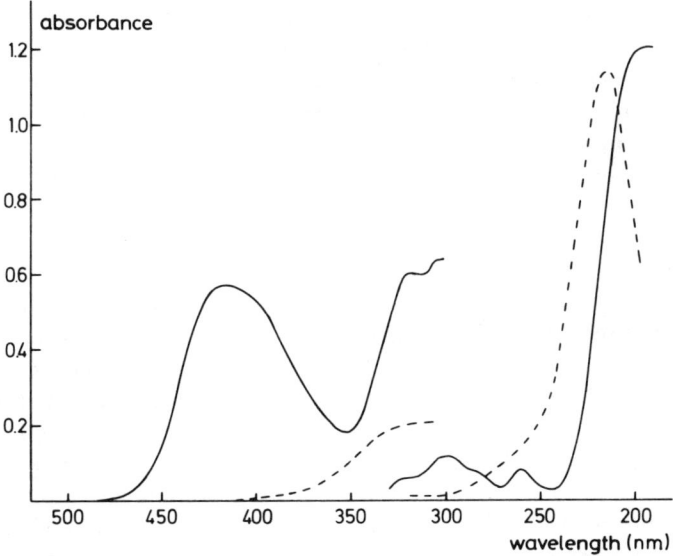

FIG. 4. Absorption spectra of reagent solution before (solid lines) and after (broken lines) freezing and thawing. For the recording from $\lambda = 320$ to $\lambda = 200$ nm, the reagent was diluted 5-fold before, and 10-fold after, freezing and thawing. Optica CF4DR grating spectrophotometer. [From Zweens et al. (78).]

decomposition on freezing and thawing does not occur. This is because transition from the liquid to the solid state then takes place abruptly after the solution has become supercooled and a phase boundary does not develop. None of these additions interferes with the formation or the spectral properties of HiCN. Turbidity, if any, is slightly increased by ethylene glycol, but diminished by ethanol and glycerol. Therefore, in circumstances where freezing of the reagent solution is likely to occur, addition of 20 ml/liter ethanol is advisable.

2.3. Handling of Blood Samples

The blood sample may be taken from a freely bleeding capillary puncture (finger; heel, in infants) or from an arterial or venous specimen. The capillary puncture must be deep enough to allow the blood to flow freely. No pressure should be exerted to obtain a sufficient volume of blood from the puncture, for the blood then becomes diluted with tissue fluid. An arterial or venous specimen may be collected into any solid anticoagulant (EDTA, heparin, mixture of ammonium and potassium oxalate). Before sampling from the tube or syringe in which the specimen has been collected, the blood should be well mixed. A tube should be gently tipped end over end 20 times; a syringe should contain a mixing ring or ball.

The optimum dilution of blood in reagent is about 250 times. A dilution of 251 times is achieved by adding 0.02 ml of blood to 5.0 ml of reagent solution. Reagent solution (5.0 ml) is pipetted into a test tube using a bulb-type or graduated-type volumetric pipet with an accuracy of ±0.5%. The accuracy of new pipets should be checked. This may be done by weighing the amount of water the pipet yields when filled to the mark. For larger series of measurements, an automatic pipetting device of the same order of accuracy is advisable.

For the transfer of blood, thick-walled capillary pipets ("Sahli type") can be used. These are one-mark, "to contain" types of pipets. The sample is drawn up to just above the mark and the outside wiped clean with a piece of tissue paper. At the same time, the excess blood is removed from the pipet by gently touching the tip with the tissue paper. The pipet is then placed with the tip at the bottom of the test tube containing the reagent solution and the blood is expelled by blowing gently. The pipet is then partly withdrawn from the solution and rinsed three times with reagent from the upper layer in the test tube. The accuracy of new pipets should be checked. This may be done by weighing the amount of mercury the pipet contains when filled to the mark, by photometry of a dye solution (e.g., patent blue V), diluted using the pipet, or by repeated determination of c_{Hb}^* of a blood sample of known concentration using the pipet. The pipets should be cleaned daily in detergent, thoroughly rinsed with distilled water, and dried. Cleaning once a week with 0.1 M HCl is also advisable.

Instead of the Sahli pipets, disposable capillary pipets can be used. These are glass capillary tubes of uniform bore, made to contain 0.02 ml when completely filled. Other types contain 0.02 ml when filled to a mark. Blood is drawn up and expelled by means of a teat which fits onto the end of the capillary. Care must be taken that no blood from the outside of the capillary gets into the reagent solution. This possible source of error can be eliminated by using break-off capillary tubes. These tubes have a greater length than necessary to contain 0.02 ml, but with a break-off point at the correct length. The capillary is filled from the noncalibrated end. When the calibrated part is filled, it is snapped off at the break-off point and dropped into the test tube with reagent solution and the tube is vigorously shaken.

When, for research purposes, the highest accuracy is to be attained in the determination of c_{Hb}^*, a 200-fold dilution should be made using a 0.5-ml Ostwald pipet and a 100-ml volumetric flask. The volume between the two marks on the Ostwald pipet can be checked by weighing the delivered amount of mercury. The volumetric flask can be checked by weighing after filling it to the mark with distilled water.

2.4. Measurement of the Diluted Hemiglobincyanide Solution

Three minutes after mixing blood and reagent solution, the absorbance of the diluted HiCN solution can be measured. The solution is stable so that the measurement may be postponed for several hours or even days if the solution is stored in a cool and dark space and evaporation is prevented. When a spectrophotometer is used, the wavelength is set at 540 nm and the slit width is adjusted for a half-intensity band width of 1 nm or less. The absorbance is measured using a 1.000-cm cuvette against a similar cuvette filled with water or reagent solution. Both can be used as a blank because the absorbance of the reagent solution at $\lambda = 540$ nm is zero. The total hemoglobin concentration is then calculated using the following equation:

$$c_{Hb}^* = A^{540} f M / \epsilon_{HiCN}^{540} l \qquad (2)$$

where c_{Hb}^* is the total hemoglobin concentration of the blood sample, A^{540} is the absorbance of the diluted HiCN solution at $\lambda = 540$ nm, f is the dilution factor, M is one-quarter of the relative molecular mass of the hemoglobin tetramer: 16114.5 d (2), ϵ_{HiCN}^{540} is the millimolar absorptivity of HiCN at $\lambda = 540$ nm: 11.0 liters mmol^{-1} cm^{-1} (Table 1), and l is the lightpath length: 1.000 cm [cf. Eq. (9) in Section 4.1].

If for f the value 251 is taken, c_{Hb}^* follows from

$$c^*_{Hb}(g/dl) = \frac{A^{540} \times 251 \times 16114.5}{11.0 \times 1.000 \times 10^4} = 36.77 \times A^{540} \qquad (3)$$

where 10^4 is the factor for the conversion of milligrams per liter to grams per deciliter.

The wavelength scale of the spectrophotometer can be checked with the help of the mercury emission line at $\lambda = 546.1$ nm, the hydrogen emission lines at $\lambda = 656.3$ and 486.1 nm, or a filter with light absorption peaks at exactly known positions (e.g., didymium glass). A very simple way to check the wavelength scale is with the aid of a solution of holmium oxide (Ho_2O_3) in perchloric acid. The absorption spectrum of such a solution is shown in Fig. 5. The most suitable peak is at $\lambda = 536$ nm, which is quite close to the absorption maximum of HiCN. The absorbance scale is checked using a filter with an exactly known absorbance, such as a carbon yellow filter (Fig. 6). This can also be done with a solution of a compound with exactly known absorptivity and concentration, such as the international HiCN reference solution (Section 2.5).

When a filter photometer is used for measuring the absorbance of the diluted HiCN solution, a filter transmitting a fairly narrow band of light around $\lambda = 540$ nm should be employed. The absorbance is measured against water or reagent solution as a blank. The hemoglobin concentration is read from a previously prepared calibration graph or table. The validity of this calibration graph or table should be checked regularly with the aid of HiCN reference solutions. As HiCN solutions strictly obey Lambert–Beer's law, a calibration line (c on abscissa, A on ordinate) can be constructed by simply connecting the coordinates of a single measurement of a reliable HiCN reference solution with the origin (cf. Fig. 5 in ref. 58). If a check on the performance of the photometer is deemed necessary, dilutions of HiCN reference solution with reagent solution are prepared, the concentrations

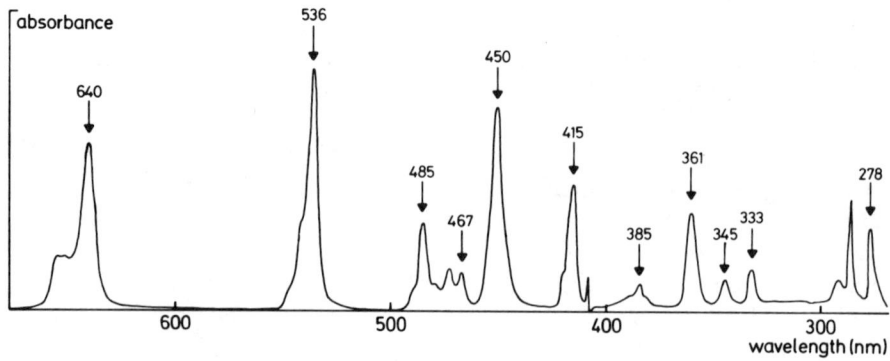

FIG. 5. Absorption spectrum of holmium oxide (Ho_2O_3) in perchloric acid.

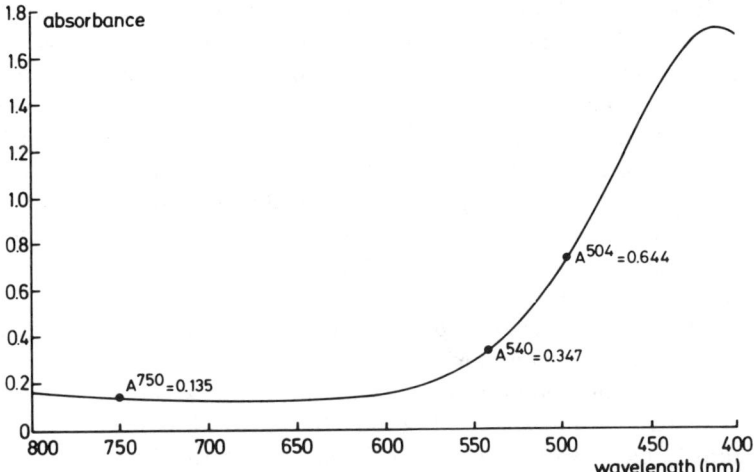

FIG. 6. Absorption spectrum of a carbon yellow filter. The absorbance curve shown is from one of a series of identical filters (Corning HT yellow 1–10) made available by the National Bureau of Standards (Washington, D.C.) to the control laboratories for the international HiCN reference solution (cf. Section 2.5). The absorbance at the three wavelengths used in the control procedure is indicated.

calculated from the diluting factors, and the absorbances measured. The results thus obtained, when plotted on a c vs A graph, should yield a straight line through the origin.

Direct-reading photoelectric hemoglobinometers are special purpose filter photometers with a scale precalibrated in hemoglobin concentration units. The diluted HiCN solution is measured with water or reagent solution as a blank and the hemoglobin concentration read from the scale. The calibration of such instruments should be checked regularly using HiCN reference solutions.

2.5. Hemiglobincyanide Reference Solutions

The international HiCN reference solution is manufactured on behalf of ICSH by the Rijks Instituut voor de Volksgezondheid, Bilthoven, The Netherlands (16), essentially following the procedure described by Zijlstra and van Kampen (72). Briefly, fresh human red cells are washed several times, hemolyzed by toluene, and centrifuged free from debris. All hemiglobin in the sample is converted to HiCN using Drabkin's solution (cf. Section 2.2). This is because the standard procedure for preparing the reference solution was established before the reagent solution containing KH_2PO_4 and nonionic detergent was developed. As the two problems—long reaction time and turbidity due to plasma protein precipitation—remedied by the modification of Drabkin's solution do not occur in the preparation of reference solutions, it was decided not to change the procedure, the more so because the reagent solution is subject to change (different brands of detergent, addition of ethanol in cold climates, etc.). The international HiCN reference solution is equivalent to a hemoglobin concentration of approximately 600 mg/liter. It is dispensed as a sterile solution (membrane filtration) in sealed 10-ml ampules of amber glass. A batch of this HiCN reference solution has been designated by WHO as International Hemiglobincyanide Reference Preparation (64).

Each batch is periodically tested by a group of laboratories nominated by ICSH (16). In these laboratories, A^{540}, A^{504}, and A^{750} are measured by means of spectrophotometers well calibrated as to wavelength and absorbance scale (cf. Section 2.4 and Fig. 6). The slit width is so chosen that the half-intensity band width is less than 1 nm. Plan-parallel glass cuvettes are used with an inner wall-to-wall distance of 1.000 cm, tolerance 0.5% (0.995–1.005 cm). The measurements are carried out at 20–25°C. The absorbance ratio A^{540}/A^{504} is calculated; it should be between 1.59 and 1.63. The measurement at $\lambda = 750$ nm is a turbidity check. The A^{750} should be less than 0.002 per cm lightpath length when measured using an appropriate blank. In addition to the above absorbance measurements, the absorption spectrum is recorded for $\lambda = 450$–750 nm with a lightpath length of 1.000 cm. Table 2 shows the results of control measurements of a batch of the international HiCN reference solution.

TABLE 2
INTERNATIONAL HEMIGLOBINCYANIDE REFERENCE PREPARATION No. 20600[a]

	A^{750}	A^{540}	A^{504}	$\dfrac{A^{540}}{A^{504}}$	λA curve
Atlanta	0.000	0.382	0.236^5	1.62	Correct
Cleveland	0.000^5	0.382	0.236^5	1.62	Correct
Groningen (vK)	0.000^5	0.383	0.238	1.61	Correct
Groningen (Z)	0.001	0.380^5	0.236^5	1.61	Correct
Kumamoto	0.000^5	0.384	0.239^5	1.60	Correct
London (Ph)	0.000^5	0.382	0.237^5	1.61	Correct
London (W)	0.000^5	0.382^5	0.238^5	1.60	Correct
Rome	0.000^5	0.382	0.237	1.61	—
Bilthoven	0.000^5	0.381^5	0.237	1.61	Correct

[a] Survey of control (mean values), August 1982. Mean = 0.382 (SD = 0.001, SEM = 0.000^5), corresponding to 560 ± 0.5 mg HiCN per liter ($n = 8$, Bilthoven not included).

Samples of all batches of the international HiCN reference solution have been stored at 4°C in one of our laboratories and are checked regularly by measuring A^{540} and A^{504}. Results of these long-term stability tests have been reported on several occasions (45, 47, 48). Table 3 shows the results up to the beginning of 1982, covering a period of 18 years. These data demonstrate that the international HiCN reference solution is amazingly stable. There is a very slight tendency for A^{540} to decrease over the years. For the seven batches which have been tested for 10 years or more (Nos. 40400– 10400; 1964–1971), this decrease over the first 10 years of shelf life was 0.5, 0, 1.0, 0.4, 1.5, 1.5, and 1.9%, respectively. The absorbance ratio A^{540}/A^{504} has in all cases been found to be within the range prescribed by ICSH (1.59–1.63), with the exception of No. 40400, after 18 years (1.58), and No. 70400, after 15 years (1.585). Thus, the present extended period of validity for batches of the international HiCN reference solution (3 years) is still very conservative.

The international HiCN reference solution is made available for reference use only to national standards committees for hematological methods or to official government-nominated holders. When there is no committee or official holder, it is distributed to an individual appointed by ICSH. The national holders take care that the international HiCN reference solution is made available to manufacturers and distributors of secondary HiCN reference solutions for the purpose of controlling the quality of their products. Secondary HiCN reference solutions are HiCN solutions made by private or governmental manufacturers, following as closely as possible the procedure used in preparing the international HiCN reference solution. The international HiCN reference solution should always be used in checking the spectral properties of the secondary reference solution.

2.6. Quality Control

After the introduction of the HiCN method in its standardized form (57, 71, 72), the routine use of this method for the determination of c^*_{Hb} steadily increased (Fig. 1). Concomitantly, the precision of the determination of c^*_{Hb} as found in three field trials held in The Netherlands improved considerably (Fig. 7). The spread, expressed as $2 \times SD$, decreased from 4 g/dl in 1960 to 2 g/dl in 1964 and to 1 g/dl in 1968. The coefficient of variation thus fell from 13 to 3.3% in a period of 8 years.

These data seemed to confirm the original concept of the standardized HiCN method: that the availability of an HiCN reference solution for providing a single calibration point would suffice to yield optimum measuring results. In actual fact, however, this did not prove to be the case. In 1973, 57 laboratories in Europe and Africa participated in an interlaboratory trial organized on behalf of the ICSH Expert Panel on Hemoglobinometry (47).

TABLE 3. Long-term Stability Test of

Age of batch (years)	40400 (1964) A^{540}	$\dfrac{A^{540}}{A^{504}}$	60400 (1966) A^{540}	$\dfrac{A^{540}}{A^{504}}$	70400 (1967) A^{540}	$\dfrac{A^{540}}{A^{504}}$	80400 (1968) A^{540}	$\dfrac{A^{540}}{A^{504}}$	90400 (1969) A^{540}	$\dfrac{A^{540}}{A^{504}}$	00400 (1970) A^{540}	$\dfrac{A^{540}}{A^{504}}$	10400 (1971) A^{540}	$\dfrac{A^{540}}{A^{504}}$
0	0.391	1.61	0.386	1.61	0.405	1.62	0.383	1.60	0.389	1.61	0.395	1.61	0.413	1.61
0.25			0.387	1.61			0.381	1.60	0.389	1.61	0.395	1.61	0.414	1.60
0.5			0.387	1.61	0.404	1.62	0.386	1.61^5	0.388	1.62	0.396	1.61	0.415	1.61^5
0.75			0.387	1.61			0.383	1.61						
1	0.389	1.61	0.388	1.61	0.405	1.59^5			0.388	1.61	0.400	1.62		
1.25					0.403	1.61							0.414	1.61
1.5	0.391	1.61			0.407	1.62	0.386	1.61^5	0.389	1.62	0.398	1.62	0.414	1.60^5
1.75					0.405	1.62	0.385	1.61	0.389	1.61	0.397	1.61	0.415	1.61
2	0.391	1.61			0.405	1.62							0.413	1.61
2.25					0.406	1.62			0.390	1.61	0.396	1.62		
2.5	0.390	1.61	0.387	1.61	0.404	1.62	0.386	1.61			0.393	1.60	0.412	1.61
2.75									0.390	1.60^5				
3	0.391	1.62			0.404	1.61	0.386	1.61			0.393	1.60	0.411	1.61
3.25														
3.5			0.388	1.61					0.387	1.61	0.393	1.61	0.412	1.61
3.75														
4	0.389	1.61	0.387	1.61	0.405	1.61	0.388	1.62	0.387	1.61	0.393	1.61	0.412	1.60
4.5					0.406	1.60^5								
5	0.393	1.61	0.386	1.61			0.384	1.61	0.387	1.60	0.393	1.60	0.411	1.61
5.5					0.404	1.62								
6	0.390	1.62					0.382	1.60	0.386	1.60	0.392	1.60	0.407^5	1.61
7					0.402	1.61	0.383	1.60	0.386	1.61	0.390	1.60		
8	0.394	1.61	0.386	1.61	0.405	1.60	0.383	1.61	0.383^5	1.60^5				
9	0.390	1.60			0.404	1.61	0.381	1.60						
10	0.389	1.60	0.386	1.61	0.401	1.61					0.389	1.59	0.405	1.61
11	0.390	1.60	0.381^5	1.60					0.378	1.62			0.405^5	1.61
12	0.389	1.59					0.382	1.60			0.387^5	1.60		
13	0.386	1.59^5			0.400	1.61			0.382^5	1.61^5				
14			0.381	1.60			0.386^5	1.60^4						
15					0.405	1.58^5								
16	0.387^5	1.59	0.380	1.60										
17														
18	0.391^5	1.58												

The material to be tested consisted of two fresh blood samples (containing EDTA as anticoagulant), two glycerol-containing hemolysates, and an HiCN reference solution. The "true" values were derived from the results obtained by members of the Expert Panel on Hemoglobinometry. Figure 8 shows that the spread in the values obtained for the blood samples and the hemolysates is considerably greater than that in the values obtained for the HiCN reference solution. A similar result was obtained in another interlaboratory trial

INTERNATIONAL REFERENCE SOLUTIONS

							Batch no. and year of preparation								
20400 (1972)		30400 (1973)		40500 (1974)		50500 (1975)		60500 (1976)		70500 (1977)		80500 (1978)		90500 (1979)	
A^{540}	A^{504}	A^{540}	A^{504}	A^{540}	A^{504}	A^{540}	A^{504}	A^{540}	A^{504}	A^{540}	A^{504}	A^{540}	A^{504}	A^{540}	A^{504}
0.408	1.61	0.403	1.61	0.382$_5$	1.60	0.379	1.61	0.392$_5$	1.61	0.394$_5$	1.60	0.396	1.60	0.392$_5$	1.61
0.408	1.61	0.402	1.61	0.380	1.61	0.381	1.61	0.394	1.61	0.396	1.61	0.393	1.61	0.392	1.61
0.408	1.61	0.403	1.62	0.381	1.60	0.381$_5$	1.61	0.392	1.61	0.395	1.61	0.394	1.60	0.393$_5$	1.61
														0.392$_5$	1.61
0.408	1.62	0.402	1.61	0.378	1.61	0.379$_5$	1.60	0.393$_5$	1.60	0.395	1.61	0.393$_5$	1.61	0.392$_5$	1.60
		0.401	1.61												
		0.403	1.60	0.379$_5$	1.62	0.381	1.60	0.393$_5$	1.61	0.394	1.60$_5$	0.395	1.61	0.392$_5$	1.61
0.406	1.61			0.380$_5$	1.61	0.382	1.60	0.392	1.61	0.394	1.61	0.393$_5$	1.61	0.393	1.60
		0.401	1.61												
0.404	1.61	0.400	1.61			0.380	1.60	0.391	1.61			0.393	1.61	0.391$_5$	1.60
										0.392	1.61	0.393	1.61		
												0.393$_5$	1.61	0.388	1.61
0.407	1.60	0.403$_5$	1.60$_5$											0.390$_5$	1.60
								0.391	1.62			0.392$_5$	1.60$_5$		
0.405	1.61	0.399	1.61												
0.402	1.60$_5$					0.379$_5$	1.61			0.391$_5$	1.61				
								0.387$_5$	1.61$_5$						
		0.400$_5$	1.60			0.376$_5$	1.62								
0.399	1.61			0.376	1.61										
		0.396	1.62												

(55) in which, besides an HiCN reference solution, two hemolysates also had to be measured (Fig. 9). It can be concluded from the distribution of the crosses in Fig. 9 that the errors made in measuring the HiCN reference solution are random errors, whereas the distribution of the dots show that, in the results obtained for the hemolysates, systematic errors play a considerable role (cf. Section 6).

Since the results obtained for the HiCN reference solutions are reasonably good (coefficient of variation < 1%), it is improbable that photometric errors have caused the unsatisfactory results for the hemolysates and the whole blood samples. Also, the absence of any significant difference between these two kinds of samples (Fig. 8) excludes nonhomogeneity of the blood samples, a possible result of insufficient mixing before the dilution with reagent solution is made, as an important source of error in the determination of c^*_{Hb} in the whole blood samples. Thus, as causes of error, there remain incomplete conversion to HiCN, resulting from the use of faulty reagent solutions (cf. Section 2.2), and incorrect dilution, due to the use of non- or falsely cali-

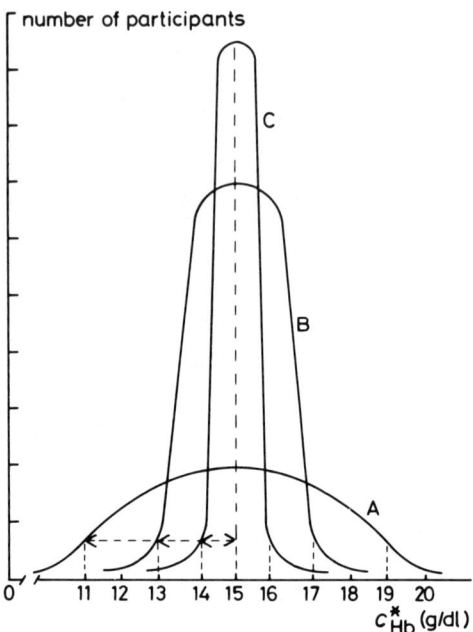

Fig. 7. Results of three field trials concerning the determination of c^*_{Hb} in Dutch hospitals. All data have been recalculated to a mean value of 15 g/dl. The horizontal arrows indicate 2 × SD. A = 1960; B = 1964; C = 1968.

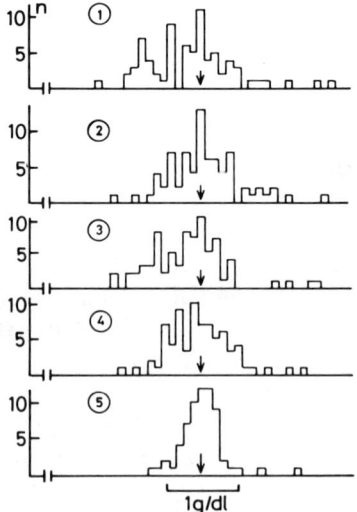

Fig. 8. Histograms of the results of determinations of c^*_{Hb} in an international trial at 67 laboratories. 1 and 2, whole blood; 3 and 4, hemolysates; 5, HiCN reference solution (A^{540} converted into c^*_{Hb} of a hypothetical blood sample). The arrows indicate the values obtained by the reference laboratories.

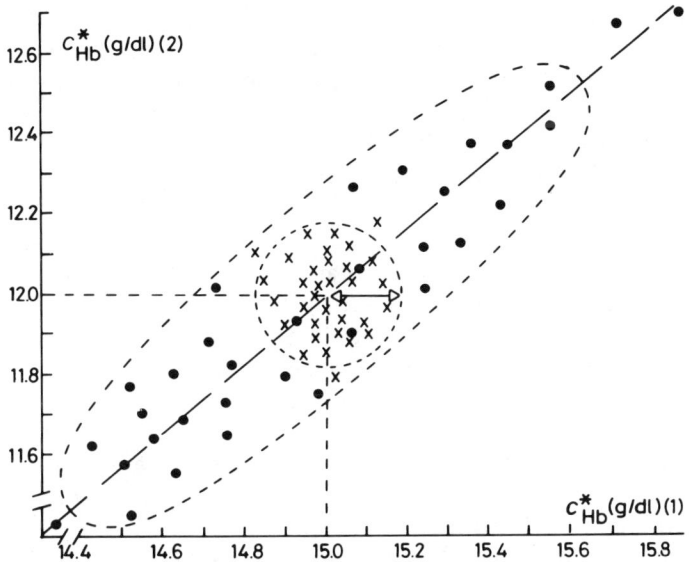

FIG. 9. Youden plot of the results of an interlaboratory trial. The crosses and dots are the result of measurements of HiCN reference solution and hemolysates, respectively (cf. Appendix).

brated pipets and/or incorrectly adjusted dilutors. It is evident from these trials that these two sources of error may play an important part in the daily practice of measuring c_{Hb}^* by the standardized HiCN method.

For the detection of conversion and dilution errors in routine hemoglobinometry, some kind of blood-like preparation with known c_{Hb}^* is necessary. Reasonably stable concentrated hemoglobin solutions of known c_{Hb}^*, prepared by means of toluene hemolysis of human blood and sterilized by membrane filtration, are now commercially available for intralaboratory quality control (47). With the aid of an HiCN reference solution to calibrate the measuring instrument and a concentrated hemoglobin solution to check the HiCN method as a whole, it should be possible for every clinical chemical laboratory to ensure that valid results are obtained.

2.7. OTHER METHODS FOR THE DETERMINATION OF TOTAL HEMOGLOBIN

Although it is generally recognized that the standardized HiCN method is the preferred method for the determination of c_{Hb}^* in all circumstances, some other procedures are also in use for routine hemoglobinometry. The reasons for using these other methods are based on considerations from outside the methodology of hemoglobinometry proper. In constructing automated analyzers, it has been considered more practical to use an HiCN method in which the absorbance is measured at a constant, short time after mixing blood and reagent solution, instead of waiting for complete conversion of all hemoglobin to HiCN. In other constructions, the use of HbO_2 instead of HiCN has been chosen for reasons of speed of operation and simplicity of design.

Another reason for using a procedure different from the standardized HiCN method has been the incorporation of a determination of c^*_{Hb} in an oximeter employing a two-wavelength method with an isobestic point (cf. Section 4.2). The absorbance measured at the isobestic wavelength is then linearly related to the total hemoglobin concentration and can easily be used for a readout of c^*_{Hb}. When, by multicomponent analysis, the concentration of all the hemoglobin derivatives in a blood sample is determined (cf. Section 4.5), the sum total of these concentrations is, of course, equal to c^*_{Hb}.

When other methods for the determination of c^*_{Hb} are used, these should be compared regularly with the standardized HiCN method. To this end, some blood samples should be measured simultaneously using the two methods. It is not a good practice to check the other method only with the concentrated hemoglobin solutions mentioned in Section 2.6.

3. Absorption Spectra and Millimolar Absorptivities of Hemoglobin and Hemoglobin Derivatives

In this section, spectral absorbance data are provided for hemoglobin and its common derivatives. Most of the data have been published previously (9, 45, 58), but some new material has been added. The importance of reliable absorption spectra and accurate values for the absorptivities of hemoglobin and its common derivatives has considerably increased since multicomponent analysis of mixtures of hemoglobin derivatives has become feasible in the clinical chemical laboratory (3, 74–76).

3.1. Sample Preparation and Measurement Procedure

Apart from the special procedures necessary to prepare the various hemoglobin derivatives, two slightly different methods were used in the preparation of the hemoglobin solutions. In most cases, the procedure as described in ref. 58 and, in more detail, in ref. 45 was used (procedure 1). This leads to quite clear hemoglobin solutions. In a later series of experiments, the solutions were prepared simply by hemolyzing the blood in a syringe with the help of a small amount of Sterox SE and pressing the hemolysate without air contact through a cotton-wool filter (75) (procedure 2). This yields hemoglobin solutions which are not as clear as those provided by the first procedure, but the absorption data obtained are very similar.

All specimens were obtained from healthy, nonsmoking humans and provided with heparin as anticoagulant. In the first procedure, the erythrocytes were washed three times with an isotonic saline solution. After the third washing, the packed cells were brought to 1.25 times the original blood volume by adding 0.7% Sterox SE solution and the hemolysate was filtered through a folded paper filter after centrifugation for 10 minutes. Finally, c^*_{Hb} was adjusted to about 8 g/dl. In order to obtain deoxygenated hemoglobin (Hb), sodium dithionite was added to the hemolysate to a molar ratio $Hb/Na_2S_2O_4$ of 1/5. To obtain oxyhemoglobin (HbO_2), the hemolysate was oxygenated in a revolving glass tonometer flushed with pure O_2. To obtain carboxyhemoglobin (HbCO), the hemolysate was saturated with carbon monoxide in a revolving glass tonometer flushed with pure CO. To obtain hemiglobin (methemoglobin; Hi), finely powdered $K_3Fe(CN)_6$ was added to the hemolysate in two steps to a molar ratio of $Hb/K_3Fe(CN)_6$ of 1/3. The pH of the resulting Hi solution was checked using a glass electrode; only solutions with pH 7.0–7.4 were accepted for the absorbance measurements. To obtain hemiglobinnitrite ($HiNO_2$), $NaNO_2$ was added to the hemolysate to a molar ratio $Hb/NaNO_2$ of 1/6 (50, 51). To obtain hemiglobinazide (HiN_3), NaN_3 was added to an Hi solution prepared as previously described to a molar ratio Hi/NaN_3 of 1/6. For the absorbance measurements at $\lambda <$ 450 nm, all solutions were diluted with water either 10 times (Hb, HbO_2, HbCO) or 20 times (Hi, $HiNO_2$, HiN_3). Small amounts of $NaNO_2$ and NaN_3 were then added to the diluted $HiNO_2$ and HiN_3 solutions, respectively.

When the second procedure, with cotton wool filtration as the only clearing procedure (75), was used, tonometry with O_2 and CO for preparing the HbO_2 and HbCO solutions was carried out with fresh whole blood. Hb-containing samples were prepared by tonometry of whole blood with a gas mixture containing 5% CO_2 and 95% N_2. No $Na_2S_2O_4$ was used.

Sulfhemoglobin-containing samples were prepared according to Siggaard-Andersen et al. (9, 37). Fresh whole blood was oxygenated by tonometry with pure O_2 and then centrifuged. The packed cells were then incubated for 30 minutes with an equal volume of a freshly prepared solution containing 40 mM Na_2S and 75 mM HCl (pH 7.5). Finally, excess H_2S was removed by tonometry with O_2. By this procedure, a blood sample is obtained containing 15–25% SHb. Sulfhemiglobin (SHi) was obtained by adding $K_3Fe(CN)_6$ to an SHb-containing solution after ultrasonic lysis of the erythrocytes; sulfhemiglobincyanide (SHiCN) was prepared by the addition of $K_3Fe(CN)_6$ and KCN.

Spectral absorbance curves of the hemoglobin solutions obtained according to the first procedure were made by means of an Optica CF4DR recording spectrophotometer. The absorbance in the regions of maximum absorption was also measured manually on an Optica CF4. Both instruments were calibrated for wavelength using mercury emission lines and for absorbance using a carbon yellow filter (cf. Section 2.4). Both instruments had been checked to ensure the absence of stray light and tested as to photometric linearity. The scale of the Optica CF4DR recorder is electronically linearized for absorbances up to 1.000. Absorption spectra of the hemoglobin solutions prepared by means of the second procedure were obtained with an HP8450A reversed-optics spectrophotometer with built-in facilities for processing and storing the absorbance data. The instrument was checked by means of a carbon yellow filter. The absorption spectra of the three sulfhemoglobin derivatives were recorded using an Aminco-Chance split-beam spectrophotometer with the reference cuvette filled with the same solution as the sample cuvette, but without the SHb derivative. The absorption spectrum of SHb was also obtained with the aid of the HP8450A, using the computing facilities of this spectrophotometer for making the necessary corrections for the absorbance of the other hemoglobin derivative (HbO_2) in the sample.

In each case, concentration and lightpath length were chosen in such a manner that the absorbances were, as far as possible, in the range of near-maximum photometric accuracy (cf. Fig. 2 in ref. 58). The smaller layer thicknesses (l = 0.013 and 0.005 cm) were attained by using 0.100-cm plan-parallel glass cuvettes with 0.087- and 0.095-cm glass inserts (Hellma Benelux, The Hague, The Netherlands) (cf. Fig. 13 in ref. 58 or Fig. 2 in ref. 75). The lightpath length of all cuvettes was determined by comparative measurements of various dilutions of a concentrated HiCN solution, taking a certified 1.000-cm plan-parallel glass cuvette as a primary standard (45, 58).

The calculation of the millimolar absorptivities and the calibration of the ordinate of the spectral absorbance curves in absorptivity units require that the concentration of the compound measured be exactly known. In order to obtain the exact concentration values, all samples, except for the SHb derivatives, were diluted with HiCN reagent solution (cf. Section 2.2) and the absorbance measured at λ = 540 nm in a certified 1.000-cm cuvette. The concentration of hemoglobin derivative X in the original sample was then calculated using the equation

$$c_X \text{ (mM)} = A_{HiCN}^{540} f / \epsilon_{HiCN}^{540} l \tag{4}$$

where f is the dilution factor, l = 1.000 cm, and ϵ_{HiCN}^{540} = 11.0 liters mmol^{-1} cm^{-1}. The millimolar absorptivity of hemoglobin derivative X at wavelength λ followed from

SPECTROPHOTOMETRY OF HEMOGLOBIN

$$\epsilon^\lambda \text{ (liters mmol}^{-1} \text{ cm}^{-1}) = A^\lambda/c_X l \tag{5}$$

where A^λ is the absorbance measured at wavelength λ for the original X-containing sample and l the lightpath length used in this measurement.

For the SHb derivatives, a special procedure had to be used, first, because these compounds must always be measured in the presence of other hemoglobin derivatives (pure solutions cannot be prepared) and, second, because they cannot be converted to HiCN. This procedure is described in Section 3.8.

Figure 10 shows the visible-range absorption spectra of the most important hemoglobin derivatives. The samples were prepared by the second procedure, except for HiCN, for which a reference solution was used (cf. Section 2.4). The measurements were carried out with the HP8450A reversed-optics spectrophotometer. The spectral absorbance curve of Hb is based on measurements of the blood of 10 donors, and that of HbO_2, HbCO, and Hi is based on the blood of 14, 8, and 16 donors, respectively. The absorption spectrum of SHb is based on a single blood sample.

3.2. DEOXYGENATED HEMOGLOBIN (Hb)

The absorption spectrum of Hb is shown in Fig. 11 and, for the visible range, also in Fig. 10. The sample preparation for the data of Fig. 11 was according to procedure 1 as described in Section 3.1. In Table 4, the milli-

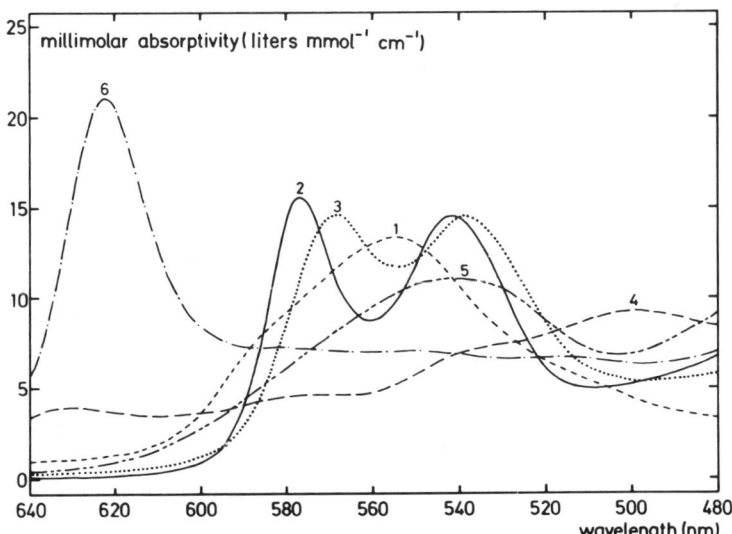

FIG. 10. Absorption spectra of (1) Hb, (2) HbO_2, (3) HbCO, (4) Hi, (5) HiCN, and (6) SHb.

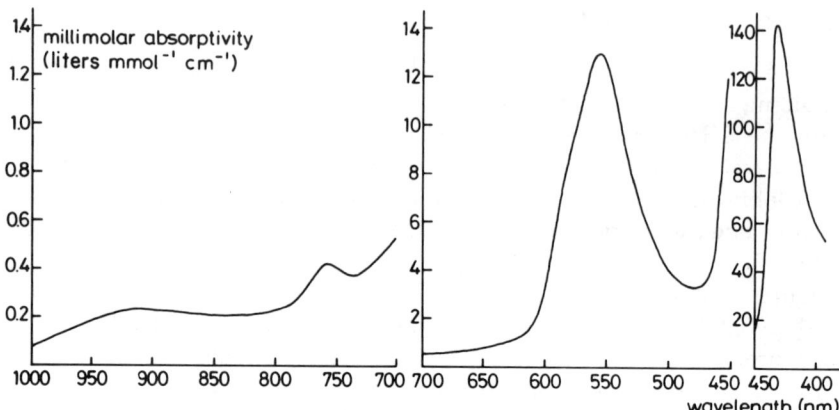

Fig. 11. Absorption spectrum of Hb. (Based on data from refs. 45 and 58.)

molar absorptivity of Hb is given for the maxima and minima in the spectral absorbance curve and for some crossover wavelengths with the spectra of various hemoglobin derivatives (isobestic points). Part of the isobestic points can also be seen in Fig. 10. The isobestic point with HbO_2 in the near infrared is found at $\lambda = 815$ nm. In the spectrophotometric determination of the oxygen saturation described in Section 4.2, an isobestic wavelength of 800 nm is used. This is the value which is consistently found when measure-

TABLE 4
MILLIMOLAR ABSORPTIVITY OF Hb

λ (nm)	ϵ^λ (liters mmol^{-1} cm^{-1})	Comments
910	0.23	Maximum
850	0.20	Minimum
815	0.22	Isobestic with HbO_2
760	0.43	Maximum
736	0.37	Minimum
600	3.20	Isobestic with Hi
586	7.23	Isobestic with HbO_2
579	8.86	Isobestic with HbCO
569	11.27	Isobestic with HbO_2
561.5	12.54	Isobestic with HbCO
555	13.04	Maximum
548.5	12.46	Isobestic with HbO_2
547.5	12.37	Isobestic with HbCO
528.5	7.71	Isobestic with Hi
522	6.42	Isobestic with HbO_2
506.5	4.81	Isobestic with HbO_2
478	3.31	Minimum
431	140.0	Maximum

ments are made on samples which have been deoxygenated by tonometry with a CO_2/N_2 mixture. When $Na_2S_2O_4$ is used for deoxygenation, a somewhat longer isobestic wavelength is found for Hb/HbO_2. The difference looks considerable, but has little practical importance because the spectral absorbance curves of Hb and HbO_2 are very flat in this region. Table 4 and Fig. 10 show that $\lambda = 548$ nm is almost an isobestic triple point for Hb, HbO_2, and HbCO. Using procedure 2 and the HP8450A spectrophotometer, a value of 13.30 liters $mmol^{-1}$ cm^{-1} was found for the millimolar absorptivity in the absorption maximum at $\lambda = 555$ nm.

3.3. Oxyhemoglobin (HbO_2)

The absorption spectrum of HbO_2 is shown in Fig. 12 and, for the visible range, also in Figs. 3 and 10. The sample preparation for the data of Fig. 12 was according to procedure 1. In Table 5, the millimolar absorptivity of HbO_2 is given for the maxima and minima in the spectral absorbance curve and for some isobestic points (cf. Fig. 10). Some comment on the position of the isobestic point for Hb/HbO_2 in the near infrared is given in Section 3.2. Using procedure 2 and the HP8450A spectrophotometer, a millimolar absorptivity of 15.42 and 14.61 liters $mmol^{-1}$ cm^{-1} was found in the maxima at $\lambda = 577$ and 542 nm and of 8.71 and 4.96 liters $mmol^{-1}$ cm^{-1} in the minima at $\lambda = 560$ and 510 nm, respectively.

3.4. Carboxyhemoglobin (HbCO)

The absorption spectrum of HbCO is shown in Fig. 13 and, for the visible range, also in Fig. 10. The sample preparation for the data of Fig. 13 was

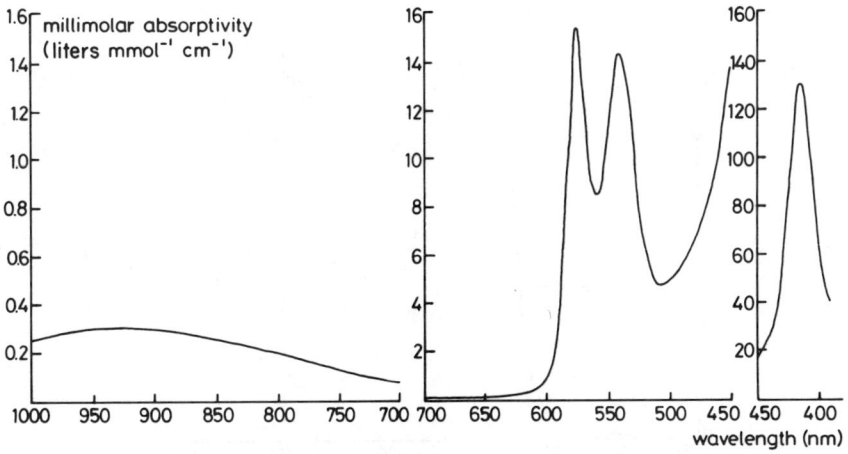

FIG. 12. Absorption spectrum of HbO_2. (Based on data from refs. 45 and 58.)

TABLE 5
Millimolar Absorptivity of HbO_2

λ (nm)	ϵ^λ (liters mmol^{-1} cm^{-1})	Comments
930	0.31	Maximum
815	0.22	Isobestic with Hb
690	0.07	Minimum
592	2.13	Isobestic with HbCO
590.5	3.62	Isobestic with Hi
586	7.23	Isobestic with Hb
577	15.37	Maximum
572.5	13.50	Isobestic with HbCO
569	11.27	Isobestic with Hb
560	8.47	Minimum
549.5	12.06	Isobestic with HbCO
548.5	12.46	Isobestic with Hb
542	14.37	Maximum
540	14.27	Isobestic with HbCO
525.5	7.72	Isobestic with Hi
522	6.42	Isobestic with Hb
510	4.76	Minimum
506.5	4.81	Isobestic with Hb
497	5.16	Isobestic with HbCO
415	131.0	Maximum

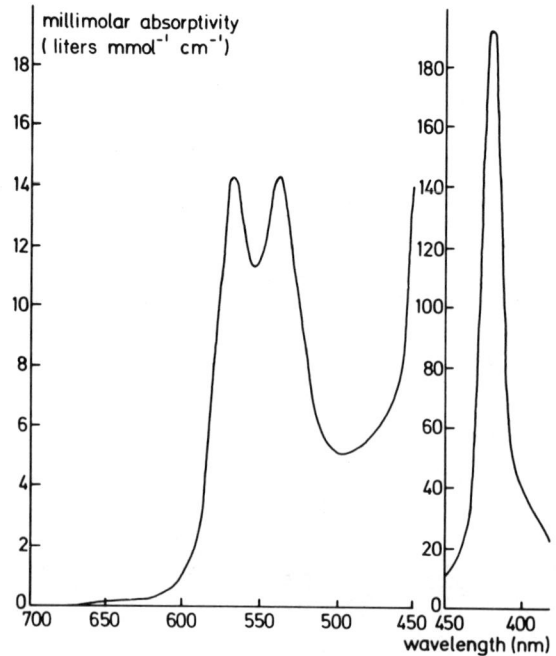

FIG. 13. Absorption spectrum of HbCO. (Based on data from refs. 45 and 58.)

TABLE 6
MILLIMOLAR ABSORPTIVITY OF HbCO

λ (nm)	ϵ^λ (liters mmol^{-1} cm^{-1})	Comments
585.5	3.79	Isobestic with Hi
579	8.86	Isobestic with Hb
572.5	13.50	Isobestic with HbO$_2$
568.5	14.31	Maximum
561.5	12.54	Isobestic with Hb
555	11.33	Minimum
549.5	12.06	Isobestic with HbO$_2$
547.5	12.37	Isobestic with Hb
540	14.27	Isobestic with HbO$_2$
539	14.36	Maximum
519.5	8.02	Isobestic with Hi
497	5.16	Isobestic with HbO$_2$
496	5.14	Minimum
420	192.0	Maximum

according to procedure 1. In Table 6, the millimolar absorptivity of HbCO is given for the maxima and minima in the spectral absorbance curve and for some isobestic points (cf. Fig. 10). Using procedure 2 and the HP8450A spectrophotometer, a millimolar absorptivity of 14.58 and 14.51 liters mmol^{-1} cm^{-1} was found in the maxima at λ = 568.6 and 539 nm and of 11.67 and 5.41 liters mmol^{-1} cm^{-1} in the minima at λ = 555 and 496 nm, respectively.

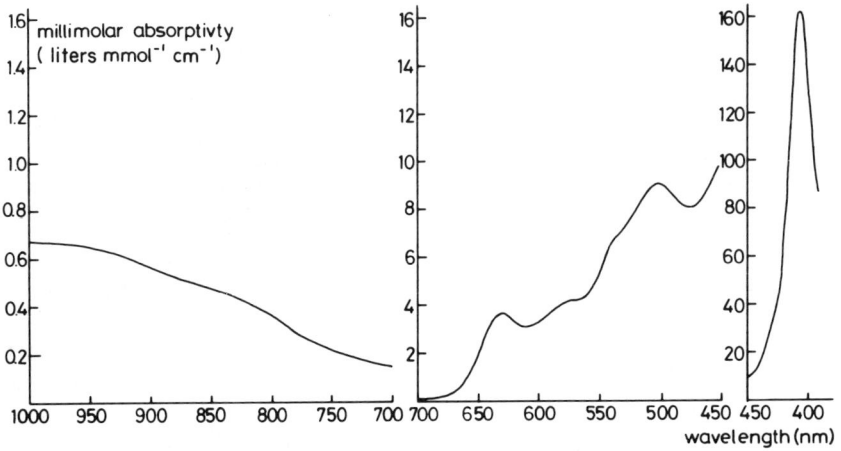

FIG. 14. Absorption spectrum of Hi. (Based on data from refs. 45 and 58.)

3.5. Hemiglobin (Methemoglobin; Hi)

The absorption spectrum of Hi is shown in Fig. 14 and, for the visible range, also in Fig. 10. The sample preparation for the data of Fig. 14 was according to procedure 1. In Table 7, the millimolar absorptivity of Hi is given for the maxima and minima in the spectral absorbance curve and for some isobestic points (cf. Fig. 10). The absorption spectrum of Hi is dependent on pH (cf. Fig. 9 in ref. 58), in contrast to those of the other hemoglobin derivatives, which are independent of pH over a wide range (pH 5.5–9.5), outside which denaturation occurs. The spectral absorptivity curves of Figs. 10 and 14 apply to pH 7.2. Using procedure 2 and the HP8450A spectrophotometer, a millimolar absorptivity of 3.90 and 9.25 liters mmol^{-1} cm^{-1} was found in the maxima at λ = 630 and 500 nm and of 3.50 and 8.38 liters mmol^{-1} cm^{-1} in the minima at λ = 608 and 476 nm, respectively.

3.6. Hemiglobincyanide (HiCN)

The spectral properties of this hemoglobin derivative have been extensively discussed in Section 2.1. The absorption spectrum is shown in Figs. 2, 3, and 10. The millimolar absorptivity in the maxima at λ = 540 and 421 nm is 10.99 (cf. Table 1) and 122.5 liters mmol^{-1} cm^{-1} (45, 58), respectively. The millimolar absorptivity in the minimum at λ = 504 nm is 6.83 liters mmol^{-1} cm^{-1}. It should be noted that $\epsilon^{540}/\epsilon^{504}$ = 10.99/6.83 = 1.609 (cf. Section 2.5 and Table 3).

HiCN is the most stable of the hemoglobin derivatives that can be formed by the addition of an anion to Hi. The absorption spectra of two other

TABLE 7
Millimolar Absorptivity of Hi (pH 7.0–7.4)

λ (nm)	ϵ^{λ} (liters mmol^{-1} cm^{-1})	Comments
690	0.13	Minimum
630	3.70	Maximum
608	3.06	Minimum
600	3.20	Isobestic with Hb
590.5	3.62	Isobestic with HbO$_2$
585.5	3.79	Isobestic with HbCO
528.5	7.71	Isobestic with Hb
525.5	7.72	Isobestic with HbO$_2$
519.5	8.02	Isobestic with HbCO
500	9.04	Maximum
476	8.04	Minimum
406	162.0	Maximum

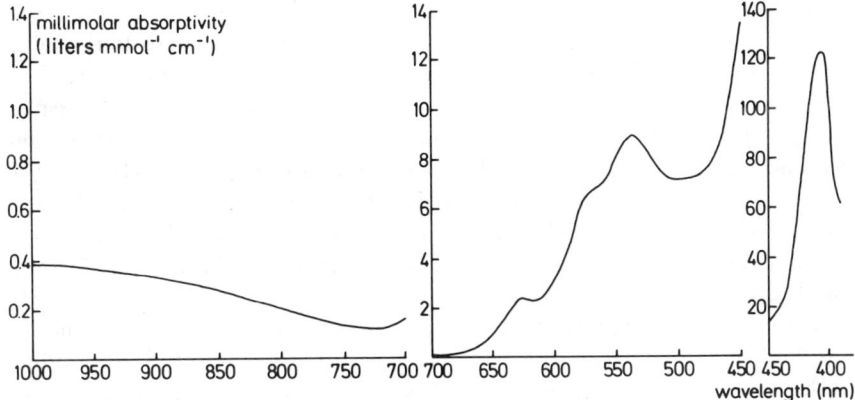

FIG. 15. Absorption spectrum of HiNO₂. (Based on data from ref. 45.)

hemoglobin derivatives of this kind (HiNO$_2$ and HiN$_3$) are dealt with in Section 3.7. A definite displacement series can be demonstrated for these compounds. The absorption spectra of HiF, HiNO$_2$, HiN$_3$, and HiCN change into each other upon the successive additon of F$^-$, NO$_2^-$, N$_3^-$, and CN$^-$ ions to a solution of Hi.

3.7. Hemiglobinnitrite (HiNO$_2$) and Hemiglobinazide (HiN$_3$)

The absorption spectra of HiNO$_2$ and HiN$_3$ are shown in Figs. 15 and 16. The sample preparation for the data in these figures was according to pro-

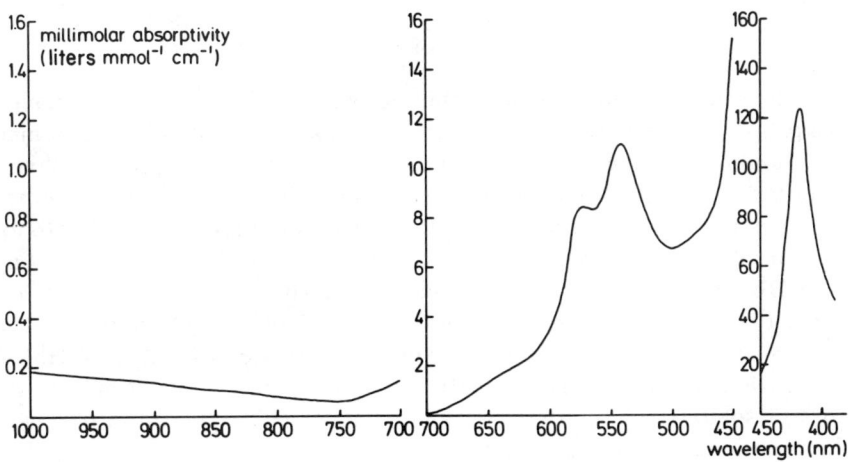

FIG. 16. Absorption spectrum of HiN₃. (Based on data from ref. 45.)

cedure 1 as described in Section 3.1. The shape of these two absorption spectra is not too different from that of HiCN. This also applies to the Soret band, where the millimolar absorptivity in the absorption maxima of $HiNO_2$, HiN_3, and HiCN is 123.5, 123.5, and 122.5 liters $mmol^{-1}$ cm^{-1}, respectively, although the wavelength of maximum absorption differs (λ = 406, 419, and 421 nm, respectively). The millimolar absorptivity of $HiNO_2$ at its two other absorption maxima, λ = 625 and 538 nm, is 2.48 and 9.02 liters $mmol^{-1}$ cm^{-1}, respectively. HiN_3 has maxima in the visible range at λ = 573 and 542 nm, with millimolar absorptivities of 8.51 and 10.95 liters $mmol^{-1}$ cm^{-1}, respectively.

It should be remembered that NO_2^- can enter the erythrocytes, whereas $Fe(CN)_6^{3-}$ cannot. Therefore, $NaNO_2$ is occasionally used to produce erythrocytes loaded with Hi. There has been some confusion in the literature as to the nature of the compound formed on the additon of $NaNO_2$ to whole blood or to a solution of hemoglobin (44). This is due to the fact that it has not always been recognized that the addition of an equimolar quantity of $NaNO_2$ leads to the formation of Hi, whereas the addition of excess $NaNO_2$ gives $HiNO_2$ (50).

Among the various reasons for which it was thought that HiN_3 could be easily substituted for HiCN in the standardized method for measuring c^*_{Hb} (cf. Section 2.2) was the near equality of the millimolar absorptivities of HiCN and HiN_3 at λ = 540 nm (ϵ^{540}_{HiCN} = 10.99; $\epsilon^{540}_{HiN_3}$ = 10.90). It has even been proposed that HiCN reference solutions be used for checking a routine HiN_3 method (60). The similarity between the two absorption spectra, however, is not strong enough to justify such a procedure (49).

3.8. Sulfhemoglobin (SHb), Sulfhemiglobin (SHi), and Sulfhemiglobincyanide (SHiCN)

In the sulfur-containing hemoglobin derivatives, the sulfur atom is bound to a pair of β-pyrrole carbon atoms at the periphery of a chlorin ring formed by saturation of the β–β double bond of the corresponding protoporphyrin IX of Hb. This structure explains the considerable departure of the absorption spectrum of SHb from that of Hb as well as the low affinity of SHb for O_2. For almost all hemoglobin derivatives which can be prepared starting from Hb, an SHb analog can be made. These compounds are of analytical rather than medical interest. The presence of SHbCO has been found once to be a source of error in the spectrophotometric determination of HbCO (56). SHiCN is formed when an SHb-containing blood sample is mixed with reagent solution for the determination of c^*_{Hb}, and is the cause of a very slight underestimation (cf. Section 2.2).

The absorption spectrum of SHb is shown in Fig. 17 and, for the visible

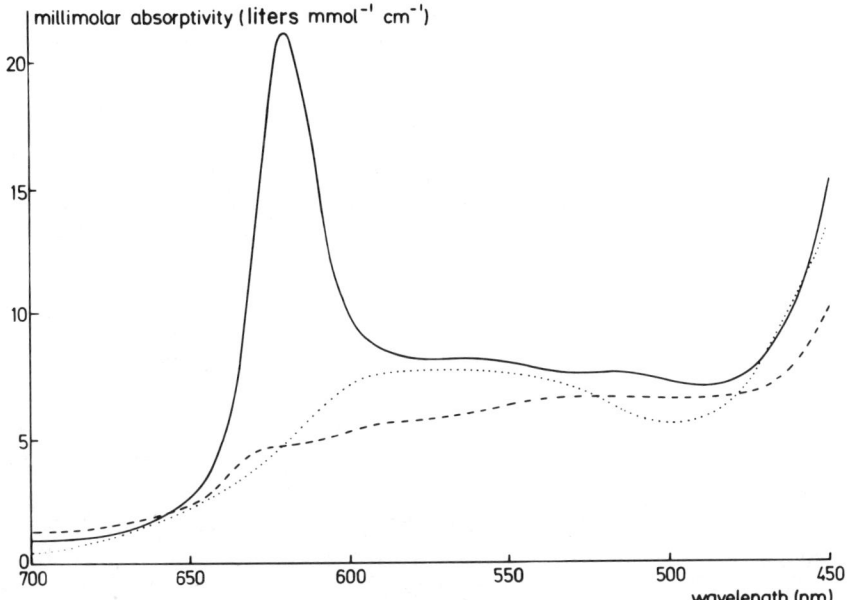

FIG. 17. Absorption spectra of SHb (—), SHi (---), and SHiCN (···). (Based on data from ref. 9.)

range, also in Fig. 10. Figure 17 shows, in addition, the absorption spectra of SHi and SHiCN. As SHb cannot be prepared in pure solution, a special procedure was necessary for the determination of its millimolar absorptivity. The SHb content of an erythrocyte suspension, prepared as described in Section 3.1, was determined by measuring the oxygen capacity before and after the formation of SHb. To this end, part of the oxygenated packed cells were mixed with the Na_2S/HCl solution of Section 3.1 (suspension A), and part with an oxygen-saturated 150 mM NaCl solution (suspension B). The erythrocyte concentrations of the two suspensions were made exactly equal. The oxygen content of the suspensions was then determined by means of the titrimetric method of Dijkhuizen et al. (10). For suspension B, c^*_{Hb} was measured using the standardized HiCN method, and O_2 capacity per g Hb (β_{O_2}) was calculated for both suspensions. The fraction of SHb in suspension A (F_{SHb}) then followed from

$$F_{SHb} = [\beta_{O_2}(B) - \beta_{O_2}(A)]/\beta_{O_2}(B) \qquad (6)$$

where $\beta_{O_2}(A)$ and $\beta_{O_2}(B)$ denote β_{O_2} for suspensions A and B, respectively.

After ultrasonic lysis of the erythrocyte suspensions, the ensuing solutions were diluted 50 times with 100 mM Tris buffer (pH 8.0) and filtered through a 0.45-μm Millipore filter. The absorbance of $\lambda = 620$ nm was measured within 5 minutes with an Optica CF4 spectrophotometer in a cuvette with a lightpath length of 1.000 cm. The absorbance by SHb (A_{SHb}^{620}) was calculated with the equation

$$A_{SHb}^{620} = A_A^{620} - A_B^{620}(1 - F_{SHb}) \tag{7}$$

where A_A^{620} is the absorbance of the SHb/HbO$_2$ solution prepared from suspension A and A_B^{620} the absorbance of the HbO$_2$ solution prepared from suspension B. The millimolar absorptivity of SHb at λ = 620 nm (ϵ_{SHb}^{620}) then followed from

$$\epsilon_{SHb}^{620} = A_{SHb}^{620} f / F_{SHb} c_{Hb}^* l \tag{8}$$

where the dilution factor f = 50, the lightpath length l = 1.000 cm, and c_{Hb}^* is expressed in millimoles per liter.

For 11 samples with F_{SHb} = 17.5–24.3% and c_{Hb}^* = 5.97–8.92 mmol/liter, a mean value of ϵ_{SHb}^{620} = 20.82 was found (SD = 1.49; SEM = 0.45) (9). These results are in fair agreement with those of Nichol and Morell (30) (ϵ_{SHb}^{620} = 21.5) and only slightly higher than those of Siggaard-Andersen et al. (37) (ϵ_{SHb}^{620} = 18.1, after recalculation to room temperature). All recent determinations give ϵ_{SHb}^{620} values which are much higher than the older ones such as those of Drabkin and Austin (12) (ϵ_{SHb}^{620} = 10.6). The value of ϵ_{SHb}^{620} being determined, the ordinate of the spectral absorbance curves of SHb, SHi, and SHiCN recorded by means of the Aminco-Chance split-beam spectrophotometer (cf. Section 3.1) could be calibrated in absorptivity units (Fig. 17).

4. Determination of Hemoglobin Derivatives

Numerous spectrophotometric methods have been developed for the determination of hemoglobin derivatives in various mixtures and are described in the literature. For many problems, several solutions have been proposed. Therefore, the methods described in this section necessarily constitute a selection from many possibilities. The selection has been made on theoretical as well as practical grounds, not the least important of the latter being that the methods concerned yielded reliable results when tested in the authors' laboratories.

4.1. Theoretical Considerations

All methods for the spectrophotometric determination of hemoglobin derivatives depend on the validity of Lambert–Beer's law and all equations used to this end can be derived from the basic equations

$$A_X^\lambda = \epsilon_X^\lambda c_X l \qquad (9)$$

and

$$A_X^\lambda = \log(I_0^\lambda/I_X^\lambda) \qquad (10)$$

where A_X^λ is the absorbance of substance X at wavelength λ, I_0^λ is the amount of light of wavelength λ impinging on X, I_X^λ is the amount of light transmitted by X, ϵ_X^λ is the absorptivity of X at wavelength λ, c_X is the concentration of X, and l is the lightpath length. A is dimensionless; thus, when c is expressed in millimoles per liter and l in centimeters, ϵ is the millimolar absorptivity expressed in liters per millimole per centimeter.

When more than one light-absorbing component is present in a solution, the absorbance measured is the sum of the absorbances of all components present:

$$A^\lambda = A_1^\lambda + A_2^\lambda \cdots A_n^\lambda = \sum_{i=1}^{n} A_i \qquad (11)$$

and

$$A^\lambda = \sum_{i=1}^{n} \epsilon_i^\lambda c_i l \qquad (12)$$

When n components are present, the absorbance should be measured (at least) at n wavelengths to obtain (at least) n equations of the type of Eq. (12), from which the concentration of each component can be calculated. This procedure presupposes that the absorptivities of all components at all the wavelengths used (n^2 in number) and the lightpath length are exactly known. A straightforward application of this method has become feasible only recently (3, 74, 75). Formerly, an attempt was always made to reduce a multicomponent system to a series of two-component systems by various interconversions of the components.

The analysis of a two-component system through absorbance measurement at two wavelengths becomes much more simplified if, for one wavelength, an isobestic point is used. The two equations ensuing from Eq. (12) with $n = 2$ are

$$A^{\lambda_1} = \epsilon_1^{\lambda_1} c_1 l + \epsilon_2^{\lambda_1} c_2 l \qquad (13)$$

and

$$A^{\lambda_2} = \epsilon_1^{\lambda_2} c_1 l + \epsilon_2^{\lambda_2} c_2 l \qquad (14)$$

If λ_2 is the isobestic wavelength, $\epsilon_1^{\lambda_2} = \epsilon_2^{\lambda_2}$. Introducing the total concentration of the two light-absorbing components ($c = c_1 + c_2$) gives, through simple transformations (58),

$$c_2/c = [\epsilon_1^{\lambda_2} A^{\lambda_1}/(\epsilon_2^{\lambda_1} - \epsilon_1^{\lambda_1})A^{\lambda_2}] - [\epsilon_1^{\lambda_1}/(\epsilon_2^{\lambda_1} - \epsilon_1^{\lambda_1})] \quad (15)$$

or

$$F_2 = c_2/c = a(A^{\lambda_1}/A^{\lambda_2}) + b \quad (16)$$

With the aid of Eq. (16), the fraction of component 2 in the solution (F_2) can be calculated from the absorbance ratio $A^{\lambda_1}/A^{\lambda_2}$ if the constants a and b are known. These constants can easily be determined by a series of measurements of the absorbance ratio of solutions exclusively containing component 1 or component 2 (58). It should be noted that neither the exact value of the absorptivities nor the exact layer thickness needs to be known. In spite of the advent of practical multicomponent methods, two-wavelength methods utilizing an isobestic point are still suitable for many purposes, e.g., for the determination of the oxygen saturation and the carbon monoxide saturation of human blood (vide infra).

Instead of selecting an isobestic point for one of the wavelengths in a method for analyzing a two-component system, one can select a pair of wavelengths at which the absorptivities of one of the components are equal ($\epsilon_1^{\lambda_1} = \epsilon_1^{\lambda_2}$). Subtracting Eq. (14) from Eq. (13) then gives

$$A^{\lambda_1} - A^{\lambda_2} = (\epsilon_2^{\lambda_1} - \epsilon_2^{\lambda_2})c_2 l \quad (17)$$

or

$$c_2 = (A^{\lambda_1} - A^{\lambda_2})/(\epsilon_2^{\lambda_1} - \epsilon_2^{\lambda_2})l = k(A^{\lambda_1} - A^{\lambda_2}) \quad (18)$$

Using Eq. (18), the concentration of component 2 in the solution (c_2) can be calculated from the absorbance difference $A^{\lambda_1} - A^{\lambda_2}$ if the constant k is known. This constant can be determined by means of a series of measurements of the absorbance difference of solutions with a known concentration of component 2. These can, of course, be pure solutions of component 2, of which the concentration can be determined after conversion to HiCN. It should be noted that k is dependent on l, so that the same cuvette or exactly equal cuvettes should be used in all measurements.

The analysis a two-component system by absorbance measurement at a single wavelength requires either an exact knowledge of the absorptivities of the two components, the lightpath length, and the sum of the concentration of the two components, or a second measurement, after conversion of the component to be measured to another hemoglobin derivative. The equation to be used with the first of these methods follows from Eq. (13) when it is taken into account that $c = c_1 + c_2$. Thus,

$$F_2 = c_2/c = A^{\lambda}/cl \, [1/(\epsilon_2^{\lambda} - \epsilon_1^{\lambda})] - [\epsilon_1^{\lambda}/(\epsilon_2^{\lambda} - \epsilon_1^{\lambda})] \quad (19)$$

where c follows from c_{Hb}^*, taking into account the dilution factor, if any, used in preparing the solution of which the absorbance is actually measured.

When a conversion method is used, two absorbance measurements are made at the same wavelength. This yields two equations, analogous to Eq. (13):

$$A_1^\lambda = \epsilon_1^\lambda c_1 l + \epsilon_2^\lambda c_2 l \tag{20}$$

$$A_2^\lambda = \epsilon_1^\lambda c_1 l + \epsilon_3^\lambda c_2 l \tag{21}$$

where ϵ_2^λ is the absorptivity of component 2 before, and ϵ_3^λ the absorptivity after, conversion to another hemoglobin derivative. Subtraction of Eq. (21) from Eq. (20) yields

$$A_1^\lambda - A_2^\lambda = (\epsilon_2^\lambda - \epsilon_3^\lambda) c_2 l \tag{22}$$

and

$$c_2 = (A_1^\lambda - A_2^\lambda)/(\epsilon_2^\lambda - \epsilon_3^\lambda) l \tag{23}$$

Dividing Eq. (23) by the sum of the concentration of the two components in the mixture gives

$$F_2 = c_2/c = (A_1^\lambda - A_2^\lambda)/(\epsilon_2^\lambda - \epsilon_3^\lambda) lc = (A_1^\lambda - A_2^\lambda)/(A_3^\lambda - A_4^\lambda) \tag{24}$$

where $(A_3^\lambda - A_4^\lambda)$ is the absorbance difference occurring when all the hemoglobin in the solution is subject to conversion. (It should be noted that the lower indices to A refer to the sequence of the absorbance measurements, whereas those to ϵ and c refer to the components.) Therefore, if the absorbance difference after first changing all the hemoglobin into component 2 can also be measured, the fraction of component 2 can be calculated from the absorbance measurement without knowing the absorptivities and the light-path length and without the necessity of determining any constants.

4.2. Determination of the Oxygen Saturation (S_{O_2})

The oxygen saturation of the blood (S_{O_2}) is defined as the amount of oxygen actually bound by the hemoglobin in the erythrocytes in a certain volume of blood divided by the oxygen capacity, i.e., the maximum amount of oxygen which can be bound by the hemoglobin in the erythrocytes in the same volume of blood. Thus

$$S_{O_2} = c_{HbO_2}/(c_{HbO_2} + c_{Hb}) \tag{25}$$

The oxygen saturation should not be confused with the oxyhemoglobin fraction (F_{HbO_2}):

$$F_{HbO_2} = c_{HbO_2}/c_{Hb}^* \tag{26}$$

Only in the theoretical case when no dyshemoglobins (hemoglobin derivatives which have permanently or temporarily lost the ability to combine reversibly with oxygen under physiological conditions) are present does S_{O_2} equal F_{HbO_2}. Usually, S_{O_2} is slightly higher than F_{HbO_2}.

For the determination of S_{O_2}, two-wavelength methods with an isobestic point are excellently suited. In our previous review (58), we described two methods in which the absorbance measurements were made at $\lambda = 560$ and 522 nm and $\lambda = 560$ and 506 nm, respectively (cf. Tables 4 and 5). The use of wavelengths at which the absorptivity is rather high, instead of the classic "oximeter band" around $\lambda = 650$ nm, was advocated on the basis of the consideration that possible turbidity due to plasma protein has less influence in regions of high hemoglobin absorptivity, and because of the wish to avoid the use of the isobestic point in the near infrared, which then appeared to be quite elusive. This isobestic point is the obvious choice for combination with a measurement in the region around $\lambda = 650$ nm because the difference in absorptivity at the two wavelengths should not be too big (Fig. 18). Only then is it possible to keep both the absorbance measurements in the range of near maximum photometric accuracy by proper selection of the lightpath length (cf. Fig. 2 in ref. 58).

A practical disadvantage of measuring at $\lambda = 560/522$ or $560/506$ nm is the necessity of using a very small lightpath length ($l = 0.01$ cm). Yet, there are more substantial reasons for preferring a method with measurement in the red/near-infrared region for the determination of S_{O_2}. First, as Mook *et al.*

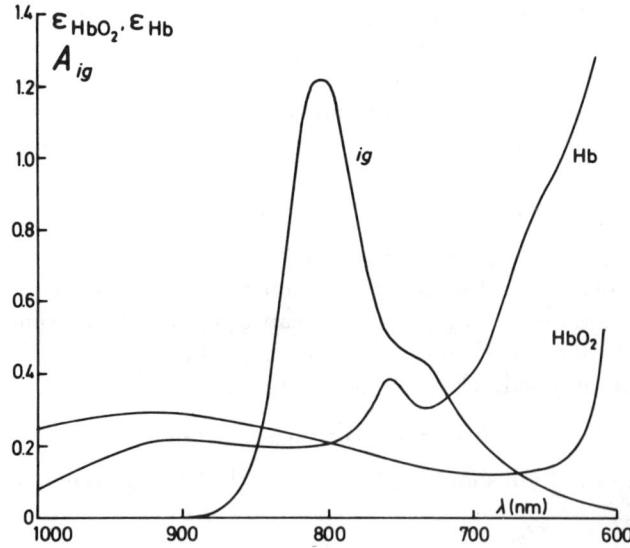

FIG. 18. Absorption spectra of HbO_2 and Hb in the red and near-infrared spectral region. Also shown is the spectral absorbance curve of a 5 mg/liter indocyanine green solution, measured in a cuvette with a lightpath length of 1.000 cm. [From Mook *et al.* (25).]

(28) have demonstrated, the sensitivity of the measurement, expressed as dA/dS_{O_2}, the change in absorbance per unit change in oxygen saturation, is much higher between $\lambda = 600$ and 730 nm than in any other region of the visible spectrum. This easily compensates for the somewhat stronger influence of possible turbidity due to plasma protein in this region than at shorter wavelengths. Second, it has been shown that the difficulty of determining the isobestic point in the near infrared is solely the result of the use of $Na_2S_2O_4$. When the blood is deoxygenated by tonometry with N_2/CO_2, the isobestic point is consistently found at $\lambda = 800$ nm, as shown in Fig. 18. For 14 blood samples of healthy human donors, Mook et al. (25) obtained $\lambda = 799.9 \pm 0.96$ nm (SD). Third, in the red/near-infrared region, the pair of wavelengths can be chosen in such a manner that the influence of the possible presence of HbCO and Hi in the blood sample on the measured value of S_{O_2} is small.

An excellent choice for the determination of S_{O_2} is the wavelength pair $\lambda = 680$ and 800 nm. The absorbance measurements are best carried out with a lightpath length of 0.2 cm. The following procedure can be recommended. An arterial or venous blood sample is collected in a glass syringe containing an amount of heparin sufficient to prevent coagulation. For an S_{O_2} measurement, 2 ml of blood is transferred anaerobically to a 2-ml syringe, which contains a glass pellet or a small metal ring for mixing and of which the dead space has been filled with a 10% solution of a nonionic detergent (cf. Section 2.2). After thoroughly mixing the sample and discarding the first three drops, the hemolysate is injected with the help of a blunt needle into a planparallel glass cuvette with a lightpath length of 0.200 cm. The cuvette is completely filled, closed with a glass cover, and placed in the spectrophotometer. The absorbance is measured at $\lambda = 680$ and 800 nm, using as a blank a similar cuvette filled with water. S_{O_2} is calculated with an equation of the type of Eq. (16). The values of the constants a and b in this equation are determined by measuring A^{680}/A^{800} for some completely oxygenated and deoxygenated blood samples. This yields absorbance ratios corresponding to $S_{O_2} = 1$ and 0, respectively. By substituting these in Eq. (16), two equations are obtained from which a and b can be solved.

This calibration was carried out by means of an Optica CF4 grating spectrophotometer, using completely oxygenated and deoxygenated blood of six humans and eight dogs. Substitution of the values for a and b in Eq. (16) results in

$$S_{O_2} = -0.3819(A^{680}/A^{800}) + 1.633 \tag{27}$$

These values of a and b are slightly different from those which would have resulted from the substitution $\epsilon_{HbO_2}^{680}$, ϵ_{Hb}^{680}, and $\epsilon_{HbO_2/Hb}^{800}$ in Eq. (15). The cause of this difference is that the absorptivities were determined for partially purified hemoglobin solutions, whereas in the direct determination of a and b, the absorbance of hemolysates is measured without any clearing procedure. The latter is obviously the correct way to determine these constants because, in a practical spectrophotometric method for the determination of S_{O_2}, no preparatory steps other than hemolysis should be necessary. Moreover, calculating a and b from the absorptivities gives quite inaccurate results in this spectral region where the absorptivities are very low and, consequently, the uncertainty in the available values is considerable.

Some data on the influence of HbCO and Hi present in a blood sample on the S_{O_2} values obtained by measuring A^{680}/A^{800} are shown in Table 8. It appears that the error, caused by such amounts of HbCO and Hi and which may go undetected for some time, is not large, except for Hi in the lower oxygen saturation range. In the latter case, an appreciable overestimation of S_{O_2} occurs. This will, however, seldom cause problems in the clinical application of the method.

The absorption maximum of indocyanine green, the dye commonly used in measuring cardiac output by means of the dye dilution method, coincides with the isobestic point of Hb and HbO_2 at $\lambda = 800$ nm, as shown in Fig. 18. This coincidence was actually the main reason for the introduction of the dye, because it allows the recording of dye dilution curves without interference due to variation in oxygen saturation. The high absorptivity of indocyanine green at $\lambda = 800$ nm (19), however, precludes the use of this wavelength for the determination of S_{O_2} when the blood contains the dye. It has been shown that by using the wavelength pair $\lambda = 660$ and 860 nm, S_{O_2} can be determined virtually independent of the presence of indocyanine green (25). The procedure is the same as with the 680/800 method. However, Eq. (16) does not hold well for the relationship between the absorbance ratio A^{680}/A^{860} and S_{O_2} since no isobestic point is used. A slightly more complicated equation can be easily derived from Eq. (12), taking $n = 2$ (25,

TABLE 8
ERRORS IN THE DETERMINATION OF OXYGEN SATURATION[a]

S_{O_2} (%)	Measured values of S_{O_2} (%)			
	$F_{HbCO} = 5\%$	$F_{HbCO} = 10\%$	$F_{Hi} = 5\%$	$F_{Hi} = 10\%$
100	99.7	99.3	99.7	99.5
80	79.7	79.3	81.5	82.8
60	59.7	59.3	63.2	66.2
40	39.7	39.3	44.9	49.5

[a] Absorbance measurement at $\lambda = 680$ and 800 nm; errors caused by presence of HbCO and Hi.

58). The 660/860 method may be useful for the determination of S_{O_2} in blood samples taken during cardiac catheterization, when dye dilution curves are being made.

4.3. Determination of the Carbon Monoxide Fraction (F_{HbCO})

In the description of the relationship between carbon monoxide and hemoglobin, the same distinction between saturation and fraction can be made as in the case of oxygen. Carbon monoxide saturation can be defined as the amount of carbon monoxide actually bound by the hemoglobin in the erythrocytes in a certain volume of blood divided by the carbon monoxide capacity, i.e., the maximum amount of carbon monoxide which can be bound by the hemoglobin in the erythrocytes in the same volume of blood. Thus

$$S_{CO} = c_{HbCO}/(c_{HbCO} + c_{HbO_2} + c_{Hb}) \tag{28}$$

This equation differs from Eq. (25) in that HbO_2 is treated as a hemoglobin derivative that can bind carbon monoxide, whereas in the definition of oxygen saturation, HbCO is treated as a hemoglobin derivative that cannot bind oxygen. From a chemical point of view this distinction may seem arbitrary but, physiologically, it does make sense. The affinity of hemoglobin for CO is so much greater than for O_2 that the P_{O_2} difference between the arterial and the venous blood has virtually no influence on the amount of HbCO in the blood. Consequently, the amount of hemoglobin that is in the form of HbCO is at least temporarily lost for the oxygen transport function. HbCO is thus correctly termed a dyshemoglobin (cf. Section 4.2) and should preferably be expressed as a fraction [cf. Eq. (29)]. That HbCO even impedes oxygen transport through its influence on the O_2 affinity of the rest of the hemoglobin in the blood is not relevant in this context.

The first choice to be made in designing a two-wavelength method for the determination of the HbCO fraction in blood is the two-component system to be used: HbCO/Hb or HbCO/ HbO_2. In our previous review (58), we described for each system a two-wavelength method with an isobestic point. The method utilizing the system HbCO/Hb has the advantage that the absorbances can be measured in a diluted solution, so that a 1.00-cm cuvette can be used. The method is reliable and accurate enough for most clinical purposes. It is carried out as described in ref. 58.

In the following, a method is described utilizing the system HbCO/HbO_2. This method has also already been dealt with in ref. 58, but it has been critically evaluated since. As the method presupposes the system HbCO/HbO_2, the fresh, heparinized blood sample is first oxygenated by rotating it for 5 minutes in a small, open, cylindrical tonometer, flushed with oxygen just before use. It has been shown that this procedure is sufficient to completely saturate the Hb present with O_2, while the influence on the HbCO in the sample is insignificant for HbCO fractions < 40% (8).

One milliliter of the tonometered blood is transferred to a 1-ml syringe, which contains a glass pellet or a small metal ring for mixing and of which the dead space has been filled with a 10% solution of a nonionic detergent (cf. Section 2.2). If less blood is available, the volume ratio of blood/detergent solution can be changed, but a correspondingly lower detergent concentration should be used. It has been shown that a twofold dilution is inconsequential for the HbCO fraction (58). After thoroughly mixing the sample and discarding the first three drops, the hemolysate is injected with the help of a blunt needle into a plan-parallel glass cuvette with a lightpath length of 0.100 cm. A 0.095-cm glass plate is inserted into the cuvette, leaving a lightpath length of 0.005 cm (cf. Fig. 13 in ref. 58 and Fig. 2 in ref. 75). The cuvette is placed in the spectrophotometer and the absorbance is measured at λ = 562 and 540 nm, using as a blank a similar cuvette filled with water. The isobestic point is at λ = 540 nm (Fig. 10; Tables 5 and 6).

For calculating the HbCO fraction from the absorbance ratio A^{562}/A^{540}, the following equation is used, which is of the type of Eq. (16):

$$F_{HbCO} = 3.215(A^{562}/A^{540}) - 1.923 \tag{29}$$

The constants a = 3.215 and b = −1.923 are based on the data of Fig. 19. These data were collected by measuring A^{562}/A^{540} of 46 human blood samples containing various amounts of HbCO. The blood was obtained from 22 healthy donors and contained heparin as anticoagulant. Part of the blood was oxygenated, part of it was equilibrated with a CO-containing gas mixture, and samples with various HbCO fractions were prepared by mixing various volumes of HbO_2- and HbCO-containing blood. The HbCO fractions were determined by a titrimetric method (8), which is a modification of the method for the determination of the oxygen content of blood described in detail in ref. 10.

The value of A^{562}/A^{540} for CO-free blood, following from the regression line of Fig. 19, is 0.598, which agrees well with the results of the direct measurement, ranging from 0.593 to 0.603. It is also in excellent agreement with the results of another series of measurements of A^{562}/A^{540} of blood after tonometry with pure O_2 for 150 minutes (7): 0.599 ± 0.004 (n = 38). The zero point of the determination of F_{HbCO} has thus been firmly established. It should be noted that the possible presence of Hi in the blood sample has practically no influence on the absorbance ratio, $(\epsilon^{562}/\epsilon^{540})_{Hi}$ being ~ 0.610. The presence of some 20% Hi would simulate less than 1% HbCO. The spread in the value of A^{562}/A^{540} (±0.004, corresponding with ±1.6% HbCO) will therefore be due to variations in the light-absorbing and -scattering properties of plasma constituents and erythrocyte stromata.

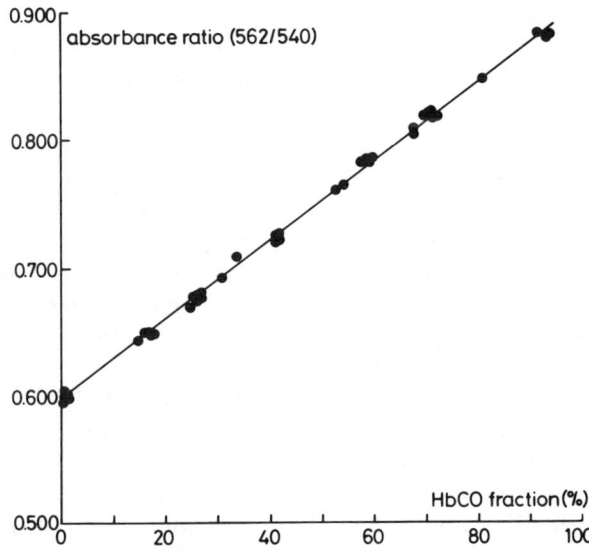

FIG. 19. Absorbance ratio A^{562}/A^{540} plotted against the corresponding HbCO fraction determined with the titrimetric method. The equation of the regression line is $A^{562}/A^{540} = 0.3118 F_{HbCO} + 0.59756$. The correlation coefficient $r = 0.9991$. [From Dijkhuizen et al. (8).]

The value of A^{562}/A^{540} for blood with 100% HbCO, following from the regression line of Fig. 19, is 0.909. Direct determination, however, yielded a slightly lower value: 0.900 ± 0.003 ($n = 84$). No cause for this discrepancy has yet been found. If the constants of Eq. (29) are calculated from the absorbance ratios directly measured for blood, with $F_{HbCO} = 0$ and 1 (0.599 and 0.900, respectively), $a = 3.322$ and $b = -1.990$. The error in the determination of F_{HbCO} caused by the use of these constants is insignificant: instead of $F_{HbCO} = 0, 10, 20, 30,$ and 40%, we get $-0.3, 10.0, 20.4, 30.7,$ and 41.0%, respectively.

4.4. Determination of the Hemiglobin Fraction (F_{Hi})

In our previous review (58), we described a two-wavelength method with an isobestic point for the determination of Hi in the system Hi/HbO$_2$ and, also, the CN$^-$ addition method according to Evelyn and Malloy (13). The latter method has the advantage that it is an absolute method in the sense that it is not dependent on constants which have been previously determined by measuring samples of known concentration. The method is invalidated by the presence of SHb (9), which is not too serious a disadvantage because of the infrequent occurrence of this hemoglobin derivative and its easy spectrophotometric detectability (cf. Fig. 10). In our experience, the

CN⁻ addition method gives reliable results for $F_{Hi} < 30\%$; with higher Hi fractions there is a progressive overestimation. Also, because of its simplicity, the method is well suited for routine application in the clinical chemical laboratory.

In the CN⁻ addition method, four diluting solutions are used. All solutions contain 0.05% of a nonionic detergent (cf. Section 2.2.), 27.50 mM Na$_2$HPO$_4$, and 13.16 mM KH$_2$PO$_4$. Solutions 2 and 3 contain, in addition, 3.84 mM KCN and 3.04 mM K$_3$Fe(CN)$_6$, respectively. Solution 4 contains, in addition, 3.84 mM KCN as well as 3.04 mM K$_3$Fe(CN)$_6$. One-half milliliter blood is added to 25 ml of each of the four solutions. In the first solution, the Hi present in the sample will remain unchanged. In the second solution, all Hi present becomes converted to HiCN. In the third solution, all hemoglobin present in the sample (except SHb) is converted to Hi, and in the fourth solution, all hemoglobin present (except SHb) is converted to HiCN. The absorbance of the four solutions is measured at $\lambda = 630$ nm, with either $l = 1.00$ cm or $l = 4.00$ cm, according to the absorbance level, using as a blank a similar cuvette filled with water. The Hi fraction is then calculated with an equation of the type of Eq. (24), in which $F_2 = F_{Hi}$, $\lambda = 630$ nm, and A_1, A_2, A_3, and A_4 are the absorbances measured for the blood diluted with solutions 1, 2, 3, and 4, respectively.

4.5. Multicomponent Analysis of Hemoglobin Derivatives

The straightforward application of a set of equations of the type of Eq. (12) for the simultaneous determination of n hemoglobin derivatives in a sample is based on the assumptions (1) that it is known which hemoglobin derivatives may be present, (2) that Lambert–Beer's law is valid for all components in the mixture, (3) that Lambert–Beer's law also applies to the actual mixture as it is presented for the absorbance measurements, (4) that the available absorptivity values are applicable to the actual measuring conditions, and (5) that the lightpath length is exactly known. Strictly speaking, the fourth assumption is simply another formulation of the third. One of the problems in the application of this method is that the measurements should preferably be carried out with simple hemolysates. The mere necessity of keeping the oxygen saturation of the sample constant precludes most clearing procedures. Therefore, the samples of which the absorbance is measured are necessarily turbid. Thus, it is a matter of semantics whether it is said that Lambert–Beer's law does not strictly apply to the solution because of the presence of light-scattering material, or that the available absorptivity values are not strictly valid for the hemoglobin derivatives as present in the hemolysate.

It has been shown that by simple anaerobic filtration through cotton wool, the hemolysate can be made clear enough for multicomponent analysis using absorptivity values as presented in Section 3. This technique has been applied in the simultaneous determination of Hb, HbO_2, HbCO, Hi, and SHb with a conventional spectrophotometer (75). In this method, measurements are made at five carefully selected wavelengths and the concentrations of the hemoglobin derivatives are calculated by solving a set of five equations of the type of Eq. (12). Such a system is called an exactly determined system, since there are just as many equations as there are unknowns. When the absorbance is measured over a considerable spectral range, either continuously or in very small steps, many more equations are obtained than there are unknowns. Such a system is called an overdetermined system. The HP8450A UV/Vis spectrophotometer (Hewlett-Packard, Palo Alto, California) is equipped with a computer program for multicomponent analysis in an overdetermined system. This procedure is now used for the simultaneous determination of the five hemoglobin derivatives previously mentioned (76).

To this end, a new set of spectral absorptivity curves is being made in our laboratories by measuring hemolyzed blood containing known concentrations of single hemoglobin derivatives (in the case of SHb, known mixtures of SHb and HbO_2). The ensuing collection of absorptivity curves are stored on magnetic tape and can be loaded into the microcomputer whenever the spectrophotometer is to be used for the analysis of a blood sample. Thus, in this procedure, the absorptivities are determined under exactly the same conditions as those under which the measurements of the unknown mixtures are made. The spectral absorptivity curves of Fig. 10 are the result of such a procedure. In the near future, it will also be possible to utilize these spectra for a similar procedure using conventional spectrophotometers, since these are being increasingly equipped with a microcomputer.

In the following, a five-wavelength method is described for the simultaneous determination of Hb, HbO_2, HbCO, Hi, and SHb, which can be carried out by means of a conventional spectrophotometer (75). The fresh, heparinized blood sample is transferred without any air contact from the syringe in which it was collected to a 2-ml glass syringe containing a mixing ball ($\phi = 5$ mm), with the dead space filled with a 5% solution of a nonionic detergent (cf. Section 2.2). After blood and detergent solution have been thoroughly mixed, a filter unit containing a piece of cotton wool is fixed onto the syringe. After flushing filter space and needle by discarding about 10 drops, two cuvettes ($l = 0.200$ and 0.100 cm) are filled with hemolysate. A 0.093-cm plan-parallel glass plate (Hellma Benelux, The Hague, The Netherlands) is inserted into the 0.100-cm cuvette. The absorbance is measured at $\lambda = 760$ and 620 nm with $l = 0.200$ cm and at $\lambda = 577$, 569, and 500 nm with $l = 0.007$ cm. In all measurements, a similar cuvette filled with water is used as a blank in the reference channel.

A spectral band width of 1 nm or less should be used. The wavelength and absorbance scales of the spectrophotometer should be checked as described in Section 2.4. The lightpath length of the cuvettes should be known exactly and can be checked as described in Section 3.1. The concentrations of the hemoglobin derivatives present in the sample are calculated from the absorbances by matrix calculation using the absorptivities of Table 9. In this calculation, the dilution of the blood with detergent solution and possible slight differences between the cuvettes in the measuring channel and the reference channels should be taken into account. An example of a program for making the calculation with a desk-top calculator (HP9845A) is given in ref. 75.

By means of an Optica CF4 spectrophotometer, the five-wavelength method has been compared for all five components with specific methods for the determination of each of these components (75). For the oxygen saturation, a comparison was made with the two-wavelength method described in Section 4.2 (λ = 680 and 800 nm), using 22 blood samples from three healthy humans. Different values of S_{O_2} were obtained by tonometry with various O_2/N_2 mixtures. The results are shown in Fig. 20. The deviation of the five-wavelength method with respect to the 680/800 method was $1.2 \pm 2.3\%$ S_{O_2} (SD).

For the carboxyhemoglobin fraction, a comparison was made with the two-wavelength method described in Section 4.3 (λ = 562 and 540 nm), using 32 blood samples from 7 healthy humans. Different values of F_{HbCO} were obtained by mixing in various proportions HbO_2-containing blood made by tonometry with O_2 and HbCO-containing blood made by tonometry with a CO-containing gas mixture. The results are shown in Fig. 21. The deviation of the five-wavelength method with respect to the 562/540 method was $1.2 \pm 1.7\%$ F_{HbCO} (SD).

For the hemiglobin fraction, a comparison was made with the method described in Section 4.4 (CN⁻ addition method), using 16 blood samples

TABLE 9
MATRIX OF MILLIMOLAR ABSORPTIVITIES

λ (nm)	ϵ^λ (liters mmol^{-1} cm^{-1})				
	Hb	HbO$_2$	HbCO	Hi	SHb
500	4.09	5.05	5.35	9.04	7.20
569	11.27	11.27	14.27	4.10	8.10
577	9.40	15.37	10.00	4.10	8.10
620	1.23	0.24	0.33	3.35	20.80
760	0.43	0.13	0.03	0.24	1.04

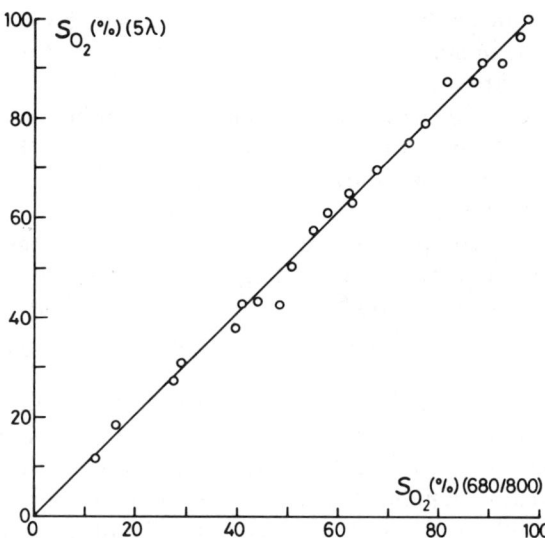

FIG. 20. Oxygen saturation (S_{O_2}) by 5 λ method plotted against S_{O_2} by 680/800 method. Equation of the regression line is S_{O_2} (5 λ) = 1.03 × S_{O_2} (680/800) − 0.32. Correlation coefficient r = 0.993. [From Zwart et al. (75).]

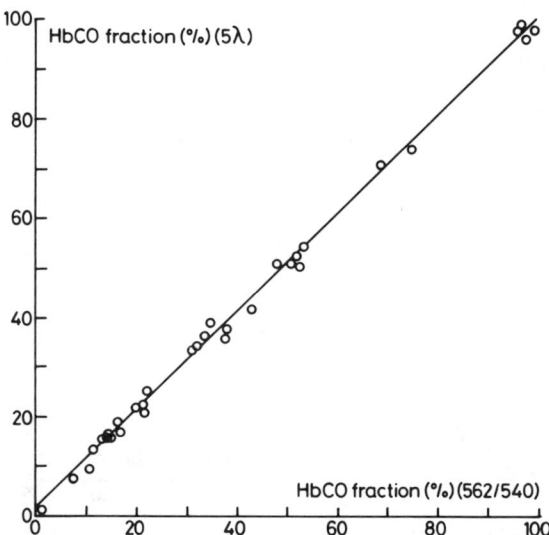

FIG. 21. Carboxyhemoglobin fraction (F_{HbCO}) by 5 λ method plotted against F_{HbCO} by 562/540 method. Equation of the regression line is F_{HbCO} (5 λ) = 0.99 × F_{HbCO} (562/540) + 1.45. Correlation coefficient r = 0.996. [From Zwart et al. (75).]

from 4 healthy humans. In order to prepare samples with various amounts of Hi, the blood was hemolyzed by the addition of a drop of undiluted Sterox SE and distributed to several test tubes to which various volumes of a 90 mM solution of hexacyanoferrate(III) were added. The tubes were kept for 90 minutes at room temperature before the absorbance of the samples was measured. The results are shown in Fig. 22. The deviation of the five-wavelength method with respect to the CN^- addition method was $-0.4 \pm 0.7\%$ F_{Hi} (SD). This comparison had to be limited to the range $F_{Hi} = 0-30\%$ because of the progressive overestimation of F_{Hi} by the CN^- addition method when $F_{Hi} > 30\%$ (cf. Section 4.4). Some additional dilution experiments indicated that the five-wavelength method gives correct values for F_{Hi} at least up to 80%.

For the sulfhemoglobin fraction, a comparison was made with a single-wavelength method utilizing an equation of the type of Eq. (19) and meeting the requirements to be fulfilled for the proper use of this equation (cf. Section 4.1). SHb-containing blood was prepared as described in Section 3.1. One-half milliliter blood was then added to 25 ml of a solution of 0.05% Sterox SE in phosphate buffer of pH 7.4 and the absorbance of this solution measured at $\lambda = 620$ nm, with $l = 1.00$ or 4.00 cm, according to the absorbance level. The results are shown in Fig. 23. The difference of the

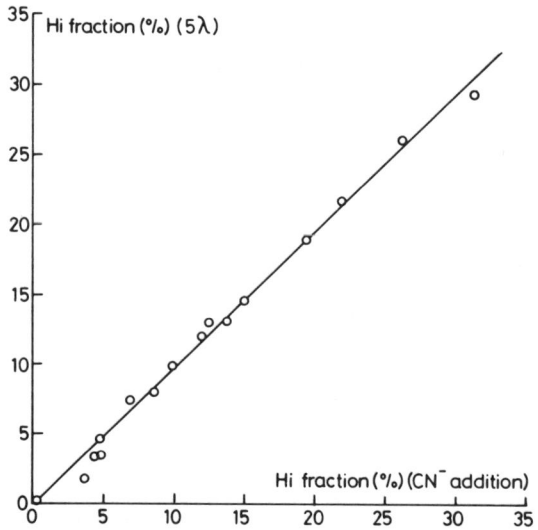

FIG. 22. Hemiglobin fraction (F_{Hi}) by 5 λ method plotted against F_{Hi} by CN^- addition method. Equation of the regression line is F_{Hi} (5 λ) = 0.99 × F_{Hi} (CN^-) − 0.31. Correlation coefficient $r = 0.994$. [From Zwart et al. (75).]

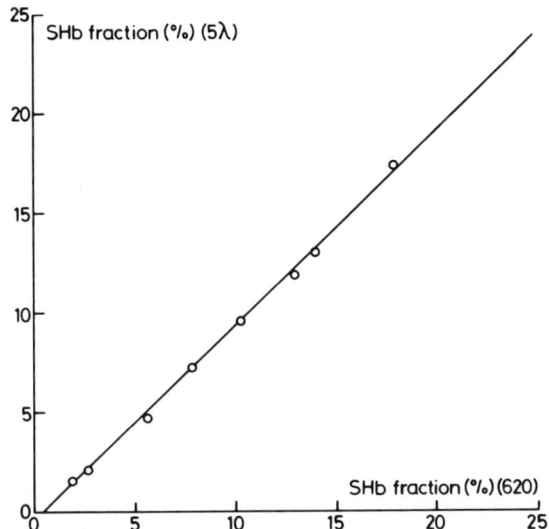

FIG. 23. Sulfhemoglobin fraction (F_{SHb}) by 5 λ method plotted against F_{SHb} by 620 method. Equation of the regression line is F_{SHb} (5 λ) = 0.99 × F_{SHb} (620) − 0.52. Correlation coefficient r = 0.998. [From Zwart et al. (75).]

five-wavelength method with respect to the 620 method was −0.6 ± 0.2% F_{SHb} (SD). The comparative measurements had to be limited to the range F_{SHb} = 0–25% because test samples with more than 25% SHb cannot be obtained by the technique used.

The total hemoglobin concentration of 22 samples from 5 healthy humans was determined with the HiCN method and the result compared with c^*_{Hb} calculated by adding the concentrations of all hemoglobin derivatives present in the sample. The results are shown in Fig. 24. The deviation of the five-wavelength method with respect to the HiCN method was −0.06 ± 0.15 g/dl (SD). The equality of c^*_{Hb} (5 λ) and c^*_{Hb} (HiCN) affords an easy opportunity for interparametric quality control (54). It is sound practice to supplement each determination of hemoglobin derivatives by means of the five-wavelength method with a determination of c^*_{Hb} as HiCN.

A four-wavelength method for the determination of Hb, HbO_2, HbCO, and Hi has been realized using an automated spectrophotometer, the IL282 CO-Oximeter (Instrumentation Laboratory Inc., Lexington, Maine) (3). The four wavelengths (λ = 535.0, 585.2, 594.5, and 626.6 nm) are obtained by means of four interference filters, each selecting a particular line from the emission spectrum of a Tl–Ne hollow cathode lamp. The advantage of this construction is that the wavelength setting is extremely stable and no regular

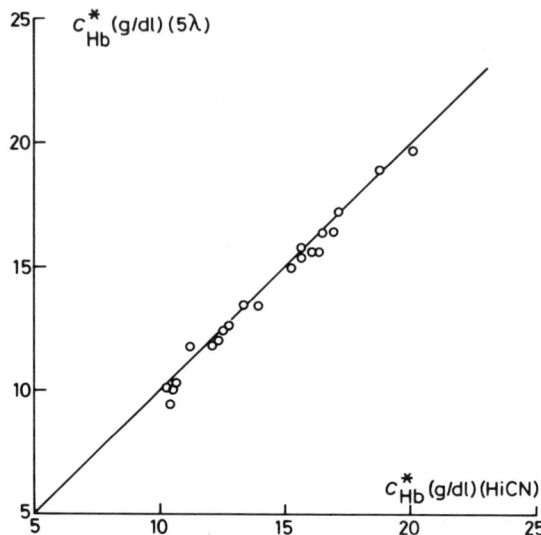

FIG. 24. Total hemoglobin concentration (c^*_{Hb}) by 5 λ method plotted against c^*_{Hb} by HiCN method. Equation of the regression line is c^*_{Hb} (5 λ) = 1.00 × c^*_{Hb} (HiCN) − 0.07. Correlation coefficient r = 0.988. [From Zwart et al. (75).]

wavelength calibration is necessary. A disadvantage is that the wavelengths to be used cannot be chosen freely at the most suitable points in the absorption spectra of the components. The IL282 CO-Oximeter was compared with the five-wavelength method described above and proved to be accurate for the determination of the HbO_2 and HbCO fractions and for c^*_{Hb}. The determination of Hi was somewhat less accurate. Even SHb could be detected, but not quantitated exactly (74).

5. Concluding Remarks

The internationally standardized HiCN method for the determination of the total hemoglobin concentration in human blood has now reached a stage of general acceptance, at least, as a reference method. The absorption spectrum of HiCN and the exact value of the millimolar absorptivity at λ = 540 nm have been established beyond any doubt, and the validity of Lambert–Beer's law for HiCN solutions has been confirmed. The production, control, and distribution of the international HiCN reference solution have become a matter of routine; no undesired incidents have occurred since the production started in 1964. A hypoosmolar $K_3Fe(CN)_6$/KCN solution, buffered with KH_2PO_4 to pH 7.2 and containing a nonionic detergent, has been proven to be the most suitable reagent solution and is widely accepted. The remaining uncertain points—the influence of the possible presence of SHb and the freeze-destruction of the reagent solution—have at last been elucidated: the influence of SHb is insignificant and the freeze-destruction of the reagent can be prevented by simple means.

The exact knowledge of the millimolar absorptivity of HiCN at $\lambda = 540$ nm constitutes a reliable reference point for the absorptivities of other hemoglobin derivatives. Because all hemoglobin derivatives except those containing sulfur can be converted into HiCN, any absorbance measured for any hemoglobin derivative can be easily related to this reference point. Thus, there remain few problems in the determination of spectral absorptivity curves of most hemoglobin derivatives.

Accurate knowledge of the spectral absorptivity curves of the common hemoglobin derivatives is the first requirement for the determination of these compounds in a mixture by multicomponent analysis. The calculating facilities necessary for the practical application of multicomponent analysis are now rapidly becoming available to the clinical chemical laboratory. Thus, it seems obvious that better and easier methods for multicomponent analysis of an increasing number of hemoglobin derivatives will be developed in the near future. Most wanted are improved tricks for suppressing turbidity in the hemolysates without changing the composition of the mixture of hemoglobin derivatives. For measuring oxygen saturation, the established two-wavelength methods utilizing an isobestic point, carried out either by means of a general purpose spectrophotometer or with a special instrument, will certainly be used for many years to come. Besides, reflection oximetry will remain an attractive alternative, especially for measuring *in vivo* (18, 26, 27, 33, 67).

In some cases, it is useful to calculate the oxygen content of the blood $(c^*_{O_2})$ after the total hemoglobin concentration (c^*_{Hb}), the oxygen saturation (S_{O_2}), and the dyshemoglobin fractions (F_{dysHb}) (cf. Section 4.2) have been determined. The following equation applies:

$$c^*_{O_2} = S_{O_2}\beta_{O_2}c^*_{Hb}(1 - F_{dysHb}) + \alpha_{O_2}P_{O_2} \tag{30}$$

where β_{O_2} is the oxygen capacity per gram hemoglobin, α_{O_2} the solubility of O_2 in whole blood, and P_{O_2} the oxygen tension. For $c^*_{Hb} = 150$ g/liter, $\alpha_{O_2} = 0.2201$ ml/liter·kPa at 37°C (6), and at an arterial P_{O_2} of 14 kPa, only 3.08 ml O_2 is in physical solution in 1 liter of blood. This is only 1.5% of the total amount of O_2 in the arterial blood and can be neglected for all practical purposes. Thus

$$c^*_{O_2} = S_{O_2}\beta_{O_2}c^*_{Hb}(1 - F_{dysHb}) = \beta_{O_2}c_{HbO_2} \tag{31}$$

Multicomponent analysis as described in Section 4.5 can give the oxyhemoglobin concentration (c_{HbO_2}) directly. As to the value of β_{O_2}, it seems obvious to use the theoretical value of $22.4 \times 10^{-3}/16114.5 = 1.390$ ml/g because, in Eq. (31), appropriate corrections for the dyshemoglobins have been made. However, in most cases this will probably give a slight overestimation. There seems to be a small, as yet unidentified, inactive hemoglobin fraction. In a series of 36 human blood samples, Dijkhuizen et al. (7) found this fraction to be $0.9 \pm 1.2\%$ (SD), but in two cases it was as high as 3.9 and 4.3%, respectively. As long as this hemoglobin fraction has not been identified and included in the multicomponent analysis, the error caused by it has to be taken for granted.

Several blood gas analyzers are equipped with a program for calculating S_{O_2} from P_{O_2}, pH, and P_{CO_2}. The calculation is based on the standard oxygen dissociation curve, and the temperature coefficient and the Bohr factors, stored in the data processor. However, there are other factors influencing the oxygen affinity of hemoglobin which cannot easily be taken into account. One of these is the 2,3-diphosphoglycerate (2,3-DPG) content of the erythrocytes. Another source of error is the presence of types of hemoglobin with an oxygen affinity different from that of HbA, of which HbF is, of

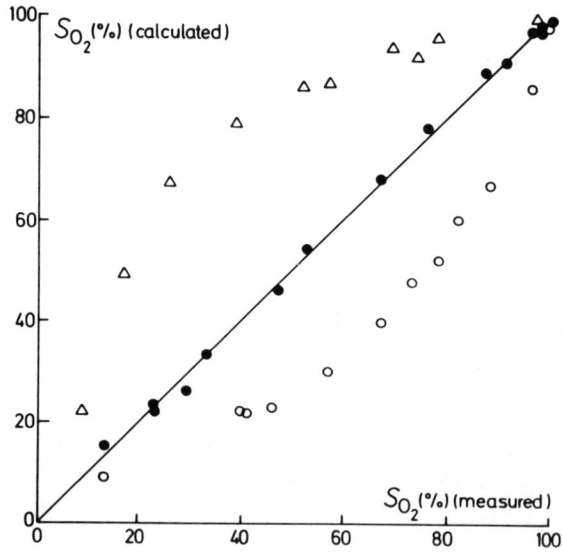

FIG. 25. Influence of 2,3-DPG content of erythrocytes on oxygen saturation (S_{O_2}) values calculated from P_{O_2} and pH (Corning 175). S_{O_2} measured with a special purpose photometer (Radiometer OSM2). Dots: fresh donor blood with normal 2,3-DPG content. Open circles: outdated bank blood, containing very little 2,3-DPG. Triangles: blood with about four times the normal 2,3-DPG content. [From Oeseburg et al. (32).]

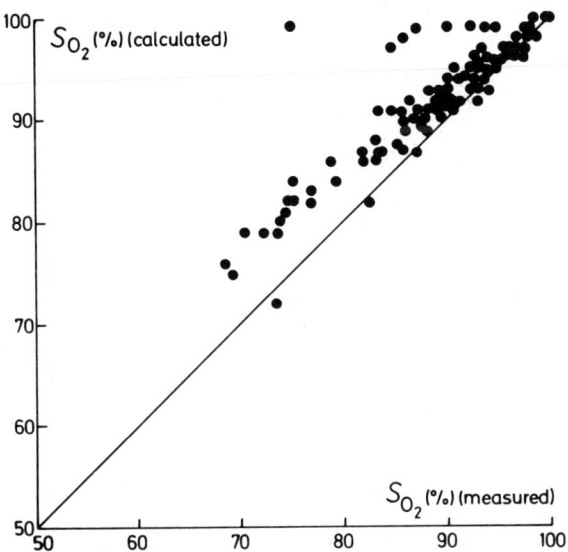

FIG. 26. Calculated (Corning 175) versus measured S_{O_2} (Radiometer OSM2) for 128 blood samples from critically ill patients. [From Oeseburg et al. (32).]

course, the most frequently occurring. Figure 25 shows the considerable influence of 2,3-DPG. For fresh donor blood with a normal 2,3-DPG content, there is an excellent agreement between the calculated and measured values of S_{O_2}. When the 2,3-DPG content is very low, the calculated values are much lower than the measured values. When the 2,3-DPG concentration is very high (about 4 times the normal [2,3-DPG]/[Hb$_4$] ratio), the calculated values are much too high. A comparison of calculated and measured values of S_{O_2} for 128 blood samples of critically ill patients (Reanimation Centre, University Hospital, Groningen, The Netherlands) showed considerable differences, the calculated values usually being too high (Fig. 26). This clearly demonstrates that the oxygen saturation should not be calculated, but should be determined by one of the several excellent methods available for this purpose.

6. Appendix

Definition of terms used in quality control.

Accuracy: agreement between the best estimate of a quantity and its true value.

Precision: agreement between results of replicate measurements in identical material; the quantitative measure is the standard deviation or the coefficient of variation.

Bias: numerical difference (+ or −) between the best estimate and the true value; equivalent term: systematic error.

As true values are essentially unknown, values from reference methods are used instead. It may happen that a method yields results with a very good reproducibility (high precision), but with a considerable systematic error (low accuracy). It should therefore be kept in mind that equality does not guarantee quality.

For obtaining more insight into the random error (standard deviation) σ_r and the systematic error (bias) σ_s, the Youden plot may be helpful. For this purpose, two samples with concentrations c_1 and c_2 are analyzed in n laboratories. The values obtained for c_1 are plotted against the values obtained for c_2 as shown in Fig. 27.

When only random errors are present, the chance of getting a deviation $+\Delta c$ is equal to the chance of getting a deviation $-\Delta c$. The points (c_1, c_2) from the various laboratories will thus be evenly spread around the point $(\bar{c}_1, \bar{c}_2)$. This result is a circular distribution. When a systematic error is present, the circular distribution changes to an elliptic one. The greater the systematic error, the narrower the ellipse.

This can be quantitated in the following way (Fig. 27). When $\sigma_s/\sigma_r = 0$, 50% of all points are found in quadrants B and D (circular distribution). When $\sigma_s/\sigma_r = 1, 2$, and 3, about 70, 80, and 90%, respectively, of all points

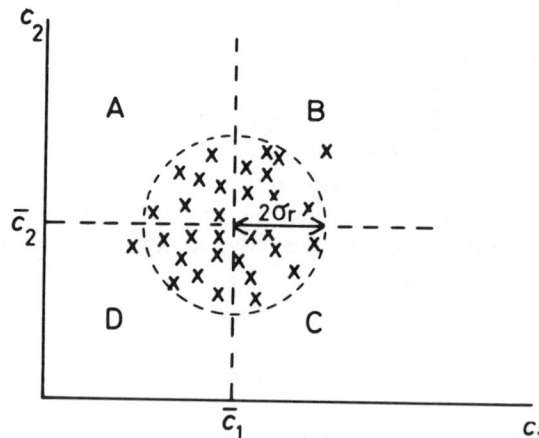

FIG. 27. Youden plot of concentration measurements c_1 and c_2 of two samples of the same compound by various laboratories. The circular distribution shown signifies the absence of systematic errors. The greater the number of laboratories making systematic errors, the more points are found in quadrants B and/or D.

are in quadrants B and D. When $\sigma_s/\sigma_r \gg 4$, all points are in quadrants B and D and the ellipse approximates a straight line.

Acknowledgments

The authors are indebted to G. A. Mook, A. Zwart, and A. Buursma for their help in the preparation of this article, to V. Sonak for language correction, to Th. Deddens for drawing the figures, and to H. Anthonio for typing the manuscript.

References

1. Bjure, J., and Nilsson, N. J., Spectrophotometric determination of oxygen saturation of hemoglobin in the presence of carboxyhemoglobin. *Scand. J. Clin. Lab. Invest.* **17**, 491–500 (1965).
2. Braunitzer, G., The molecular weight of human haemoglobin. *Bibl. Haematol.* **18**, 59–60 (1964).
3. Brown, L. J., A new instrument of the simultaneous determination of total hemoglobin, % oxyhaemoglobin, % carboxyhaemoglobin, % methaemoglobin, and oxygen content in whole blood. *IEEE Trans. Biomed. Eng.* **BME-27**, 132–138 (1980).
4. Burkhard, O., and Barnikol, W. K. R., Dependence of visible spectrum [$\epsilon(\lambda)$] of fully oxygenated hemoglobin on concentration of hemoglobin. *J. Appl. Physiol.: Respir., Environ. Exercise Physiol.* **52**, 124–130 (1982).
5. Chilcote, M. E., and O'Dea, A. E., Lyophilized carbonylhemoglobin as a colorimetric hemoglobin standard. *J. Biol. Chem.* **200**, 117–124 (1953).
6. Christoforides, C., and Hedley-Whyte, J., Effect of temperature and hemoglobin concentration on solubility of O_2 in blood. *J. Appl. Physiol.* **27**, 592–596 (1969).
7. Dijkhuizen, P., Buursma, A. Fongers, T. M. E., Gerding, A. M., Oeseburg, B., and Zijlstra, W. G., The oxygen binding capacity of human haemoglobin. Hüfner's factor redetermined. *Pfluegers Arch.* **369**, 223–231 (1977).
8. Dijkhuizen, P., Buursma, A., Gerding, A. M., van Kampen, E. J., and Zijlstra, W. G., Carboxyhemoglobin. Spectrophotometric determination tested and calibrated using a new reference method for measuring carbon monoxide in blood. *Clin. Chim. Acta* **80**, 95–104 (1977).
9. Dijkhuizen, P., Buursma, A., Gerding, A. M., and Zijlstra, W. G., Sulfhaemoglobin. Absorption spectrum, millimolar extinction coefficient at $\lambda = 620$ nm, and interference with the determination of haemiglobin and of haemiglobincyanide. *Clin. Chim. Acta* **78**, 479–487 (1977).
10. Dijkhuizen, P., Kwant, G., and Zijlstra, W. G., A new reference method for the determination of the oxygen content of blood. *Clin. Chim. Acta* **68**, 79–85 (1976).
11. Drabkin, D. L., Spectrophotometric studies. XIV. The crystallographic and optical properties of the hemoglobin of man in comparison with those of other species. *J. Biol. Chem.* **164**, 703–723 (1946).
12. Drabkin, D. L., and Austin, J. H., Preparation from washed blood cells; nitric oxide hemoglobin and sulfhemoglobin. *J. Biol. Chem.* **112**, 51–65 (1935).
13. Evelyn, K. A., and Malloy, H. T., Microdetermination of oxyhemoglobin, methemoglobin, and sulfhemoglobin in a single sample of blood. *J. Biol. Chem.* **126**, 655–662 (1938).
14. Hoek, W., Kamphuis, N., and Gast, R., Haemiglobincyanide: Molar lineic absorbance and stability constant. *J. Clin. Chem. Clin. Biochem.* **19**, 1209–1210 (1981).

15. International Committee for Standardization in Haematology, Recommendations for haemoglobinometry in human blood. *Br. J. Haematol.* **13**, Suppl., 71–75 (1967).
16. International Committee for Standardization in Haematology, Recommendations for reference method for haemoglobinometry in human blood (ICSH Standard EP 6/2: 1977) and specifications for international haemiglobincyanide reference preparation (ICSH) Standard EP 6/3: 1977). *J. Clin. Pathol.* **31**, 139–143 (1978).
17. Itano, H. A., Molar extinction coefficients cyanmethemoglobin (hemiglobincyanide) at 540 and 281 nm. *In* "Modern Concepts in Hematology" (G. Izak and S. M. Lewis, eds.), pp. 26–28. Academic Press, New York, 1972.
18. Landsman, M. L. J., Knop, N., Kwant, G., Mook, G. A., and Zijlstra, W. G., Fiber optic reflection oximeter. *Pfluegers Arch.* **373**, 273–282 (1978).
19. Landsman, M. L. J., Kwant, G., Mook, G. A., and Zijlstra, W. G. Light absorbing properties, stability, and spectral stabilization of indocyanine green. *J. Appl. Physiol.* **40**, 575–583 (1976).
20. Matsubara, T., Okuzono, H., and Senba, U., A modification of van Kampen-Zijlstra's reagent for the hemiglobincyanide method. *Clin. Chim. Acta* **93**, 163–164 (1979).
21. Matsubara, T., Okuzono, H., and Tamagawa, S., Proposal for an improved reagent in the hemiglobincyanide method. *In* "Modern Concepts in Hematology" (G. Izak and S. M. Lewis, eds.), pp. 29–43. Academic Press, New York, 1972.
22. Meyer-Wilmes, J., and Remmer, H., Die Standardisierung des roten Blutfarbstoffes durch Hämiglobincyanid. I. Mitteilung: Bestimmung der spezifischen Extinktion von Hämiglobincyanid. *Naunyn-Schmiedebergs Arch. Exp. Pathol. Pharmakol.* **229**, 441–449 (1965).
23. Michelsen, O., Woolard, H., and Ness, A. T., Decolorization on freezing of ferricyanide solution used for hemoglobin determination *Clin. Chem. (Winston-Salem, N.D.)* **10**, 611–618 (1964).
24. Minkowski, A., and Swierczewski, E., The oxygen capacity of the human foetal blood. *In* "Oxygen Supply to the Human Foetus" (J. Walker and A. C. Turnbull, eds.), pp. 237–253. Blackwell, Oxford, 1959.
25. Mook, G. A., Buursma, A., Gerding, A., Kwant, G., and Zijlstra, W. G., Spectrophotometric determination of oxygen saturation of blood independent of the presence of indocyanine green. *Cardiovasc. Res.* **13**, 233–237 (1979).
26. Mook, G. A., Osypka, P., Sturm, R. E., and Wood, E. H., Fibre optic reflection photometry on blood. *Cardiovasc. Res.* **2**, 199–209 (1968).
27. Mook, G. A., van Assendelft, O. W., Kwant, G., Landsman, M. L. J., and Zijlstra, W. G., Two-colour reflection oximetry. *Proc. K. Ned. Akad. Wet., Ser. C* **79**, 472–490. (1976).
28. Mook, G. A., van Assendelft, O. W., and Zijlstra, W. G., Wavelength dependency of the spectrophotometric determination of blood oxygen saturation. *Clin. Chim. Acta* **26**, 170–173 (1969).
29. Morningstar, D. A., Williams, G. Z., and Suutarinen, P., The millimolar extinction coefficient of cyanmethemoglobin from direct measurements of hemoglobin iron by X-ray emission spectrography. *Am. J. Clin. Pathol.* **46**, 603–607 (1966).
30. Nichol. A. W., and Morell, D. B., Spectrophotometric determination of mixtures of sulphhaemoglobin and methaemoglobin in blood. *Clin. Chim. Acta* **22**, 157–160 (1968).
31. Nilsson, N. J., Oximetry. *Physiol. Rev.* **40**, 1–26 (1960).
32. Oeseburg, B., Kwant, G., and Zwart, A., Determination of factors modulating hemoglobin oxygen affinity and the significance of these factors for the calculation of oxygen saturation by automated blood gas machines. *In* "Blood pH, Carbon Dioxide, Oxygen and Calcium-ion" (O. Siggaard-Andersen, ed.), pp. 139–150, 1981.
33. Polanyi, M. L., and Hehir, R. M., In vivo oximeter with fast dynamic response. *Rev. Sci. Instrum.* **33**, 1050–1054 (1962).

34. Refsum, H. E., Spectrophotometric determination of hemoglobin oxygen saturation in hemolyzed whole blood by means of various wavelength combinations. *Scand. J. Clin. Lab. Invest.* **9,** 190–193 (1957).
35. Remmer, A., Die Standardisierung des roten Blutfarbstoffes durch Hämiglobincyanid. II. Mitteilung: Eisengehalt und O_2-Bindungsvermögen von menschlichem Blut. *Naunyn-Schmiedebergs Arch. Exp. Pathol. Pharmakol.* **229,** 450–462 (1956).
36. Salvati, A. M., Tentori, L., and Vivaldi, G., The extinction coefficient of human hemiglobincyanide. *Clin. Chim. Acta* **11,** 477–479 (1965).
37. Siggaard-Andersen, O., Nørgaard-Pedersen, B., and Rem, J., Hemoglobin pigments. Spectrophotometric determination of oxy-, carboxy-, met-, and sulfhemoglobin in capillary blood. *Clin. Chim. Acta* **42,** 85–100 (1972).
38. Small, K. A., Radford, E. P., Frazier, J. M., Rodkey, F. L., and Collison, H. A., A rapid method for simultaneous measurement of carboxy- and methemoglobin in blood. *J. Appl. Physiol.* **31,** 154–160 (1971).
39. Standardizing Committee of the European Society of Haematology, Decision concerning haemoglobinometry. *Bibl. Haematol.* **18,** 110–111 (1964).
40. Standardizing Committee of the European Society of Haematology: Recommendations and requirements for haemoglobinometry in human blood (preparation and use of a haemoglobin reference standard). *Bibl. Haematol.* **21,** 213–216 (1965).
41. Stigbrand, T., Molar absorbancy of cyanmethaemoglobin. *Scand. J. Clin. Lab. Invest.* **20,** 252–254 (1967).
42. Taylor, J. D., and Miller, J. D. M., A source of error in the cyanmethemoglobin method of determination of hemoglobin concentration in blood containing carbon monoxide *Am. J. Clin. Pathol.* **43,** 265–271 (1965).
43. Tentori, L., Vivaldi, G., and Salvati, A. M., The extinction coefficient of human hemiglobincyanide as determined by nitrogen analysis. *Clin. Chim. Acta* **14,** 276–277 (1966).
44. Theil, G. B., and Auer, J. E., Hemiglobin sodium nitrite vs. hemiglobin potassium ferricyanide. *Clin. Chem. (Winston-Salem, N.C.)* **13,** 1010–1013 (1967).
45. Van Assendelft, O. W., "Spectrophotometry of Haemoglobin Derivatives." Van Gorcum & Co., Assen, The Netherlands, 1970.
46. Van Assendelft, O. W., Photometry and the standardized method for the determination of haemoglobin. *Schweiz, Med. Wochenschr.* **101,** 1649–1652 (1971).
47. Van Assendelft, O. W., Buursma, A., Holtz, A. H., van Kampen, E. J., and Zijlstra, W. G., Quality control in haemoglobinometry with special reference to the stability of haemiglobincyanide reference solutions. *Clin. Chim. Acta* **70,** 161–169 (1976).
48. Van Assendelft, O. W., Holtz, A. H., van Kampen, E. J., and Zijlstra, W. G., Control data of international haemiglobincyanide reference solutions. *Clin. Chim. Acta* **18,** 78–81 (1967).
49. Van Assendelft, O. W., van Kampen, E. J., and Zijlstra, W. G., Standardization of haemoglobinometry. The use of haemiglobinazide instead of haemiglobincyanide for the determination of haemoglobin. *Proc. K. Ned. Akad. Wet. Ser. C* **72,** 249–253 (1969).
50. Van Assendelft, O. W., and Zijlstra, W. G., The formation of haemiglobin using nitrites. *Clin. Chim. Acta* **11,** 571–577 (1965).
51. Van Assendelft, O. W., and Zijlstra, W. G., Hemiglobin and hemiglobin nitrite. *Clin. Chem. (Winston-Salem, N.C.)* **14,** 918–919 (1968).
52. Van Assendelft, O. W., and Zijlstra, W. G., Extinction coefficients for use in equations for the spectrophotometric analysis of haemoglobin mixtures. *Anal. Biochem.* **69,** 43–48 (1975).
53. Van Assendelft, O. W., Zijlstra, W. G., and van Kampen, E. J., Haemoglobinometry. Challenges and pittfalls. *Proc. K. Ned. Akad. Wet. Ser. C* **73,** 104–112 (1970).

54. van Kampen, E. J., Finding the path? A critical evaluation of quality control. In "New Pathways in Laboratory Medicine' (S. B. Rosalki, ed.), pp. 132–140. Huber, Bern, 1978.
55. van Kampen, E. J., and Van Assendelft, O. W., Quality control and hematology. I. *Qual. Control Clin. Chem. Trans. Int. Symp. 6th, 1975* pp. 325–333.
56. van Kampen, E. J., Volger, H. C., and Zijlstra, W. G., Spectrophotometric determination of carboxyhemoglobin in human blood and the application of this method to the estimation of carbonmonoxide in gas mixtures. *Proc. K. Ned. Akad. Wet., Ser. C* **57**, 320–331 (1954).
57. van Kampen, E. J., and Zijlstra, W. G., Standardization of hemoglobinometry. II. The hemiglobincyanide method. *Clin. Chim. Acta* **6**, 538–544 (1961).
58. van Kampen, E. J., and Zijlstra, W. G., Determination of hemoglobin and its derivatives. *Adv. Clin. Chem.* **8**, 141–187 (1965).
59. Van Oudheusden, A. P. M., van de Heuvel, J. M., van Stekelenburg, G. J., Siertsema, L. H., and Wadman, S. K., De ijking van de hemoglobinebepaling op basis van ijzer. *Ned. Tijdschr. Geneeskd.* **108**, 265–267 (1964).
60. Vanzetti, G., An azidemethemoglobin method for hemoglobin determination in blood. *J. Lab. Clin. Med.* **67**, 116–126 (1966).
61. Vanzetti, G., and Franzini, C., Improved reagents and concentrated reference solutions in hemoglobinometry. *In* "Modern Concepts in Hematology" (G. Izak and S. M. Lewis, eds.), pp. 44–53. Academic Press, New York, 1972.
62. Weatherburn, M. V., and Logan, J. E., The effect of freezing on potassium ferricyanide-potassium cyanide reagent used in the cyanmethemoglobin procedure. *Clin. chim. Acta* **9**, 581–584 (1964).
63. Wootton, I. D. P., and Blevin, W. R., The extinction coefficient of cyanmethemoglobin. *Lancet* **1**, 434–436 (1964).
64. World Health Organization, WHO Expert Committee on Biological Standardization, 20th Report. *W. H. O. Tech. Rep. Ser.* **384**, 12–13 and 86–87 (1968).
65. Zijlstra, W. G., The photometric determination of the oxygen saturation of human blood by transmission and reflection techniques. *Proc. K. Ned. Akad. Wet., Ser. C* **56**, 598–608 (1953).
66. Zijlstra, W. G., Buursma, A., and Zwart, A., Molar absorptivities of human hemoglobin in the visible spectral range are independent of total hemoglobin concentration. *J. Appl. Physiol.: Respir., Environ. Exercise Physiol.* **54**, 1287–1291 (1983).
67. Zijlstra, W. G., and Mook, G. A., "Medical Reflection Photometry." Van Gorcum & Co., Assen, The Netherlands, 1962.
68. Zijlstra, W. G., and Muller, C. J., Spectrophotometry of solutions containing three components, with special reference to the simultaneous determination of carboxyhemoglobin and methemoglobin in human blood. *Clin. Chim. Acta* **2**, 237–245 (1957).
69. Zijlstra, W. G., Van Assendelft, O. W., Buursma, A., and van Kampen, E. J., The use of an ion-selective electrode for checking the CN^- content of reagent solutions used in the HiCN method. *In* "Modern Concept in Hematology" (G. Izak and S. M. Lewis, eds.), pp. 54–57. Academic Press, New York, 1972.
70. Zijlstra, W. G., Van Assendelft, O. W., and van Kampen, E. J., Standardization of haemoglobinometry. Spectral characteristics of the international haemiglobincyanide reference solution. *Proc. K. Ned. Akad. Wet., Ser. C* **72**, 238–241 (1969).
71. Zijlstra, W. G., and van Kampen, E. J., Standardization of hemoglobinometry. I. The extinction coefficient of hemiglobincyanide. *Clin. Chim. Acta* **5**, 719–726 (1960).
72. Zijlstra, W. G., and van Kampen, E. J., Standardization of hemoglobinometry. III. Preparation and use of a stable hemiglobincyanide standard. *Clin. Chim. Acta* **7**, 96–99 (1962).
73. Zijlstra, W. G., van Kampen, E. J., and Van Assendelft, O. W., Standardization of haemoglobinometry. Establishing the reference point. *Proc. K. Ned. Akad. Wet., Ser. C* **72**, 231–237 (1969).

74. Zwart, A., Buursma, A., Oeseburg, B., and Zijlstra, W. G., Determination of hemoglobin derivatives with the IL282 CO-Oximeter as compared with a manual spectrophotometric five-wavelength method. *Clin. Chem. (Winston-Salem, N.C.)* **27**, 1903–1907 (1981).
75. Zwart, A., Buursma, A., van Kampen, E. J., Oeseburg, B., van der Ploeg, P. H. W., and Zijlstra, W. G., A multi-wavelength spectrophotometric method for the simultaneous determination of five haemoglobin derivatives. *J. Clin. Chem. Clin. Biochem.* **19**, 457–463 (1981).
76. Zwart, A., Buursma, A., and Zijlstra, W. G., Determination of haemoglobin derivatives with the HP8450 A UV/VIS spectrophotometer. *In* "Methodology and Physiology of Blood Gases and pH" (B. Oeseburg and W. G. Zijlstra, eds.), pp. 127–138. Private Press, Groningen, 1982.
77. Zwart, A., Kwant, G., Oeseburg, B., and Zijlstra, W. G., Oxygen dissoiation curves for whole blood, recorded with an instrument that continuously measures pO_2 and S_{O_2} independently at constant t, pCO_2 and pH. *Clin. Chem. (Winston-Salem, N.C.)* **28**, 1287–1292 (1982).
78. Zweens, J., Frankena, H., and Zijlstra, W. G., Decomposition on freezing of reagents used in the ICSH-recommended method for determination of total haemoglobin in blood; its nature, cause and prevention. *Clin. Chim. Acta* **91**, 337–352 (1979).

THE CLINICAL CHEMISTRY OF OXALATE METABOLISM

M. F. Laker

Department of Clinical Biochemistry and Metabolic Medicine, Royal Victoria Infirmary, Newcastle upon Tyne, England

1. Introduction 259
2. Quantitative Analytical Methods 260
   2.1. Titration 261
   2.2. Colorimetry 261
   2.3. Fluorometry 262
   2.4. Enzymic Techniques 262
   2.5. Chromatography 266
   2.6. Isotachophoresis 267
   2.7. Indirect Methods 267
   2.8. *In Vivo* Isotope Dilution 267
   2.9. Specimen Collection 268
   2.10. Reference Ranges 268
3. Oxalate Metabolism 269
   3.1. Biosynthesis 269
   3.2. Oxalate Absorption 270
   3.3. Distribution and Excretion 271
4. Disorders of Oxalate Metabolism 271
   4.1. Primary Hyperoxaluria 271
   4.2. Increased Intake of Oxalate 276
   4.3. Excessive Intake of Oxalate Precursors 277
   4.4. Pyridoxine Deficiency 278
   4.5. Oxalate Metabolism in Gastrointestinal Disorders 278
   4.6. Oxalosis in Renal Failure 283
   4.7. Localized Oxalosis 284
   4.8. Oxalate Gallstones 284
5. Conclusions 285
   References 285

1. Introduction

Disorders of oxalate metabolism (Table I) which cause tissue deposition of insoluble calcium oxalate (oxalosis) or excessive excretion in urine (hyperox-

TABLE 1
Disorders of Oxalate Metabolism

Hyperoxaluria	Oxalosis
Congenital	Generalized
Primary hyperoxaluria types I and II	Hyperoxaluria
Acquired	Renal failure
Increased intake of oxalate	Localized
Increased intake of an oxalate precursor	*Aspergillus* infections
Ascorbic acid	Ocular disease
Ethylene glycol	
Methoxyflurane	
Xylitol	
Pyridoxine deficiency	
Enteric hyperoxaluria	

aluria) have been increasingly recognized in recent years. Primary hyperoxaluria was characterized as a distinct entity in the 1950s and the enzyme deficiencies responsible for the disorder were described in the 1960s. During the 1970s, hyperoxaluria complicating fat malabsorption was described and its pathogenesis investigated. It also became apparent that oxalosis was a complication of end-stage renal disease. Ascorbic acid, ethylene glycol, xylitol, and the anesthetic agent methoxyflurane have been shown to be precursors of oxalic acid in man. Localized oxalosis has been described as a complication of *Aspergillus* infections.

Studies of oxalate metabolism depend on reliable methodology. Numerous methods for measuring oxalate in biological fluids are available. However, many are technically demanding, time consuming, and do not appear suitable for the busy clinical laboratory. A review of these analytical methods together with a consideration of clinical disorders of oxalate metabolism forms the basis for the present article.

2. Quantitative Analytical Methods

Many methods have been described for estimating oxalate directly or for measuring a derivative of oxalic acid, including titration, colorimetric, fluorometric, enzymic, atomic absorption, and isotope dilution procedures. Recently, gas–liquid chromatography, high-performance liquid chromatography, and isotachophoresis have been utilized for oxalate measurement and enzymes have been immobilized in continuous-flow and electrode systems. Preliminary separation of oxalate from interfering substances is undertaken in many techniques, ion exchange, solvent extraction, precipitation, or a combination of these being used. This increases the complexity of the methods and [^{14}C]oxalic acid is often used as a recovery marker.

2.1. Titration

Titration with potassium permanganate or cerium(IV) sulfate has been used for determining oxalic acid concentrations.

Preliminary extraction with ether has been combined with permanganate titration to overcome poor analytical recovery caused by the presence of inorganic ions (P8). However, the technique is time consuming, and an 18-hour extraction with peroxide-free ether is necessary for complete recovery (Y1). Precipitation of calcium oxalate has been used as an alternative to ether extraction, the resulting precipitate being dissolved in sulfuric acid prior to titration (A10). Urinary values of 0.10–0.41 mmol/24 hours were obtained in 60 collections from six subjects. The procedure has been further simplified by precipitating at a higher pH and using fewer washings of the precipitate (D12).

Cerate titration with nitroferroin as the indicator has been adapted to urine analysis, the technique having been previously used for measuring oxalate in beer (K7, K8). Incomplete recovery was overcome by adding [^{14}C]oxalate with a standard amount of unlabeled oxalate and correcting for the amount not precipitated.

Disadvantages of titration include the need to isolate oxalate and the inability to do batch analysis.

2.2. Colorimetry

Several different colorimetric reactions have been adapted to oxalate analysis in biological fluids. The most widely used have been based on the reduction of oxalate to glycolate and its subsequent conversion to formaldehyde by heating with concentrated sulfuric acid. Formaldehyde reacts with 2,7-dihydroxynaphthalene (C1) or the more specific chromotropic acid (M2) to form a purple complex. The technique has been adapted to urine analysis by extracting continuously with ether, precipitating oxalate as its calcium salt, and using zinc rather than magnesium for reducing to glycolate (D8). Improved recovery and sensitivity were reported by omitting heating the urine with acid, extracting the filtrate obtained after acidification, with the addition of ammonium sulfate to urine, and shortening the period of reduction with zinc and sulfuric acid (H16).

Several modifications of this technique have been described, including the use of [^{14}C]oxalic acid to compensate for losses in extraction and precipitation (D5, E2, G9, H10, J5), omitting extraction (H10), extracting with tri-n-butyl phosphate (G9), precipitating oxalate with calcium sulfate and ethanol (H15), and isolating glycolate (H10) or oxalate (D1, J5, O4) by ion-exchange chromatography. Improved sensitivity may be obtained by modifying the zinc reduction procedure (H15), and using ion-exchange chromatography to isolate oxalate has enabled analysis in plasma to be undertaken (D1).

Dihydroxynaphthalene methods may be made highly specific, but at the cost of complexity. One such method involves extraction of urine with ether/ethanol, back extraction into an alkalinized aqueous phase, precipitation of calcium oxalate, reduction to glycolate, isolation of glycolate by ion-exchange chromatography, and colorimetric determination of glycolate after its elution from the column. Losses are compensated for by using [^{14}C]oxalic acid as a recovery marker (E2).

Methods based on other colorimetric reactions have been described, but these have not been as widely used as dihydroxynaphthalene procedures. Conversion of oxalate to formate and the reaction of formate with indole results in a colored complex and this method has been applied to urine (H6). Glycolate reacts with phenylhydrazine to form a product which may be oxidized to a red complex (P5). An automated procedure has been described utilizing the inhibitory effect of oxalate on the development of the red uranium(IV)–4-(2-pyridylazo)resorcinol complex (B1). Isotopically labeled oxalic acid was used to compensate for losses in precipitation.

2.3. Fluorometry

Oxalic acid has been estimated in blood and urine by extracting with tri-n-butyl phosphate, precipitating with calcium sulfate, and reducing to glyoxylic acid, which, when coupled with resorcinol, forms a highly fluorescent complex (Z2). Recoveries of 95–98% were obtained and the method was sensitive, detecting 0.9 μmol oxalic acid. However, the values for serum or plasma were considerably higher than those reported with isotope dilution, enzymic, or some gas–liquid chromatographic methods (Table 2).

2.4. Enzymic Techniques

Two enzymes have been used in analytical methods, oxalate decarboxylase (EC 4.1.1.2) and oxalate oxidase (EC 1.2.3.4). Oxalate decarboxylase, the more widely used enzyme, catalyzes the reaction:

$$(COOH)_2 \rightarrow CO_2 + HCOOH$$

TABLE 2
REFERENCE RANGES FOR PLASMA OXALATE

References	Number of subjects	Method	Reference range (μmol/liter)
Zarembski and Hodgkinson (Z2)	15	Fluorometric	22–36
Knowles and Hodgkinson (K4)	20	Enzymic	9–16
Hatch et al. (H4)	40	Enzymic	8–52
Ackay and Rose (A3)	7	Enzymic	0–5.4
Sugiura et al. (S21)	50	Enzymic	7–46
Kohlbecker and Butz (K11)	10	Enzymic	17–45
Bennett et al. (B10)	8	Radioenzymic	4–15
Chambers and Russell (C9)	20	GLC	62–260
Dagneaux et al. (D1)	20	GLC	13–28
Nuret and Offner (N2)	20	GLC	12–21
Gelot et al. (G3)	40	GLC	9–41
Wolthers and Hayer (W14)	22	GLC	0.6–5.0
Hodgkinson and Wilkinson (H14)	3	In vivo isotope dilution	1.3–1.6
Constable et al. (C13)	5	In vivo isotope dilution	0.8–1.4
Prenen et al. (P10)	10	In vivo isotope dilution	1.0–1.8

Oxalate oxidase, which has recently become commercially available, catalyzes the reaction:

$$(COOH)_2 + O_2 \rightarrow 2CO_2 + H_2O_2$$

2.4.1. Oxalate Decarboxylase

The first enzymic methods for determining oxalate in biological fluids used oxalate decarboxylase with measurement of liberated carbon dioxide. Crawhall and Watts (C17) described a method for plasma analysis which involved manometric estimation of carbon dioxide. However, it was not sufficiently sensitive to allow quantification of oxalate present in blood from normal subjects (less than 90 μmol/liter). A similar manometric determination of carbon dioxide has been described for urine analysis after first precipitating calcium oxalate and redissolving the precipitate in a citrate buffer. The presence of enzyme inhibitors was overcome by adding EDTA to the reaction mixture (M9). A more direct method has been described involving ultrafiltration of urine followed by concentration of the ultrafiltrate. Precipitation was omitted and the mean recovery for the technique was 99.7% (R4). Inhibition due to the presence of phosphate and sulfate in urine has been overcome by using increased amounts of enzyme, the carbon dioxide released being quantified by measuring the pH change in a trapping buffer after incubating for 18 hours (H2). This method has been adapted to plasma analyses, although the volumes of blood required are too large for it to be used routinely for this purpose (A3).

An automated method for determining oxalate in serum has been described which involved ultrafiltration, removal of endogenous carbon dioxide, and sampling by an AutoAnalyzer. The sample stream was mixed with buffer and enzyme and the liberated carbon dioxide was passed into an alkaline solution containing phenolphthalein, the color change being recorded automatically (K4). The method was also suitable for urine. Inhibition of the enzyme by sulfate and phosphate was corrected for by adding these ions to the standard solutions.

Enzyme-liberated carbon dioxide has been quantified by a conductometric technique (S1). Urine samples could be assayed directly after adjusting the pH to 3.0 and centrifuging to remove debris, although the method was more satisfactory if specimens were first extracted with chloroform. The procedure was linear over the range 0.5 to 3.0 μmol and was sensitive enough to detect 0.1 μmol.

Carbon dioxide released by oxalate decarboxylase has been quantified radioenzymically (B9, B10). Under the conditions used, a constant amount of $^{14}CO_2$ was evolved from a standard amount of [^{14}C]oxalic acid, trapped, and counted. If unlabeled oxalate from urine (B9) or plasma (B10) was present, a proportionately decreased amount of $^{14}CO_2$ was released. The initial preparation of samples was relatively complex, oxalate being precipitated, redissolved, extracted into ether, dried, and the residue dissolved in an acetate buffer.

An alternative approach is to measure formate produced by oxalate decarboxylase using a second enzyme, either formyltetrahydrofolate synthetase (EC 6.3.4.3) or formate dehydrogenase (EC 1.2.1.2). Formyltetrahydrofolate synthetase catalyzes the formation of N^{10}-formyltetrahydrofolic acid which is converted to 5,10-methenyltetrahydrofolate under acidic conditions. This sequence has been used to determine urinary oxalate by measuring the increase in absorbance at 350 nm (J3). Formate dehydrogenase catalyzes the conversion of formic acid to carbon dioxide with the reduction of NAD^+ to NADH. This has been adapted to the analysis of urine (C14) and serum (H4) by initially incubating specimens with oxalate decarboxylase at pH 3.0, then with formate dehydrogenase at pH 7.0. Purification of NAD^+ was necessary, as the commercially available coenzyme was contaminated with formate. Urine analysis was undertaken with a citrate extract, although direct estimation was possible. Ultrafiltration followed by precipitation and citrate extraction of the precipitate was used for plasma, with [^{14}C]oxalate being used as a recovery marker. The urine method has been adapted to a semiautomated procedure involving the addition of [^{14}C]oxalate, precipitation, extraction of the precipitate with an EDTA/citrate buffer, analysis in a continuous-flow system, and correction for the losses which occurred in preparation (Y5).

2.4.2. Oxalate Oxidase

The first analytical application of oxalate oxidase described the measurement of released carbon dioxide by the pH change which occurred after trapping the gas in a buffer (K12). The method was direct and, since 2 mol carbon dioxide per mol oxalate are produced by this enzyme, the method was more sensitive than decarboxylase procedures, the limit of detection being 10 nmol oxalate. However, the method was time consuming since it involved incubating for 18 hours.

The enzyme has been coupled with horseradish peroxidase (EC 1.11.1.7) and peroxide estimation undertaken by measuring oxidation of a chromophore (L2). In this method, interfering substances were removed by ion-exchange chromatography and adsorption with charcoal. A method using the same chromophore (3-methyl-2-benzothiazoline hydrazine and N,N-dimethylaniline) has been described for plasma (S21). The enzyme has been linked with catalase (EC 1.11.1.6) and aldehyde dehydrogenase (EC 1.2.1.5), $NADP^+$ reduction being measured. This method was rapid, sensitive, and suitable for use with plasma or urine specimens (K11).

2.4.3. Immobilized Enzymes

Techniques using immobilized enzymes have the great advantage of being economical in the use of expensive reagents. Oxalate oxidase has been immobilized in glutaraldehyde-activated tubing which was included in a continuous-flow system (B2). When the method was applied to urine, it was found that a preliminary precipitation stage was necessary to overcome the effects of an inhibitor which could not be removed by other means (P7).

Potentiometric sensors have been used to quantify carbon dioxide production with oxalate decarboxylase in free solution (Y2). More recently, an electrode has been described in which a thin layer of oxalate decarboxylase was chemically immobilized over a carbon dioxide sensor (K5). However, calibration curves were nonlinear and relatively large amounts of enzyme were required. A similar method has been reported in which linear calibration curves were obtained using less enzyme by altering the immobilization technique to entrapment of the enzyme within polyacrylamide gel and optimizing the analytical procedure (V1). Both electrode techniques were direct and a minimum of sample preparation was required for satisfactory results.

2.5. CHROMATOGRAPHY

Chromatographic procedures have the potential advantage of great specificity since the analytical procedures themselves involve separation of compounds. In addition, very sensitive detectors are available. Three types of chromatographic methods have been used to estimate oxalic acid concentrations: gas-phase chromatography (GC), high-performance liquid chromatography (HPLC), and ion-exchange chromatography.

2.5.1. *Gas-Phase Chromatography*

Oxalic acid is a polar, nonvolatile compound and thus derivatives must be formed for GC techniques. Methyl (C9, C11, D14, F4, M11, P2), ethyl (C10, D17, N2), propyl (D7, D20, G3), the electron capturing dichloroethyl (M17, T3), and trimethylsilyl (C8, V5, W14) derivatives of oxalic acid have been described. Many of these methods require preliminary preparative stages, and extraction, precipitation, and ion-exchange techniques are used for this purpose. However, several procedures have been described in which drying is the only step prior to derivatization other than the addition of internal standard (C10, D7, M11, M17). Packed columns using silicone or ethylene glycol stationary phases have been most frequently used, but porous polymer stationary phases (C9, C11, P2) and capillary columns (D17, W14) have also been utilized. Selective ion monitoring by gas chromatography–mass spectrometry has been described for the determination of oxalate pool size following the intravenous injection of [^{13}C]oxalic acid (D20). Several methods are sufficiently sensitive to enable blood levels to be estimated (C9, G3, M11, N2, W14).

2.5.2. *High-Performance Liquid Chromatography*

An HPLC method has been described which involved initial precipitation followed by separation of the redissolved oxalate on a strong cation-exchange column with acetic acid–sodium acetate–tetrabutyl ammonium–tetrafluoroborate as the mobile phase (M10). An electrochemical detector was used and quantitative data were obtained by using the standard addition technique. Other methods have involved using UV detectors after forming the strongly UV-absorbing 2,3-dihydroxyquinoxaline derivative (H21, M18). Oxalate was separated by reverse-phase chromatography either with (H21) or without (M18) gradient elution. Minimal sample preparation was required and oxalate was eluted within 10 minutes. A simple procedure has been described in which the sample was initially passed through a minicolumn followed by direct injection of the eluate onto the chromatograph (L4). Oxalate was separated by reverse-phase chromatography with tetrabutyl ammonium as the counterion. Gradient elution was not necessary, oxalate was eluted within 3 minutes, and the turnround time was 20 minutes/sample.

2.5.3. Ion-Exchange Chromatography

Oxalate may be measured by applying urine directly to an ion-exchange column and determining the conductivity, in 50-μl aliquots, of the eluate, taken every minute. The background conductivity due to the eluting electrolyte was compensated for by running a "suppressor" column in addition to the oxalate column (M3). Oxalate was eluted within 30 minutes.

2.6. Isotachophoresis

It has been demonstrated that oxalate may be separated from other ions by isotachophoresis in PTFE capillaries and detected by UV absorption at 254 nm (S8). Preliminary precipitation was used, with [^{14}C]oxalic acid being included as a recovery marker. The method appeared sensitive, precise, and free from interference, although the quantitative characteristics of the procedure were not fully evaluated.

2.7. Indirect Methods

Indirect methods involve precipitating oxalate with a calcium salt and then measuring either the calcium remaining in the supernate or the calcium present in the precipitate. Supernate calcium has been determined by titration (G11), colorimetrically by continuous-flow analysis (S4), or by atomic absorption spectroscopy (K10, M12). Precipitated calcium has been measured by atomic absorption spectroscopy (F11). In the latter study, low recovery was overcome by adding a known amount of sodium oxalate before precipitating with calcium.

2.8. In Vivo Isotope Dilution

The use of isotopes as recovery markers and in gas chromatography–mass spectrometry methods has been referred to. Isotope dilution techniques have also been used to determine plasma concentrations of oxalic acid following intravenous infusion of [^{14}C]oxalate (C13, H14, P9, P10, W7). In these studies, ^{14}C activities have been measured in plasma and urine, the concentration of oxalate in urine determined by a colorimetric, titrimetric, or enzymic technique, and the concentration of plasma oxalate calculated from the specific activity of oxalate in urine.

2.9. Specimen Collection

An investigation of the conditions necessary to avoid loss of calcium oxalate in urine collections has been described (H13). An acid concentration of 0.1 mol/liter was required to ensure the dissolution of all calcium oxalate crystals present in urine. The conditions for taking blood specimens have also been investigated (A3), plasma oxalate levels being determined with and without the addition to the samples of inhibitors of glyoxylate metabolism. Values without added inhibitors were higher than when inhibitors had been added, and glyoxylate added to blood resulted in high oxalate levels. It was concluded that spuriously high plasma oxalate concentrations would occur if *in vitro* glyoxylate metabolism was not inhibited. Other investigators have found blood specimens to be stable if plasma was separated from the cells within 1 hour. Plasma was stable for 3 hours, but generation of oxalate occurred if plasma was left standing at room temperature for 24 hours (W14).

2.10. Reference Ranges

Published reference ranges for urinary and plasma oxalate concentrations are given in Tables 2 and 3. Excluding indirect techniques, most methods give similar reference ranges for urine; the mean values for each type of method are colorimetric, 128–469 μmol/24 hours; enzymic, 138–451 μmol/24 hours; GLC, 109–500 μmol/24 hours; and indirect, 15.5–324 μmol/24 hours. Urinary excretion of oxalate increases during childhood, reaching adult values at about the age of 14 (G9, H15). If urinary excretion in children is expressed per unit of body surface area and corrected to adult standard body surface area (1.73 m^2), the results for children lie in the same range as those obtained for adults (G9).

Although there is a measure of agreement for urinary reference ranges, there are very wide discrepencies in ranges for plasma. The lowest concentrations (approximately 1 μmol/liter) have been obtained with *in vivo* isotopic methods and, in some studies with other techniques, concentrations 100 times greater than this have been reported. Since the values in plasma are very much lower than those in urine, specificity and recovery are more critical. The greatest range in plasma levels has been obtained with GLC methods, where the discrepancies can be attributed, in part, to poorly resolving columns operating under insensitive conditions. When capillary columns with greater resolution and higher sensitivity have been used, reference ranges of the same order as *in vivo* isotopic techniques have been obtained (W14). Similar values have also been obtained with an enzymic method in which sensitivity was increased by using large quantities of blood (A3).

TABLE 3
REFERENCE RANGES FOR URINARY OXALATE[a]

References	Number of subjects	Method	Reference range (μmol/24 hours)
Dempsey et al. (D8)	33	Colorimetric	167–555
Hodgkinson and Zarembski (H16)	39	Colorimetric	100–264
Hodgkinson and Williams (H15)	22	Colorimetric	189–522
Olthuis et al. (O4)	25	Colorimetric	126–505
Dobbins and Binder (D16)	15	Colorimetric	56–500
Zarembski and Hodgkinson (Z2)	60	Fluorometric	100–317
Mayer et al. (M9)	25	Enzymic	52–407
Hallson and Rose (H2)	22	Enzymic	143–459
Costello et al. (C14)	22	Enzymic	156–418
Yriberri and Posen (Y5)	22	Enzymic	200–520
Farrington and Chambers (F4)	150	GLC	80–500
Gelot et al. (G3)	121	GLC	100–488
Decoux et al. (D7)	18	GLC	44–577
Di Corcia et al. (D14)	15	GLC	188–859
Charransol et al. (C10)	34	GLC	111–500
Park and Gregory (P2)	18	GLC	100–255
Wolthers and Heyer (W14)	20	GLC	140–320
Hughes et al. (H21)	23	HPLC	184–616
Fraser and Campbell (F10)	11	Indirect; atomic absorption	40–143
Menarche (M12)	30	Indirect; atomic absorption	0–444
Kochl and Abecassus (K10)	30	Indirect; atomic absorption	22–422
Giterson et al. (G11)	25	Indirect; calcium titration	0–286

[a] Studies with values on more than 10 healthy subjects have been included.

The effect of inhibitors of glyoxylate metabolism on plasma levels of oxalate is discussed in Section 2.9.

3. Oxalate Metabolism

3.1. BIOSYNTHESIS

The two major endogenous precursors of oxalate in man are glycine and ascorbate (Fig. 1). Although less than 0.5% of glycine is converted to oxalate (E4), experiments with isotopically labeled glycine have shown that 40% of urinary oxalate may originate from this source (C16). Glycine is converted to oxalate via glyoxylate, either directly by oxidative deamination or transamination, or indirectly, via serine and glycolate (H11). Glyoxylate is an active intermediate which is involved in several metabolic pathways. It is

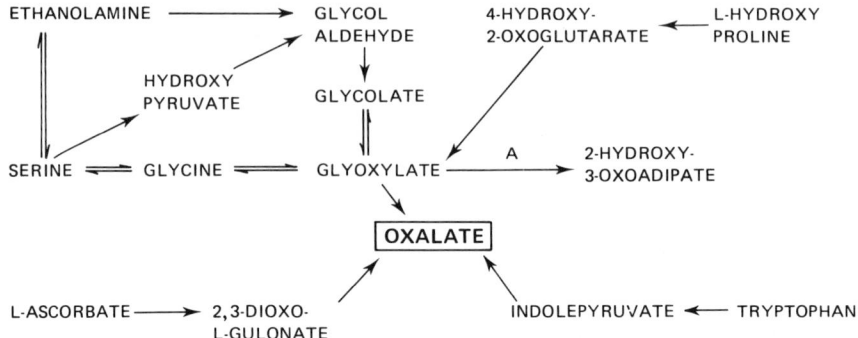

FIG. 1. Outline of oxalate metabolism in man. The reaction (A), the synergistic decarboxylation of glyoxalate and 2-oxoglutarate, is catalyzed by 2-oxoglutarate:glyoxylate carboligase, the cytosolic form of which is deficient in type I hyperoxaluria. This leads to increased conversion of glyoxylate to oxalate and glycolate.

converted to oxalate by three enzymes, glycolic acid oxidase (EC 1.1.3.1), lactate dehydrogenase (EC 1.1.1.27), and xanthine oxidase (EC 1.2.3.2), which is probably the least important of the three (G8). Conversion to glycine and carbon dioxide are the major routes of metabolism of glyoxylate.

Ascorbate metabolism is normally responsible for 20–40% of urinary oxalate (A12, B3) and 25–65% of ascorbate is metabolized to oxalate (B4, M8). Oxalate is derived from C-1 and C-2 of ascorbic acid (B7, H8). Other quantitatively less important precursors of oxalate include tryptophan, phenylalanine (F14), and hydroxyproline (R3).

A fuller account of the biosynthesis of oxalate is given in the reviews by Hodgkinson (H11) and Williams and Smith (W11).

3.2. Oxalate Absorption

The oxalate content of foods varies considerably. Spinach, rhubarb, and beetroot have a very high content (greater than 2.5 mmol/100 g) and moderate amounts are found in coffee, tea, chocolate, nuts, and parsley (0.5–2.5 mmol/100 g). Dairy produce, meat, fish, cereals, and, surprisingly, strawberries contain relatively little oxalate (K1, Z1). Many factors affect the oxalate content of food, including the season, soil conditions, method of preparing, part of the plant, and its age. Various dietary factors, including calcium content of the food, will affect bioavailability of oxalate (B19, M6).

The amount of dietary oxalate appearing in the urine is 2–7% of that ingested (A10, C5, E2), although absorption is greater if test meals are taken without additional food (C5). Oxalate absorption increases linearly with increasing concentration and is not inhibited by dinitrophenol or ouabain; thus, absorption is thought to be by passive, nonmediated diffusion. Absorption can occur throughout the small bowel and in the colon (B11). However, it has recently been suggested that the results of these transport studies may have been distorted by an omission of calcium from perfusing fluid, and the possibility of an active component to oxalate absorption has not been ruled out (F12).

3.3. Distribution and Excretion

If [^{14}C]oxalate is injected intravenously, more than 95% is recovered in the urine within 36 hours and no radioactivity is detected in breath carbon dioxide. Thus, oxalic acid is a metabolic end product in man and is excreted unchanged (E4, W7). Miscible pool size in normal adults is 20–75 μmol and the volume of distribution is approximately 40–45% of body weight (C13, E4, H7, H14). Oxalate clearance is greater than inulin or creatinine clearance, indicating that tubular secretion occurs (C13, H7, H14, W7). In animal studies, tubular secretion can be inhibited by various drugs (C4, K3).

4. Disorders of Oxalate Metabolism

Oxalic acid is a relatively strong acid ($pK_a' = 1.23$) and, if taken in large amounts, is locally corrosive. Other pathological effects are caused by the limited solubility of calcium oxalate, with deposits occurring in tissues in oxalosis and the urinary tract in hyperoxaluria. A detailed consideration of the factors involved in urinary tract stone formation is beyond the scope of the present article, since it would involve inhibitors of stone formation such as pyrophosphate and glycoproteins in addition to the constituents of calculi such as calcium and oxalate. However, in addition to stones occurring in patients with overt hyperoxaluria, it is recognized that urinary oxalate concentrations are somewhat higher in patients with idiopathic calcium oxalate stone formation than in subjects without stones, and that increased oxalate excretion is more common than increased calcium excretion (R5, W2). Increased absorption of dietary oxalate has been implicated in some of these patients (H12, M5).

Oxalate has been reported to interfere with glucose metabolism (J6, Y4) and to inhibit several enzymes, including pyruvate kinase (B20, R2) and lactate dehydrogenase (E6). The significance of these effects in intact animals is unclear.

4.1. Primary Hyperoxaluria

There are two recognized genetic variants of glyoxylate metabolism which cause the clinical syndrome of primary hyperoxaluria. Both are characterized

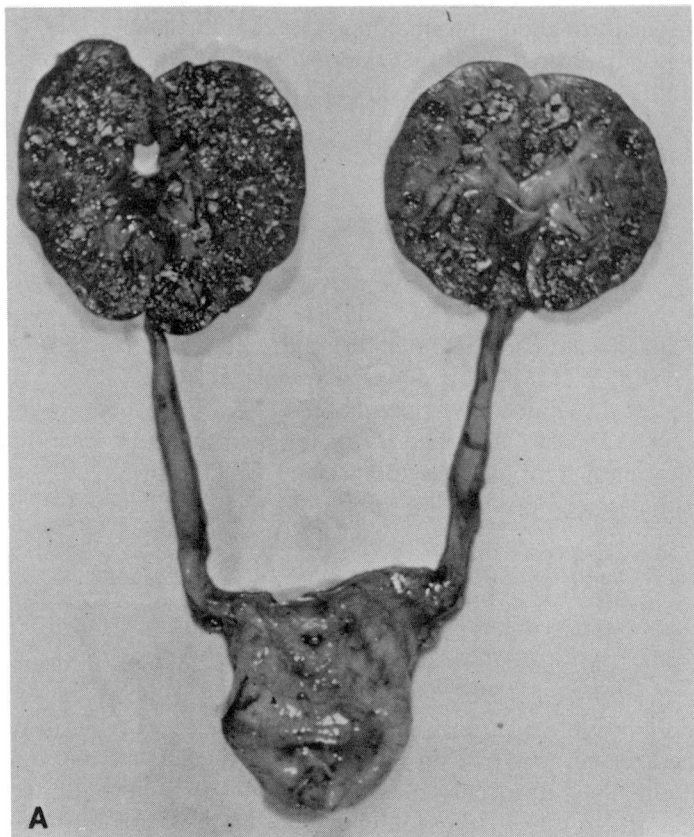

Fig. 2. (A) Kidneys and bladder from a patient with primary hyperoxaluria. The kidneys are small, scarred, and were gritty when cut. (B) Postmortem X ray of the cut left kidney showing extensive nephrocalcinosis. Chemical analysis of the kidney demonstrated oxalate deposits. [Case 1 of Walls et al. (W3), reproduced with permission.]

by overproduction of oxalic acid, leading to hyperoxaluria and oxalosis (H9, W8, W11). The first reported case was probably described in 1925 by Lepontre (L5), who described a 4½-year-old child with urolithiasis and nephrocalcinosis due to calcium oxalate. The disease has been increasingly recognized since the first detailed report appeared in 1950 (D4, A8).

4.1.1. Clinical Features

The cases reported between 1950 and 1964 have been reviewed by Hockaday et al. (H9). Patients were divided into three groups: those who showed the clinical, biochemical, and pathological features of primary hyperoxaluria (63 cases), sibs of typical cases who were also thought to have inherited the

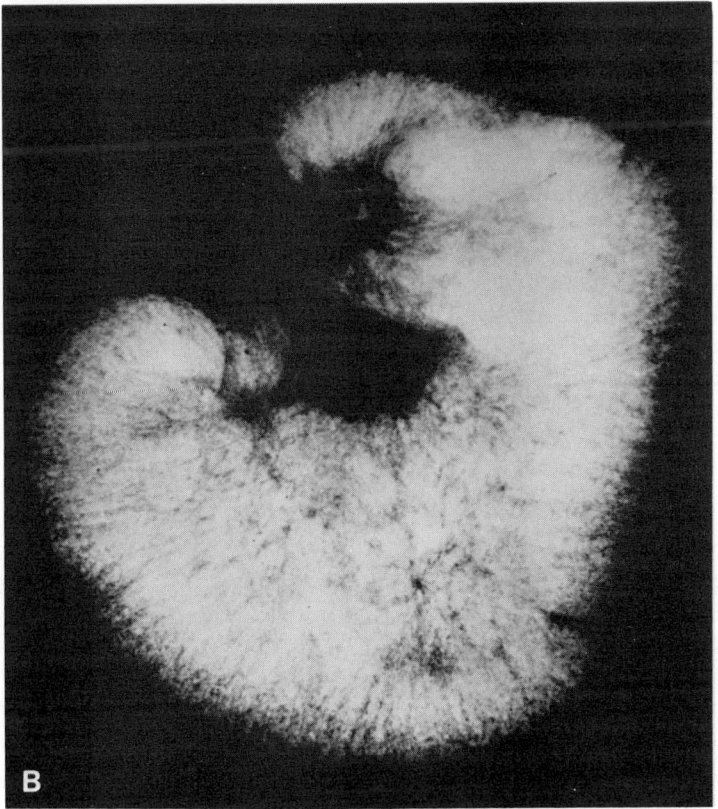

Fig. 2B.

disease (14), and atypical cases (42). Typically, the disease presents in early childhood, although atypical patients have been described who presented in the sixth decade of life. Most present with symptoms attributable to urinary tract stones, either renal colic or hematuria. There is usually progression to chronic renal failure and over 80% of patients die of this complication before they reach the age of 20. However, some patients with typical features of the disease may survive for more than 20 years. Occasionally, patients may present with acute renal failure (B16, O2). Other features include heart block (C12, M7, W5), peripheral neuropathy (H1, M15, S16), Raynaud's phenomenon (W3), and peripheral gangrene (A9, B12). Bone disease is being increasingly recognized and pathological fractures may occur, particularly in patients being treated by dialysis (A1, B14, B15). Urinary excretion of oxalate is high, often exceeding 3.0 mmol/24 hours. However, hyperoxaluria may diminish as chronic renal failure develops, causing oxalate retention to occur (W11).

4.1.2. Pathological Features

Renal damage results from a combination of nephrolithiasis, nephrocalcinosis, and interstitial nephritis. At postmortem, the kidneys are typically small, scarred, and have a granular appearance (Fig. 2). They are gritty when cut. Crystals of oxalic acid occur in the tubules, causing tubular atrophy, and may also be seen in the interstitial tissue. Interstitial fibrosis and inflammation are common (H9, W8). It has been suggested that deposition of oxalate in the kidneys may occur relatively late in the course of the disease, since crystals were not demonstrated in renal biopsy specimens from two patients 6 and $1\frac{1}{2}$ years, respectively, prior to autopsy, which revealed extensive deposits (S7, W3).

Extrarenal deposits have been noted in many tissues, including the myofibrils and conducting tissue of the heart (C12, L7, W3), blood vessels (S7, W3), central nervous system (H3, S7), retina (F6), and many other tissues (W8).

4.1.3. Pathogenesis

Two genetic variants of primary hyperoxaluria are recognized, types I and II. Both are thought to be inherited as autosomal recessive traits.

It has been demonstrated by *in vivo* isotope techniques that the metabolism of glyoxylate is abnormal in type I hyperoxaluria (F11, H9). Conversion of glyoxylate to $^{14}CO_2$ is reduced, while there is increased conversion to oxalic and glycolic acids. This is thought to be due to decreased conversion of glyoxylate to 2-hydroxy-3-oxoadipic acid, caused by a deficiency of cytosolic 2-oxoglutarate:glyoxylate carboligase (Fig. 1). This enzyme deficiency has been demonstrated in liver, spleen, and kidney tissue obtained at operation from four patients and postmortem tissue from one patient (K9). The activity of the mitochondrial form of the enzyme is normal (C18, K9). There may be heterogeneity of this variant of hyperoxaluria, since normal cytosolic and mitochondrial 2-oxoglutarate:glyoxylate carboligase activities have been reported in skeletal muscle from a 10-year-old girl with increased glycolate excretion and primary hyperoxaluria (B13).

Patients with type II primary hyperoxaluria have increased excretion of L-glyceric acid in addition to oxalic acid (W9). They have a deficiency of D-glycerate dehydrogenase (EC 1.1.1.29) which has been demonstrated in leukocytes (W8). This enzyme catalyzes the interconversion of hydroxypyruvate and D-glycerate; in its absence, hydroxypyruvate is thought to stimulate the oxidation of glyoxylate to oxalate by a coupled reaction catalyzed by lactate dehydrogenase (W10). This effect of hydroxypyruvate apparently results from the increased oxidation of the LDH–NADH complex secondary to the reduction of hydroxypyruvate to L-glycerate (Fig. 3).

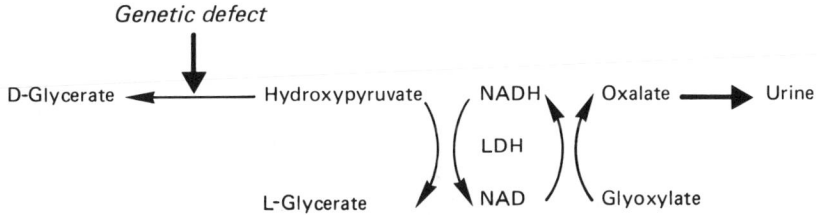

FIG. 3. Biochemical defect in type II primary hyperoxaluria. In the absence of D-glycerate dehydrogenase, hydroxypyruvate is converted to L-glycerate with the linked reduction of glyoxylate to oxalate.

Other aspects of oxalate metabolism have been investigated in primary hyperoxaluria, but without demonstrating abnormalities. Thus, the fractional conversion of ascorbic acid (A12) and glycine (C15) to oxalate are normal, as are the absorption (A11) and renal clearance (C13, W7) of oxalic acid.

4.1.4. *Treatment*

There is no specific treatment for primary hyperoxaluria and therapeutic measures are directed at reducing oxalate synthesis, preventing renal tract stone formation and growth, and treating chronic renal failure.

Pyridoxine therapy has been widely used to reduce oxalate synthesis, since hyperoxaluria accompanies vitamin B_6 deficiency in experimental animals (G5) and man (F1). Doses of 100–150 mg/day have been reported to reduce the daily mean oxalate excretion in four patients (S14). Not all patients respond to these amounts and up to 600 mg/day has been used to induce a response by Gibbs and Watts (G7). Although only half of their patients responded to these high doses, a therapeutic response was maintained for long periods in those who did. In these studies, urinary oxalate was not normalized by pyridoxine, but doses of up to 1 g/day have been reported to reduce oxalate excretion by 60–70%, the value falling to within the reference range for one of the two patients studied (W10). In the latter study, a therapeutic response was maintained for 20 and 30 months, respectively, in the two subjects. Although pyridoxine therapy may be effective, the typical response of urinary metabolites to oral tryptophan loading, as seen in vitamin B_6 deficiency, is absent from patients with primary hyperoxaluria (G10).

Sodium benzoate has been administered to trap glycine as hippurate and prevent its conversion to oxalate. However, less than 0.5% of glycine is converted to oxalate in patients with primary hyperoxaluria (E4), and although responses have been reported, they are temporary, escape soon occurring (A11, D2).

Various enzyme inhibitors have been investigated. Calcium carbimide, an inhibitor of aldehyde dehydrogenase, has been reported to reduce oxalate excretion to normal in one subject (S15). However, this has not been confirmed and it was suggested that the apparent decrease in oxalate excretion was due to inhibition of oxalate decarboxylase, which was used for oxalate analysis (Z3). Disulfiram (Antabuse), also an inhibitor of aldehyde dehydrogenase, and allopurinol, an inhibitor of xanthine oxidase, do not affect oxalate excretion (G8). Several *in vitro* inhibitors of oxalate production from glyoxylate, such as sulfonates and succinamide, have been investigated, but have not been shown to be effective *in vivo* (W4, W11). The monoamine oxidase inhibitor isocarboxazid has been reported to reduce oxalate synthesis by 40% in one patient (B13) and tyrosine has also been reported to have a therapeutic effect (Z7).

Magnesium oxide and phosphates are sometimes used in an attempt to prevent urinary tract stone formation (D9, L8). Methylene blue is also administered since it has been shown to inhibit oxalate crystal formation in urine (S22).

Oxalate is readily dialyzable, but the rate of removal by peritoneal dialysis and hemodialysis does not keep pace with production and fails to halt the progress of the disease (B12, J2, K6, W3, Z5). However, survival up to 102 months has been reported in patients treated with hemodialysis (J1). Early experience with transplantation was not encouraging since deposition of oxalate crystals occurred in the transplanted kidney as early as the seventeenth postoperative day (D10, S6, S15). Recently, more encouraging reports have appeared, particularly where transplantation has been combined with pyridoxine, magnesium oxide, and methylene blue therapy, and dietary control of oxalate-rich foods has been exerted (L6, M16, O3). In three cases, graft function has remained stable for over 2 years.

4.2. Increased Intake of Oxalate

The effects of ingesting oxalic acid have been reviewed by Jeghers and Murphy (J4). Symptoms of acute poisoning include sour taste and burning of the mouth, pharynx, and stomach. Retching and vomiting may occur and can be severe. Systemic effects include lowering of serum calcium and renal impairment. Acute poisoning develops within a few minutes to several hours after ingesting the acid. The range of lethal doses is extremely wide, values between 2 and 30 g of oxalic acid having been reported. Although oxalic acid poisoning used to be relatively common, its incidence has declined since its use in cleaning fluids has decreased (H11). Ingestion of foods with high oxalate contents causes urinary oxalate excretion to increase significantly (F7).

4.3. Excessive Intake of Oxalate Precursors

Excessive intake of several oxalate precursors may lead to hyperoxaluria; ascorbic acid, ethylene glycol, methoxyflurane, and xylitol have all been implicated. Oral ingestion of glycine does not cause increased urinary oxalate (B21), presumably because of the small fraction metabolized by this pathway. However, a report has recently appeared of hyperoxaluria occurring in three patients following bladder irrigation with a glycine-containing solution during or shortly after transurethral prostatectomy (F8).

4.3.1. Ascorbic Acid

Two types of response to ingesting large amounts of ascorbic acid have been described: a modest enhancement in urinary oxalate affecting many subjects and a marked increase which affects relatively few. Thus, the ingestion of an extra 4 g/day vitamin C caused an average increase of 0.13 mmol/day in urinary oxalate, while an extra 9 g/day resulted in an average daily increase of 0.75 mmol oxalate excretion (L3). This latter figure may be excessive, since in a more recent study, the administration of 10 g ascorbate to four subjects caused the mean oxalate excretion to rise from 0.55 to 1.00 mmol/day (S9). Oxalate excretion returned to pretreatment levels within 24 hours of reverting to normal vitamin C intakes.

The administration of ascorbic acid may induce a 10-fold increase in urinary oxalate in susceptible subjects (B17, B18). In these studies, three subjects, including two members of one family, responded in this manner, suggesting there may be an inherited variant of ascorbic acid metabolism.

4.3.2. Ethylene Glycol

Ethylene glycol is widely used as an antifreeze agent and is sometimes ingested as a substitute for ethanol or for attempted suicide. It is metabolized to carbon dioxide, is excreted unchanged in the urine, and a variable amount is converted to oxalic acid via glycol aldehyde and glyoxylic acid. The proportion converted to oxalic acid varies from species to species (G6). Ethylene glycol is nontoxic and the main clinical features of ingestion, metabolic acidosis, central nervous system disorders, and acute renal failure, result from its metabolism to oxalic acid (P3). Since the initial stage in the conversion of ethylene glycol to oxalate is catalyzed by alcohol dehydrogenase, ethanol has been administered in cases of poisoning as an alternative substrate for the enzyme (P3, W1). If the metabolism of ethylene glycol is thus blocked, more is excreted unchanged.

4.3.3. Methoxyflurane

Impaired renal function and oxalate deposits in the kidney have been noted after administration of the anesthetic agent methoxyflurane, but not following halothane (F9, M8, P1, S10). Urinary oxalate excretion has been noted to increase 10-fold on the first postoperative day and to fall progressively over the next 7 days (F9). Methoxyflurane is metabolized to oxalic acid and fluoride is released: both factors have been implicated in methoxyflurane-induced oxalosis and oxaluria (M8, W12).

4.3.4. *Xylitol*

Several adverse reactions have been reported in patients receiving intravenous xylitol as an energy source in parenteral nutrition regimes, including disturbances of renal function and tissue deposits of calcium oxalate, particularly in the kidney and arteries of the brain (S7, T1). Three factors have been implicated in the pathogenesis of xylitol-induced oxalosis, namely, pyridoxine deficiency, barbiturate administration, and impaired renal function. Thus, in investigations of oxalate metabolism in vitamin B_6-deficient rats, urinary excretion of oxalic and glyoxylic acids was increased following infusion with xylitol. Correction of vitamin B_6 deficiency resulted in a return to normal of urinary oxalate excretion and of the percentage of [^{14}C]xylitol converted to [^{14}C]oxalate (T2). Phenobarbitone pretreatment has been shown to result in a two- to threefold increase in urinary [^{14}C]oxalate excretion following the intraperitoneal injection of [^{14}C]xylitol or [^{14}C]glycolate in rats (R6). However, in studies on postoperative patients receiving intravenous xylitol, some of whom showed evidence of pyridoxine deficiency, Watts and colleagues (C7, H5) were unable to demonstrate greater than normal urinary oxalate excretion. Examination of urinary organic acids by gas chromatography–mass spectrometry demonstrated abnormal glycolic aciduria. They concluded that another factor, possibly preexisting impaired renal function, was necessary for xylitol-induced oxalosis.

4.4. PYRIDOXINE DEFICIENCY

Hyperoxaluria occurs in pyridoxine deficiency in experimental animals (G5) and man (F1). The mechanism is unclear, although altered metabolism of a number of precursors of oxalic acid has been reported (R8, T2).

Pyridoxine supplements are effective in reducing oxalate excretion in patients with primary hyperoxaluria, although the mode of action of the vitamin is not understood (see Section 4.1.4).

4.5. OXALATE METABOLISM IN GASTROINTESTINAL DISORDERS

4.5.1. *Enteric (Absorptive) Hyperoxaluria*

It has been recognized for many years that urolithiasis occurs as a complication of small intestinal disease. Thus, Deren *et al.* (D11), in reviewing 583 patients with ulcerative colitis or regional enteritis, found that urolithiasis was twice as frequent in patients with Crohn's disease (6.3%) as in those with ulcerative colitis (3.4%). In another large series of patients with inflammatory bowel disease, it was noted that stones were radiopaque in 81% of patients in whom they occurred (G4). An extremely high incidence of calcium oxalate stones (33%) has been reported in patients within 6 months of undergoing jejunoileal bypass for the treatment of morbid obesity (D13, O1).

Hofmann, Smith, and colleagues at the Mayo Clinic first noted the association between hyperoxaluria, ileal disease, and urolithiasis (H19, S13), and this association has been confirmed by others (A2, D18). In addition to Crohn's disease and jejunoileal bypass, hyperoxaluria may occur in other diseases, including radiation enteritis (D18), celiac disease, bacterial overgrowth, hepatobiliary disorders (S12), and in patients with ileal resection (A2, D18, S12), in whom the risk of developing hyperoxaluria is related to the length of bowel removed (E2, S19). Increased urinary excretion of oxalate also occurs in children following ileal resection (V3).

Enteric hyperoxaluria was initially thought to result from disturbed bile acid metabolism. It was hypothesized that the malabsorption of bile acids which occurs in these conditions leads to increased deconjugation with release of large amounts of glycine. Enteric bacteria then convert glycine to glyoxylate, which is absorbed, converted to oxalate, and excreted by the kidney, leading to hyperoxaluria (H19, S13). This hypothesis and variants of it were disproved by the same group when they failed to demonstrate that the label from cholyl[1-^{14}C]glycine was incorporated into urinary oxalate in patients with enteric hyperoxaluria (H18). Similar results were reported by Chadwick et al. (C5), who also measured [^{14}C]oxalate excretion following oral administration of isotopically labeled oxalate with either unlabeled sodium oxalate or with a meal. If sodium oxalate was given to fasting subjects, mean oxalate absorption in patients with ileal disease and hyperoxaluria was approximately twice that found in control subjects and patients with ileal resection but without hyperoxaluria. When the test meal was given with food, hyperoxaluric patients absorbed approximately four times as much oxalate as nonhyperoxaluric subjects. Excretion decreased to normal within 24 hours of starting a low-oxalate diet and returned to high values when a normal diet was resumed. Thus, enteric hyperoxaluria is due to enhanced absorption, a finding that has been confirmed by other groups (E2, S19).

The colon has been identified as the site of enhanced oxalate absorption. In one study, hyperoxaluria was demonstrated in patients with ileal resection and an intact colon, but urinary oxalate excretion was normal in patients with ileal resection and a colectomy (E2). These findings were confirmed and extended by Dobbins and Binder (D16), who found hyperoxaluria in eight patients with steatorrhea and intact colons, but normal oxalate excretion in five patients with steatorrhea and ileostomies. Absorption of [^{14}C]oxalate was enhanced in the former group, but normal in the latter. The amount absorbed has been shown to be similar if oxalate is administered orally or instilled into the cecum of bypass patients (E1).

Two factors have been implicated in enhanced colonic absorption of oxalate, namely, malabsorbed fatty acids and bile acids. Fatty acids have been thought to be involved since it was observed that the degree of hyperoxaluria directly correlated with the severity of steatorrhea (E2). Addition of fat to an otherwise constant diet increased urinary oxalate excretion in patients with ileal resection who were already hyperoxaluric, and the effect was most pronounced in patients with extensive ileal resection. Under these conditions, mean oxalate excretion increased from 2.2 to 3.5 mmol/day. It was suggested that malabsorbed fatty acids decrease the amount of intraluminal calcium available in the colon for binding oxalate, which thus remains in solution (Fig. 4). In this soluble form, oxalate is absorbed and excreted by the kidney, causing hyperoxaluria. The solubility hypothesis has received support from *in vitro* studies in which adding sodium oleate to solutions of sodium oxalate reduced precipitation of oxalate or, conversely, increased oxalate solubility when calcium chloride was added (B11).

The role of fat in the pathogenesis of hyperoxaluria has been supported by several clinical studies. Reducing the saturated-fat content of the diet reduces urinary oxalate excretion in patients with ileal disease (A7). Oxalate excretion in patients with steatorrhea due to celiac disease is reduced if fat excretion is decreased by treatment with a gluten-free diet (M1). Hyperoxaluria occurs in steatorrhea in the absence of intestinal disease in pancreatic disorders (A6, S17). Indeed, so predictable is the relationship between steatorrhea and hyperoxaluria in patients with an intact colon that measuring urinary oxalate excretion after an oral oxalate load has been suggested as a screening test for fat malabsorption (A6, R1).

It has been demonstrated that different fatty acids vary in the extent to which they enhance oxalate absorption in experimental animals (D15). Ricinoleic acid caused a greater enhancement than oleic acid of oxalate absorption by the rat colon; octanoic acid had no effect. Those compounds that increased oxalate absorption were also associated with increased water secretion, suggesting that the effect of fatty acids in enhancing oxalate absorption could, at least in part, be due to an effect on colonic permeability.

The effect of bile acids on oxalate absorption has been studied in rat and rhesus monkey (C6). Bile acids caused increased oxalate absorption in the isolated, perfused colon in rat and after infusion directly into the colon of monkeys with colostomies. These findings have been confirmed in rats and it has been suggested that bile acids exert their effects by inducing increased colonic permeability to oxalate (D15). Bile acids have been shown to have an effect in man by Fairclough *et al.* (F2), who investigated oxalate absorption

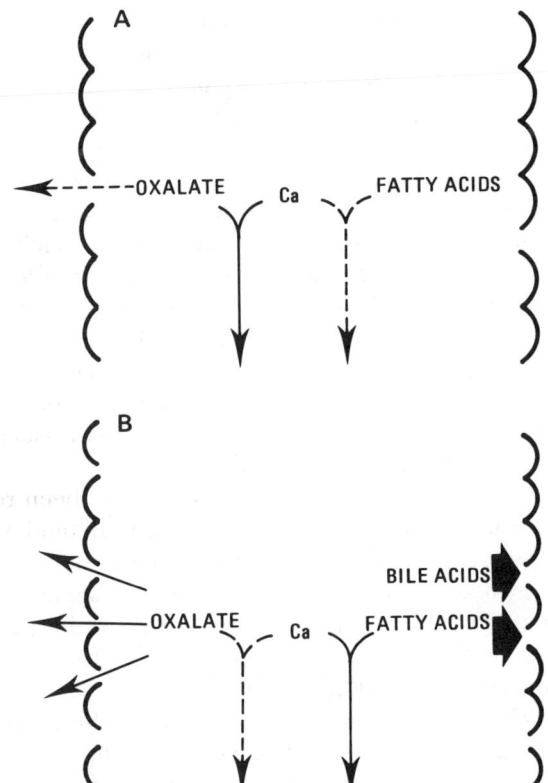

FIG. 4. Oxalate absorption from the colon. (A) Normal. There is little fat in the lumen of the colon and free calcium binds to oxalate, rendering it insoluble and preventing its absorption. (B) Fat malabsorption causing enteric hyperoxaluria. Possible mechanisms include (1) fatty acids and bile acids affect the colonic mucosa, rendering it more permeable to oxalate, and (2) excess fatty acids in the colonic lumen preferentially bind calcium, thus enhancing oxalate solubility. As a result, increased amounts of oxalate are absorbed.

from surgically excluded colons in two patients with chronic liver disease. Oxalate absorption was increased fivefold when sodium chenodeoxycholate was incorporated into the perfusion solutions.

Thus, bile acids and fatty acids have both been implicated in the pathogenesis of hyperoxaluria. The solubility and permeability theories of enhanced absorption are not mutually incompatible and may both be implicated at the same time (S18).

Treatment of enteric hyperoxaluria has been directed at dietary control and the administration of agents which sequester oxalate in the lumen of the gut. Reduction in urinary oxalate excretion with low-oxalate diets has been demonstrated (C5, E2, S19). However, such treatment is not altogether satisfactory, partly because it is difficult to define a palatable low-oxalate diet and partly because such diets do not always abolish hyperoxaluria (H17, H20). This latter finding suggests that an endogenous component or a dietary precursor of oxalic acid may contribute to enteric hyperoxaluria. Reduction in dietary fat has been suggested as treatment (A5, A7), but it may be difficult to abolish steatorrhea by this means in these patients.

An alternative approach is to increase the intraluminal concentration of calcium in the colon and thus sequester oxalate as the calcium salt. It has been noted that urinary oxalate excretion decreased as dietary intake of calcium was increased from 250 to 3000 mg/day (S17). Other studies have similarly shown reduction in oxalate excretion using calcium therapy (C3, H22, P4, S20). However, even at high levels of calcium intake, oxalate excretion is not normalized, possibly because delivery of calcium to the colon is inadequate. Direct perfusion of the colon with calcium has been shown to greatly reduce enteric hyperoxaluria (M14).

The bile acid-sequestering resin cholestyramine has been reported to be an effective therapeutic agent (S12, S19). The resin will bind oxalate *in vitro* (S19) but not in the presence of physiological concentrations of chloride (L1). It is possible that cholestyramine is effective not by sequestering oxalate, but because it binds bile acids and fatty acids, thus reducing intraluminal binding of calcium or decreasing colonic permeability to oxalate. A study of various potential oxalate-binding agents has been reported (L1), aluminium hydroxide being suggested as worthy for trial. This reduced oxalate absorption when *in vivo* studies were performed, although the effect was not marked (H20). Oral DEAE-cellulose will also reduce urinary oxalate excretion (P6).

It has recently been demonstrated that many patients with enteric hyperoxaluria continue to excrete greater amounts of oxalate than normal subjects when on a very low-oxalate, normal protein diet (H17). Urinary oxalate remained high on a low-oxalate, hydrolyzed protein diet, but fell to normal if the protein content was reduced. It was suggested that tissue or bacterial production of oxalate or an oxalate precursor from dietary constituents may occur and would explain why it may be difficult to reduce urinary oxalate on a low-oxalate diet (A4).

Risk factors other than urinary oxalate concentrations may be important in urolithiasis complicating fat malabsorption. The solubility of calcium oxalate is reduced in these patients, possibly because the excretion of citric acid, an inhibitor of calcium oxalate precipitation, is reduced (E5, R7).

4.5.2. *Renal Oxalosis in Intestinal Disease*

Patients with jejunoileal bypass may develop oxalate deposits in the kidneys. Cryer *et al.* (C19) described a patient who developed renal impairment and hyperchloremic metabolic acidosis; hyperoxaluria was present and oxalate crystals were seen in the renal biopsy, which also showed interstitial nephritis. Renal function improved when the bypass was reversed and, postoperatively, oxalate excretion fell to normal. Several other patients have been reported in whom renal oxalosis was noted, the distribution of oxalate deposits being mainly interstitial and tubular (C2, D3, E3, G2, V2). Two patients died and, at postmortem, extrarenal oxalate crystals were minimal (C2, G2). All of the patients had jejunoileal bypass except one with longstanding Crohn's disease and small bowel resection who developed acute renal failure (M4). This patient was treated with hemodialysis, and hyperoxaluria developed in the recovery phase.

Renal function and renal biopsies have been studied in 18 patients with jejunoileal bypass, only two of whom had significantly impaired renal function (D19). Hyperoxaluria (0.77–2.78 mmol/24 hours) was present in all 16 patients in whom oxalate excretion was determined. All of the renal biopsies showed interstitial nephritis and tubular atrophy, although in 10 patients, oxalate crystals were not seen, and in the other 8, most of the damaged interstitial areas were remote from crystal deposits. One of the patients in this study has been found in further investigations to show staining of tubular basement membranes with antisera to IgG and C3 (Z6). It was suggested that immunologically mediated tubular damage occurred due to antitubular basement membrane antibodies. These were thought to be produced either because oxalate induced renal damage, leading to the release of basement membrane antigens, or because bacterial overgrowth in the bypass gut caused mucosal damage, with the release of basement membrane that shared structural characteristics with renal basement membrane.

An association between renal oxalosis and renal failure complicating small intestinal disease has thus been established. However, it has not been demonstrated that oxalosis is responsible for the impaired renal function.

4.6. Oxalosis in Renal Failure

Deposits of calcium oxalate have been found in tissues from uremic patients, particularly kidney (F3, Y3), heart (B8, S2), bone (M13), and thyroid gland (F5). Renal deposits occur mainly in the tubules and only rarely in interstitial tissue (F3, S3). In the heart, oxalate crystals have been noted in myocardial fibers and in interstitial tissue (S3). In this study, congestive cardiac failure or varying degrees of heart block had occurred in most patients in whom extensive cardiac deposits were found, and crystal deposition in the conduction system of a patient with complete heart block was noted. It was suggested that extensive cardiac oxalosis was a clinically important complication of prolonged renal insufficiency. Deposits in both kidney and heart occur more extensively in chronic rather than acute renal failure (S3) and, also, in patients dying after long-term hemodialysis, compared to short-term treatment (F5). A clinical syndrome of muscular weakness, peripheral ulceration, and multiple shunt complications has been attributed to obliterative vasculitis caused by extensive deposition of oxalate in the peripheral arteries of three patients being treated by hemodialysis (O5).

Oxalosis in renal insufficiency is most probably due to oxalate retention with subsequent tissue deposition. Increases in blood oxalate levels have been described in renal failure (B5, W14, Z4) and, as mentioned above, the extent of tissue deposits is related to the duration of disease. The degree of oxalemia has been shown to correlate linearly with plasma creatinine (B6). Paradoxically, marked hyperoxaluria has been described in nine patients with renal failure complicating spinal cord injury (V4). It was hypothesized that increased absorption of dietary oxalate might occur in these patients due to disturbed colonic function. However, tests of oxalate absorption were not performed.

Hemodialysis reduces but does not normalize blood oxalate levels (B5, W14, Z4). Oral and intravenous vitamin B_6 treatment has also been shown to reduce blood oxalate but it has not been established if this is due to B_6 deficiency occurring in these patients (B6).

4.7. Localized Oxalosis

Localized deposits of oxalate crystals have been described in patients in whom there has not been any identifiable disorder of oxalate metabolism. Many microorganisms, including bacteria, algae, and fungi such as *Aspergillus* species are known to produce oxalate (H11). Nime and Hutchins (N1) noted that large amounts of oxalate were present in the infected tissues of a patient with aspergillosis and when 68 other cases were reviewed, a further 10 cases were found. Various species of *Aspergillus* were responsible for the infection in these cases. In addition to occurring in association with aspergillosis, localized oxalosis has been reported in patients with ocular disease, in diabetics (F13), and in one patient following a retinal detachment (G1).

4.8. Oxalate Gallstones

A unique case of calcium oxalate–calcium phosphate gallstones has recently been described (W13). The patient was a female infant with bronchopulmonary dysplasia who died at 6 months of age after periods of ventilation and hyperalimentation. Drug treatment included fursemide, theophylline, spironolactone, and digoxin. Infections with *Candida albicans* and cytomegalovirus had occurred and been treated with amphotericin B, fluorocytosine, and acyclovir. Abdominal calcifications in the upper right quadrant had been noted at 4 months. At postmortem, gallstones were found which were shown to contain calcium oxalate and calcium phosphate. Although the cause was obscure, it was suggested that either the previous drug treatment or hyperalimentation could be important factors in the development of calcium oxalate gallstones. Other cases may have been overlooked if the possibility of such stones occurring was not considered.

5. Conclusions

Some disorders of oxalate metabolism, such as primary hyperoxaluria, are rare, and others, such as methoxyflurane-induced hyperoxaluria, are curiosities. However, enteric hyperoxaluria, the commonest cause of increased oxalate excretion, is not uncommon and it is possible that oxalosis may prove to be an important cause of morbidity and mortality in patients with end-stage renal disease. It is also possible that many patients with idiopathic renal tract stones may have a disorder of oxalate metabolism. Thus, it is apparent that there are several clinically important disorders of oxalate metabolism. To facilitate investigations in these and other areas, reliable and robust methods for determining oxalate in urine and plasma are required. Although many are available for urine analysis, some appear more suitable for research laboratories than for a busy service service laboratory. In addition, there is a need for a rapid, straightforward plasma assay.

REFERENCES

A1. Adams, N. D., Carrera, G. F., Johnson, R. P., Latorraca, R., and Lemann, J., Calcium-oxalate-crystal-induced bone disease. *Am. J. Kidney Dis.* **1**, 294–299 (1982).

A2. Admirand, W. H., Earnest, D. L., and Williams, H. E., Hyperoxaluria and bowel disease. *Trans. Assoc. Am. Physicians* **84**, 307–312 (1971).

A3. Akcay, T., and Rose, G. A., The real and apparent plasma oxalate. *Clin. Chim. Acta* **101**, 305–311 (1980).

A4. Andersson H., and Bosaeus, I., Hyperoxaluria in malabsorptive states. *Urol. Int.* **36**, 1–9 (1981).

A5. Andersson, H., Filipsson, S., and Hultén, L., Urinary oxalate excretion related to ileocolic surgery in patients with Crohn's disease. *Scand. J. Gastroenterol.* **13**, 465–469 (1978).

A6. Andersson, H., and Gillberg, R., Urinary oxalate on a high-oxalate diet as a clinical test of malabsorption. *Lancet* **2**, 677–679 (1977).

A7. Andersson, H., and Jagenburg, R., Fat-reduced diet in the treatment of hyperoxaluria in patients with ileopathy. *Gut* **15**, 360–366 (1974).

A8. Aponte, G. E., and Fetter, T. R., Familial idiopathic oxalate nephrocalcinosis. *Am. J. Clin. Pathol.* **24**, 1363–1373 (1954).

A9. Arbus, G. S., and Sniderman, S., Oxalosis with peripheral gangrene. *Arch. Pathol.* **97**, 107–110 (1974).

A10. Archer, H. E., Dormer, A. E., Scowen, E. F., and Watts, R. W. E., Studies on the urinary excretion of oxalate by normal subjects. *Clin. Sci.* **16**, 405–411 (1957).

A11. Archer, H. E., Dormer, A. E., Scowen, E. F., and Watts, R. W. E., The Aetiology of primary hyperoxaluria. *Br. Med. J.* **1**, 175–181 (1958).

A12. Atkins, G. L., Dean, B. M., Griffin, W. J., Scowen, E. F., and Watts, R. W. E., Quantitative aspects of ascorbic acid metabolism in patients with primary hyperoxaluria. *Clin. Sci.* **29**, 305–314 (1965).

B1. Baadenhuisen, H., and Jansen, A. P., Colorimetric determination of urinary oxalate recovered as calcium oxalate. *Clin. Chim. Acta* **62**, 315–324 (1975).

B2. Bais, R., Potezny, N., Edwards, J. B., Rofe, A. M., and Conyers, R. A. J., Oxalate determination by immobilized oxalate oxidase in a continuous flow system. *Anal. Chem.* **52**, 508–511 (1980).

B3. Baker, E. M., Saari, J. C., and Tolbert, B. M., Ascorbic acid metabolism in man. *Am. J. Clin. Nutr.*, **19**, 371–378 (1966).

B4. Baker, E. M., Sauberlich, H. E., Wolfskill, S. J., Wallace, W. T., and Dean, E. E., Tracer studies of vitamin C utilization in men: Metabolism of D-glucuronolactone-6-C^{14}, D-glucuronic-6-C^{14} acid and L-ascorbic-1-C^{14} acid. *Proc. Soc. Exp. Biol. Med.* **109**, 737–741 (1962).

B5. Balcke, P., Schmidt, P., Zazgornik, J., Kopsa, H., and Deutsch, E., Secondary hyperoxalemia in chronic renal failure. *Int. J. Artif. Organs* **5**, 141–143 (1982).

B6. Balcke, P., Schmidt, P., Zazgornik, J., Kopsa, H., and Deutsch, E., Effects of vitamin B6 administration on elevated plasma oxalic acid levels in haemodialysed patients. *Eur. J. Clin. Invest.* **12**, 481–483 (1982).

B7. Banay, M., and Dimant, E., On the metabolism of L-ascorbic acid in the scorbutic guinea-pig. *Biochim. Biophys. Acta* **59**, 313–319 (1962).

B8. Bennett, B., and Rosenblum, C., Identification of calcium oxalate crystals in the myocardium in patients with uremia. *Lab. Invest.* **10**, 947–955 (1961).

B9. Bennett, D. J., Cole, F. E., Frohlich, E. D., and Erwin, D. T., Radioenzymatic procedure for urinary oxalate determination. *J. Lab. Clin. Med.* **91**, 822–830 (1978).

B10. Bennett, D. J., Cole, F. E., Frohlich, E. D., and Erwin, D. T., A radioenzymatic isotope-dilution assay for oxalate in serum or plasma. *Clin. Chem.* **25**, 1810–1813 (1979).

B11. Binder, H. J., Intestinal oxalate absorption. *Gastroenterology* **67**, 441–446 (1974).

B12. Boquist, L., Lindqvist, B., Östberg, Y., and Steen, L., Primary oxalosis. *Am. J. Med.* **54**, 673–681 (1973).

B13. Bourke, E., Frindt, G., Flynn, P., and Schreiner, G. E., Primary hyperoxaluria with normal alpha-ketoglutarate: Glyoxylate carboligase activity. *Ann. Intern. Med.* **76**, 279–284 (1972).

B14. Brancaccio, D., Poggi, A., Ciccarelli, C., Bellini, F., Galmozzi, C., Poletti, I., and Maggiore, Q., Bone changes in end-stage oxalosis. *Am. J. Roentgenol.* **136**, 935–939 (1981).

B15. Breed, A., Chesney, R., Friedman, A., Gilbert, E., Langer, L., and Latorraca, R., Oxalosis induced bone disease: A complication of transplantation and prolonged survival in primary hyperoxaluria. *J. Bone Jt. Surg.* **63-A**, 310–316 (1981).

B16. Brennan, J. N., Diwan, R. V., Makkar, S. P., Cromer, S. P., and Bellon, E. M., Ultrasonic diagnosis of primary hyperoxaluria in infancy. *Radiology* **145**, 147–148 (1982).

B17. Briggs, M., Vitamin-C-induced hyperoxaluria. *Lancet* **1**, 154 (1976).

B18. Briggs, M. H., Garcia-Webb, P., and Davies, P., Urinary oxalate and vitamin-C supplements. *Lancet* **2**, 201 (1973).

B19. Brinkley, L., McGuire, J., Gregory, J., and Pak, C. Y. C., Bioavailability of oxalate in foods. *Urology* **17**, 534–538 (1981).

B20. Buc. H., Demaugre, F., and Leroux, J. P., The kinetic effects of oxalate on liver and erythrocyte pyruvate kinases. *Biochem. Biophys. Res. Commun.* **85**, 774–779 (1978).

B21. Butz, M., Hoffman, H., and Kohlbecker, G., Dietary influence on serum and urinary oxalate in healthy subjects and oxalate stone formers. *Urol. Int.* **35**, 309–315 (1980).

C1. Calkins, V. P., Microdetermination of glycolic and oxalic acids. *Ind. Eng. Chem. Anal. Ed.* **15**, 762–763 (1943).

C2. Canos, H. J., Hogg, G. A., and Jeffrey, J. R., Oxalate nephropathy due to gastrointestinal disorders. *Can. Med. Assoc. J.* **124**, 729–733 (1981).

C3. Caspary, W. F., Tönissen, J., and Lankisch, P. G., 'Enteral' hyperoxaluria. Effect of cholestyramine, calcium, neomycin, and bile acids on intestinal oxalate absorption in man. *Acta Hepato-Gastroenterol.* **24**, 193–200 (1977).

C4. Cattell, W. R., Spencer, A. G., Taylor, G. W., and Watts, R. W. E., The mechanism of the renal excretion of oxalate in the dog. *Clin. Sci.* **23**, 43–52 (1962).

C5. Chadwick, V. S., Modha, K., and Dowling, R. H., Mechanism for hyperoxaluria in patients with ileal dysfunction. *N. Engl. J. Med.* **289**, 172–176 (1973).

C6. Chadwick, V. S., Elias, E., Bell, G. D., and Dowling, R. H., *In* "Advances in Bile Acid Research" (S. Matern, J. Hackenschmidt, P. Back, and W. Gerok, eds.), pp. 435–440. Schattaur, Stuttgart, 1974.

C7. Chalmers, R. A., Lawson, A. M., Hauschildt, S., and Watts, R. W. E., The urinary excretion of glycollic acid and threonic acid by xylitol-infused patients and their relationship to the possible role of 'active glycolaldehyde' in the transketolase reaction in vivo. *Biochem. Soc. Trans.* **3**, 518–521 (1975).

C8. Chalmers, R. A., and Watts, R. W. E., The quantitative extraction and gas-liquid chromatographic determination of organic acids in urine. *Analyst* **97**, 958–967 (1972).

C9. Chambers, M. M., and Russell, J. C., A specific assay for plasma oxalate. *Clin. Biochem.* **6**, 22–28 (1973).

C10. Charransol, G., Barthelemy, Ch., and Desgrez, P., Rapid determination of urinary oxalic acid by gas-liquid chromatography without extraction. *J. Chromatogr.* **145**, 452–455 (1978).

C11. Charransol, G., and Desgrez, P., Utilisation de la chromatographie en phase gazeuse pour la determination quantitative de L'acide oxalique. *J. Chromatogr.* **48**, 530–532 (1970).

C12. Coltart, D. J., and Hudson, R. E. B., Primary oxalosis of the heart: A cause of heart block. *Br. Heart J.* **33**, 315–319 (1971).

C13. Constable, A. R., Joekes, A. M., Kasidas, G. P., O'Regan, P., and Rose, G. A., Plasma level and renal clearance of oxalate in normal subjects and in patients with primary hyperoxaluria or chronic renal failure or both. *Clin. Sci.* **56**, 299–304 (1979).

C14. Costello, J., Hatch, M., and Bourke, E., An enzymatic method for the spectrophotometric determination of oxalic acid. *J. Lab. Clin. Med.* **87**, 903–908 (1976).

C15. Crawhall, J. C., Scowen, E. F., and Watts, R. W. E., Conversion of glycine to oxalate in primary hyperoxaluria. *Lancet* **2**, 806–809 (1959).

C16. Crawhall, J. C., De Mowbray, R. R., Scowen, E. F., and Watts, R. W. E., Conversion of glycine to oxalate in normal subjects. *Lancet* **2**, 810 (1959).

C17. Crawhall, J. C., and Watts, R. W. E., The oxalate content of human plasma. *Clin. Sci.* **20**, 357–366 (1961).

C18. Crawhall, J. C., and Watts, R. W. E., The metabolism of 1-^{14}C glyoxylate by the liver mitochondria of patients with primary hyperoxaluria and non-hyperoxaluric subjects. *Clin. Sci.* **23**, 163–168 (1962).

C19. Cryer, P. E., Garber, A. J., Lucas, B., and Wise, L., Renal failure after small intestinal bypass for obesity. *Arch. Intern. Med.* **135**, 1610–1612 (1975).
D1. Dagneaux, P. G. L. C. K., Elhorst, J. T. K., and Olthius, F. M. F. G., Oxalic acid determination in plasma. *Clin. Chim. Acta* **71**, 319–325 (1976).
D2. Daniels, R. A., Michels, R., Aisen, P., and Goldstein, G., Familial hyperoxaluria. *Am. J. Med.* **29**, 820–831 (1960).
D3. Das, S., Joseph, B., and Dilk, A. L., Renal failure owing to oxalate nephrosis after jejunoileal bypass. *J. Urol.* **121**, 506–509 (1979).
D4. Davis, J. S., Klingberg, W. G., and Stowell, R. E., Nephrolithiasis and nephrocalcinosis with calcium oxalate crystals in kidneys and bones. *J. Pediatrics* **36**, 323–334 (1950).
D5. Dean, B. M., and Griffin, W. J., Estimation of urinary oxalate by the method of isotope dilution. *Nature (London)* **205**, 598–599 (1965).
D6. Dean, B. M., Watts, R. W. E., and Westwick, W. J., Metabolism of [1-^{14}C]glyoxylate, [1-^{14}C]glycollate, [1-^{14}C]glycine and [2-^{14}C]glycine by homogenates of kidney and liver tissue from hyperoxaluric and control subjects. *Biochem. J.* **105**, 701–707 (1967).
D7. Decoux, G., Boujet, C., Rual, M., Drouet, C., Lafond, J. L., Favier, A., and Agnius-Delord, C., Methode rapide de dosage de L'acide oxalique urinaire par chromatographie en phase gazeuse. *Clin. Chim. Acta* **120**, 207–217 (1982).
D8. Dempsey, E. F., Forbes, A. P., Melick, R. A., and Henneman, P. H., Urinary oxalate excretion. *Metabolism* **9**, 52–58 (1960).
D9. Dent, C. E., and Stamp, T. C. B., Treatment of primary hyperoxaluria. *Arch. Dis. Child.* **45**, 735–745 (1970).
D10. Deodhar, S. D., Tung, K. S. K., Zühlke, V., and Nakamoto, S., Renal homotransplantation in a patient with primary familial oxalosis. *Arch. Pathol.* **87**, 118–124 (1969).
D11. Deren, J. J., Porush, J. G., Levitt, M. F., and Khilani, M. T., Nephrolithiasis as a complication of ulcerative colitis and regional enteritis. *Ann. Intern. Med.* **56**, 843–853 (1962).
D12. Dick, M., A simplified urinary oxalate method. *Proc. Assoc. Clin. Biochem.* **4**, 186–187 (1967).
D13. Dickstein, S. S., and Frame, B., Urinary tract calculi after intestinal shunt operations for the treatment of obesity. *Surg. Gynecol. Obstet.* **136**, 257–260 (1973).
D14. Di Corcia, A., Samperi, R., Vinci, G., and D'Ascenzo, G., Simple, reliable chromatographic measurement of oxalate in urine. *Clin. Chem.* **28**, 1457–1460 (1982).
D15. Dobbins, J. W., and Binder, H. J., Effect of bile salts and fatty acids on the colonic absorption of oxalate. *Gastroenterology* **70**, 1096–1100 (1976).
D16. Dobbins, J. W., and Binder, H. J., Importance of the colon in enteric hyperoxaluria. *N. Engl. J. Med.* **296**, 298–301 (1977).
D17. Dosch, W., Rapid and direct gas chromatographic determination of oxalic acid in urine. *Urol. Res.* **7**, 227–234 (1979).
D18. Dowling, R. H., Rose, G. A., and Sutor, D. J., Hyperoxaluria and renal calculi in ileal disease. *Lancet* **1**, 1103–1106 (1971).
D19. Drenick, E. J., Stanley, T. M., Border, W. A., Zawada, E. T., Dornfield, L. P., Upham, T., and Llach, F., Renal damage with intestinal bypass. *Ann. Intern. Med.* **89**, 594–599 (1978).
D20. Duggan, D. E., Walker, R. W., Noll, R. M., and Vandenheuvel, W. J. A., Determination of urinary oxalic acid and of oxalate pool size by stable isotope dilution. *Anal. Biochem.* **94**, 477–482 (1979).
E1. Earnest, D. L., Perspectives on incidence, etiology, and treatment of enteric hyperoxaluria. *Am. J. Clin. Nutr.* **30**, 72–75 (1977).

E2. Earnest, D. L., Johnson, G., Williams, H. E., and Admirand, W. E., Hyperoxaluria in patients with ileal resection: An abnormality in dietary oxalate absorption. *Gastroenterology* **66**, 1114–1122 (1974).

E3. Ehlers, S. M., Posalaky, Z., Strate, R. G., and Quattlebaum, F. W., Acute reversible renal failure following jejunoileal bypass for morbid obesity: A clinical and pathological (EM) study of a case. *Surgery* **82**, 629–634 (1977).

E4. Elder, T. D., and Wyngaarden, J. B., The biosynthesis and turnover of oxalate in normal and hyperoxaluric subjects. *J. Clin. Invest.* **39**, 1337–1344 (1960).

E5. Elliot, J. S., and Soles, W. P., Excretion of calcium and citric acid in patients with small bowel disease. *J. Urol.* **111**, 810–812 (1974).

E6. Emerson, P. M., and Wilkinson, J. H., Urea and oxalate inhibition of the serum lactate dehydrogenase. *J. Clin. Pathol.* **18**, 803–807 (1965).

E7. Evans, G. W., Phillips, G., Mukherjee, T. M., Snow, M. R., Lawrence, J. R., and Thomas, D. W., Identification of crystals deposited in brain and kidney after xylitol administration by biochemical, histochemical and electron diffraction methods. *J. Clin. Pathol.* **26**, 32–36 (1973).

F1. Faber, S. R., Feitler, W. W., Bleiler, R. E., Ohlson, M. A., and Hodges, R. E., The effects of an induced pyridoxine and pantothenic acid deficiency on excretions of oxalic and xanthurenic acids in urine. *Am. J. Clin. Nutr.* **12**, 406–412 (1963).

F2. Fairclough, P. D., Feest, T. G., Chadwick, V. S., and Clark, M. S., Effect of sodium chenodeoxycholate on oxalate absorption from the excluded human colon—A mechanism for 'enteric' hyperoxaluria. *Gut* **18**, 240–244 (1977).

F3. Fanger, H., and Esparza, A., Crystals of calcium oxalate in kidneys in uremia. *Am. J. Clin. Pathol.* **41**, 597–603 (1964).

F4. Farrington, C. J., and Chalmers, A. H., Gas-chromatographic estimation of urinary oxalate and its comparison with a colorimetric method. *Clin. Chem.* **25**, 1993–1996 (1979).

F5. Fayemi, A. O., Ali, M., and Braun, E. V., Oxalosis in hemodialysis patients. *Arch. Pathol. Lab. Med.* **103**, 58–62 (1979).

F6. Fielder, A. R., Garner, A., and Chambers, T. L., Opthalmic manifestations of primary oxalosis. *Br. J. Ophthalmol.* **64**, 782–788 (1980).

F7. Finch, A. M., Kasidas, G. P., and Rose, G. A., Urine composition in normal subjects after oral ingestion of oxalate-rich foods. *Clin. Sci.* **60**, 411–418 (1981).

F8. Fitzpatrick, J. M., Kasidas, G. P., and Rose, G. A., Hyperoxaluria following glycine irrigation for transurethral prostatectomy. *Br. J. Urol.* **53**, 250–252 (1981).

F9. Frascino, J. A., Vanamee, P., and Rosen, P. P., Renal oxalosis and azotemia after methoxyflurane anesthesia. *N. Engl. J. Med.* **283**, 676–679 (1970).

F10. Fraser, J., and Campbell, D. J., Indirect measure of oxalic acid in urine by atomic absorption spectrophotometry. *Clin. Biochem.* **5**, 99–103 (1972).

F11. Frederick, E. W., Rabkin, M. T., Richie, R. H., and Smith, L. H., Studies on primary hyperoxaluria: In vivo demonstration of a defect in oxalate metabolism. *N. Engl. J. Med.* **269**, 821–829 (1963).

F12. Freel, R. W., Hatch, M., Earnest, D. L., and Goldner, A., Oxalate transport across the isolated rat colon. *Biochim. Biophys. Acta* **600**, 838–843 (1980).

F13. Friedman, A. H., and Charles, N. C., Retinal oxalosis in two diabetic patients. *Am. J. Ophthalmol.* **78**, 189–195 (1974).

F14. Gambardella, R. L., and Richardson, K. E., The pathways of oxalate formation from phenylalanine, tyrosine, tryptophan and ascorbic acid in the rat. *Biochim. Biophys. Acta* **499**, 156–168 (1977).

G1. Garner, A., Retinal oxalosis. *Br. J. Ophthalmol.* **58**, 613–619 (1974).
G2. Gelbart, D. R., Brewer, L. L., and Weinstein, A. B., Oxalosis and chronic renal failure after intestinal bypass. *Arch. Intern. Med.* **137**, 239–243 (1977).
G3. Gelot, M. A., Lavoue, G., Belleville, F., and Nabet, P., Determination of oxalates in plasma and urine using gas chromatography. *Clin. Chim. Acta* **106**, 279–285 (1980).
G4. Gelzayd, E. A., Breuer, R. I., and Kirsner, J. B., Nephrolithiasis in inflammatory bowel disease. *Am. J. Dig. Dis.* **13**, 1027–1034 (1968).
G5. Gershoff, S. N., Faragalla, F. F., Nelson, D. A., and Andrus, S. B., Vitamin B6 deficiency and oxalate nephrocalcinosis in the cat. *Am. J. Med.* **27**, 72–80 (1959).
G6. Gessner, P. K., Parke, D. V., and Williams, R. T., Studies in detoxication. 86. The metabolism of ^{14}C-labelled ethylene glycol. *Biochem. J.* **79**, 482–489 (1962).
G7. Gibbs, D. A., Hauschildt, S., and Watts, R. W. E., Glyoxalate oxidation in rat liver and kidney. *J. Biochem.* **82**, 221–230 (1977).
G8. Gibbs, D. A., and Watts, R. W. E., An investigation of the possible role of xanthine oxidase in the oxidation of glyoxylate to oxalate. *Clin. Sci.* **31**, 285–297 (1966).
G9. Gibbs, D. A., and Watts, R. W. E., The variation of urinary oxalate excretion with age. *J. Lab. Clin. Med.* **73**, 901–908 (1969).
G10. Gibbs, D. A., and Watts, R. W. E., The action of pyridoxine in primary hyperoxaluria. *Clin. Sci.* **38**, 277–286 (1970).
G11. Giterson, A. L. Slooff, P. A. M., and Schouten, H., Oxalate in urine. *Clin. Chim. Acta* **29**, 342–343 (1970).
H1. Hall, B. M., Walsh, J. C., Horuath, J. S., and Lytton, D. G., Peripheral neuropathy complicating primary hyperoxaluria. *J. Neurol. Sci.* **29**, 343–349 (1976).
H2. Hallson, P. C., and Rose, G. A., A simplified and rapid enzymatic method for determination of urinary oxalate. *Clin. Chim. Acta* **55**, 29–39 (1974).
H3. Haqqani, M. T., Crystals in brain and meninges in primary hyperoxaluria and oxalosis. *J. Clin. Pathol.* **30**, 16–18 (1977).
H4. Hatch, M., Bourke, E., and Costello, J., New enzymatic method for serum oxalate determination. *Clin. Chem.* **23**, 76–78 (1977).
H5. Hauschildt, S., Chalmers, R. A., Lawson, A. M., Schultis, K., and Watts, R. W. E., Metabolic investigations after xylitol infusion in human subjects. *Am. J. Clin. Nutr.* **29**, 258–273 (1976).
H6. Hausman, E. R., McAnally, J. S., and Lewis, G. T., Determination of oxalate in urine. *Clin. Chem.* **2**, 439–444 (1956).
H7. Hautmann, R., and Osswald, H., Pharmacokinetic studies of oxalate in man. *Invest. Urol.* **16**, 395–398 (1979).
H8. Hellman, L., and Burns, J. J., Metabolism of L-ascorbic acid-1-C^{14} in man. *J. Biol. Chem.* **230**, 923–930 (1958).
H9. Hockaday, T. D. R., Clayton, J. E., Frederick, E. W., and Smith, L. H., Primary hyperoxaluria. *Medicine* **43**, 315–345 (1964).
H10. Hockaday, T. D. R., Frederick, E. W., Clayton, J. E., and Smith, L. H., Studies on primary hyperoxaluria. 11. Urinary oxalate, glycolate, and glyoxylate measurement by isotope dilution methods. *J. Lab. Clin. Med.* **65**, 677–687 (1965).
H11. Hodgkinson, A., "Oxalic Acid in Biology and Medicine." Academic Press, New York, 1977.
H12. Hodgkinson, A., Evidence of increased oxalate absorption in patients with calcium-containing stones. *Clin. Sci. Mol. Med.* **54**, 291–294 (1978).
H13. Hodgkinson, A., Sampling errors in the determination of urine calcium and oxalate: Solubility of calcium oxalate in HCl-urine mixtures. *Clin. Chim. Acta* **109**, 239–244 (1981).

H14. Hodgkinson, A., and Wilkinson, R., Plasma oxalate concentration and renal excretion of oxalate in man. *Clin. Sci. Mol. Med.* **46**, 61–73 (1974).
H15. Hodgkinson, A., and Williams, A., An improved colorimetric procedure for urine oxalate. *Clin. Chim. Acta* **36**, 127–132 (1972).
H16. Hodgkinson, A., and Zarembski, P. M., The determination of oxalic acid in urine. *Analyst* **86**, 16–21 (1961).
H17. Hofmann, A. F., Laker, M. F., Dharmsathaphorn, K., Sherr, H. P., and Lorenzo, D., Complex pathogenesis of hyperoxaluria after jejunoileal bypass surgery. *Gastroenterology* **84**, 293–300 (1983).
H18. Hofmann, A. F., Tacker, M., Fromm, H., Thomas, P. J., and Smith, L. H., Acquired hyperoxaluria and intestinal disease. *Mayo Clin. Proc.* **48**, 35–42 (1973).
H19. Hofmann, A. F., Thomas, P. J., Smith, L. H., and McCall, J. T., Pathogenesis of secondary hyperoxaluria in patients with ileal resection and diarrhea. *Gastroenterology* **58**, 960 (1970).
H20. Hofmann, A. F., Schmuck, G., Scopinaro, N., Laker, M. F., Sherr, H. P., Lorenzo, D., and Meeuse, B. J. D., Hyperoxaluria associated with intestinal bypass surgery for morbid obesity: Occurrence, pathogenesis and approaches to treatment. *Int. J. Obes.* **5**, 513–518 (1981).
H21. Hughes, H., Hagen, L., and Sutton, R. A. L., Determination of urinary oxalate by high-performance liquid chromatography. *Anal. Biochem.* **119**, 1–3 (1982).
H22. Hylander, E., Jarnum, S., and Neilsen, K., Calcium treatment of enteric hyperoxaluria after jejunoileal bypass for morbid obesity. *Scand. J. Gastroenterol.* **15**, 349–352 (1980).
J1. Jacobs, C., Rottembourgh, J., Reach, I., and Legrain, M., Terminal renal failure due to oxalosis in 14 patients. *Proc. Eur. Dial. Transplant Assoc.* **11**, 359–366 (1975).
J2. Jacobsen, E., and Mosback, N., Primary hyperoxaluria, treated with haemodialysis and kidney transplantation. *Dan. Med. Bull.* **21**, 72–76 (1974).
J3. Jakoby, W. B., *In* "Methods of Enzymatic Analysis" (H. U. Bergmeyer, ed.), 2nd Ed., Vol. 3, pp. 1542–1545. Academic Press, New York, 1974.
J4. Jeghers, H., and Murphy, R., Practical aspects of oxalate metabolism. *N. Engl. J. Med.* **233**, 208–215 (1945).
J5. Johansson, S., and Tabuva, R., Determination of oxalic and glycolic acid with isotope dilution methods and studies on the determination of glyoxylic acid. *Biochem. Med.* **11**, 1–9 (1974).
J6. Jones, A. R., Gadiel, P., and Stevenson, D., The fate of oxalic acid in the wistar rat. *Xenobiotica* **11**, 385–390 (1981).
K1. Kasidas, G. P., and Rose, G. A., Oxalate content of some common foods: Determination by an enzymatic method. *J. Hum. Nutr.* **34**, 255–266 (1980).
K2. Klauwers, J., Wolf, P. L., and Cohn, R., Failure of renal transplantation in primary oxalosis. *J. Am. Med. Assoc.* **209**, 551 (1969).
K3. Knight, T. F., Sansom, S. C., Senekjian, H. O., and Weinman, E. J., Oxalate secretion in the rat proximal tubule. *Am. J. Physiol.* **240**, F295–F298 (1981).
K4. Knowles, C. F., and Hodgkinson, A., Automated enzymic determination of oxalic acid in human serum. *Analyst* **97**, 474–481.
K5. Kobos, R. K., and Ramsey, T. A., Enzyme electrode system for oxalate determination utilizing oxalate decarboxylase immobilized on a carbon dioxide sensor. *Anal. Chim. Acta* **121**, 111–118 (1980).
K6. Koch, B., Irvine, A. H., Barr, J. R., and Poznanski, W. J., Three kidney transplantations in a patient with primary hereditary hyperoxaluria. *Can. Med. Assoc. J.* **106**, 1323–1331 (1972).
K7. Koch, G. H., and Strong, F. M., Determination of oxalate in beer by cerate oxidimetry

and the application of isotopically-derived correction factor. *Anal. Chem.* **37**, 1092–1095 (1965).
K8. Koch, G. H., and Strong, F. M., Determination of oxalate in urine. *Anal. Biochem.* **27**, 162–171 (1969).
K9. Koch, J., Stokstad, R., Williams, H. E., and Smith, L. H., Deficiency of 2-oxo-glutarate:glyoxylate carboligase activity in primary hyperoxaluria. *Proc. Natl. Acad. Sci. U.S.A.* **57**, 1123–1129 (1967).
K10. Koehl, C., and Abecassis, J., Determination of oxalic acid in urine by atomic absorption spectrophotometry. *Clin. Chim. Acta* **70**, 71–77 (1976).
K11. Kohlbecker, G., and Butz, M., Direct spectrophotometric determination of serum and urinary oxalate with oxalate oxidase. *J. Clin. Chem. Clin. Biochem.* **19**, 1103–1106 (1981).
K12. Kohlbecker, G., Richter, L., and Butz, M., Determination of oxalate in urine using oxalate oxidase: Comparison with oxalate decarboxylase. *J. Clin. Chem. Clin. Biochem.* **17**, 309–313 (1979).
L1. Laker, M. F., and Hofmann, A. F., Effective therapy of enteric hyperoxaluria: In vitro binding of oxalate by anion exchange resins and aluminium hydroxide. *J. Pharm. Sci.* **70**, 1065–1067 (1981).
L2. Laker, M. F., Hofmann, A. F., and Meeuse, B. J. D., Spectrophotometric determination of urinary oxalate with oxalate oxidase from moss. *Clin. Chem.* **26**, 827–830 (1980).
L3. Lamden, M. P., and Chrystowski, G. A., Urinary oxalate excretion by man following ascorbic acid ingestion. *Proc. Soc. Exp. Biol. Med.* **85**, 190–192 (1954).
L4. Larsson, L., Libert, B., and Asperud, M., Determination of urinary oxalate by reversed-phase ion-pair "high-performance" liquid chromatography. *Clin. Chem.*, **28**, 2272–2274 (1982).
L5. Lepoutre, C., Calculs multiples chez un enfant: Infiltration du parenchyme renal par des depots cristallins. *J. Urol.* **20**, 424 (1925).
L6. Leumann, E. P., Wegmann, W., and Largiader, F., Prolonged survival after renal transplantation in primary hyperoxaluria of childhood. *Clin. Nephrol.* **9**, 29–34 (1978).
L7. Lewis, R. D., Lowenstam, H. A., and Rossman, G. R., Oxalate nephrosis and crystalline myocarditis. *Arch. Pathol.* **98**, 149–155 (1974).
L8. Lyon, E. S., Borden, T. A., Ellis, J. E., and Vermegulen, C. W., Calcium oxalate lithiasis produced by pyridoxine deficiency and inhibition with high magnesium diets. *Invest. Urol.* **4**, 133–142 (1966).
M1. McDonald, G. B., Earnest, D. L., and Admirand, W. H., Hyperoxaluria correlates with fat malabsorption in patients with sprue. *Gut* **18**, 561–566 (1977).
M2. MacFadyen, D. A., Estimation of formaldehyde in biological mixtures. *J. Biol. Chem.* **158**, 107–133 (1945).
M3. Mahle, C. J., and Menon, M., Determination of urinary oxalate by ion chromatography: Preliminary observation. *J. Urol.* **127**, 159–162 (1982).
M4. Mandell, I., Krauss, E., and Millan, J. C., Oxalate-induced acute renal failure in Crohn's disease. *Am. J. Med.* **69**, 628–632 (1980).
M5. Marangella, M., Fruttero, B., Bruno, M., and Linari, F., Hyperoxaluria in idiopathic calcium stone disease: Further evidence of intestinal hyperabsorption of oxalate. *Clin. Sci.* **63**, 381–385 (1982).
M6. Marshall, R. W., Cochran, M., and Hodgkinson, A., Relationships between calcium and oxalic acid intake in the diet and their excretion in the urine of normal and renal stone forming subjects. *Clin. Sci.* **43**, 91–99 (1972).
M7. Massie, B. M., Bharati, S., Scheinman, M. M., Lev, M., Desai, J., Rubeson, E., and

Schmidt, W., Primary oxalosis with pan conduction cardiac disease: Electrophysiologic and anatomic correlation. *Circulation* **64**, 845–852 (1981).

M8. Mazze, R. I., Trudell, J. R., and Cousins, M. J., Methoxyflurane metabolism and renal dysfunction. *Anesthesiology* **35**, 247–252 (1971).

M9. Mayer, G. G., Markow, D., and Karp, F., Enzymatic oxalate determination in urine. *Clin. Chem.* **9**, 334–339 (1963).

M10. Mayer, W. J., McCarthy, J. P., and Greenberg, M. S., The determination of oxalic acid in urine by high performance liquid chromatography with electrochemical detection. *J. Chromatogr. Sci.* **17**, 656–660 (1979).

M11. Mee, J. M. L., and Stanley, R. W., A rapid gas liquid chromatographic method for determining oxalic acid in biological materials. *J. Chromatogr.* **76**, 242–243 (1973).

M12. Menarchè, R., Routine micromethod for determination of oxalic acid in urine by atomic absorption spectrophotometry. *Clin. Chem.* **20**, 1444–1445 (1974).

M13. Milgram, J. W., and Salyer, W. R., Secondary oxalosis of bone in chronic renal failure. *J. Bone Jt. Surg.* **56-A**, 387–395 (1974).

M14. Modigliani, R., Labayle, D., Aymes, C., and Denvil, R., Evidence for excessive absorption of oxalate by the colon in enteric hyperoxaluria. *Scand. J. Gastroenterol.* **13**, 187–192 (1978).

M15. Moorhead, P. J., Cooper, D. J., and Timperley, W. R., Progressive peripheral neuropathy in a patient with primary hyperoxaluria. *Br. Med. J.* **2**, 312–313 (1975).

M16. Morgan, J. M., Hartley, M. W., Miller, A. C., and Diethelm, A. G., Successful renal transplantation in hyperoxaluria. *Arch. Surg.* **109**, 430–433 (1974).

M17. Moye, H. A., Malagodi, M. H., Clarke, D. H., and Miles, C. J., A rapid gas chromatographic procedure for the analysis of oxalate ion in urine. *Clin. Chim. Acta* **114**, 173–185 (1981).

M18. Murray, J. F., Nolen, H. W., Gordon, R., and Peters, J. H., The measurement of urinary oxalic acid by derivatization coupled with liquid chromatography. *Anal. Biochem.* **121**, 301–309 (1982).

N1. Nime, F. A., and Hutchins, G. M., Oxalosis caused by *Aspergillus* infections. *Johns Hopkins Med. J.* **133**, 183–194 (1973).

N2. Nuret, P. and Offner, M., A new method for determination of oxalate in blood serum by gas chromatography. *Clin. Chim. Acta* **82**, 9–12 (1978).

O1. O'Leary, J. P., Thomas, W. C., and Woodward, E. R., Urinary tract stone after small bowel bypass for morbid obesity. *Am. J. Surg.* **127**, 142–147 (1974).

O2. Oli, H., and Davison, A. M., Adult systemic oxalosis presenting as acute renal failure. *Post Grad. Med. J.* **55**, 44–45 (1979).

O3. O'Regan, P., Constable, A. R., Joekes, A. M., Kasidas, G. P., and Rose, G. A., Successful renal transplantation in primary hyperoxaluria. *Postgrad. Med. J.* **56**, 288–293 (1980).

O4. Olthuis, F. M. F. G., Markscag, A. M. G., Elhorst, J. T. K., and Dagneaux, P. G. L. C. K., Urinary oxalate estimation. *Clin. Chim. Acta* **75**, 123–128 (1977).

O5. Op De Hoek, C. T., Diderich, P. P. N. M., Gratama, S., Weijs-V Hofwegen, E. J. M., Oxalosis in chronic renal failure. *Proc. Eur. Dial. Tranplant Assoc.* **17**, 730–735 (1980).

P1. Paddock, R. B., Parker, J. W., and Guadini, N. P., The effects of methoxyflurane on renal function. *Anesthesiology* **25**, 707–708 (1964).

P2. Park, K. Y., and Gregory, J., Gas-chromatographic determination of urinary oxalate. *Clin. Chem.* **26**, 1170–1172 (1980).

P3. Parry, M. F., and Wallach, R., Ethylene glycol poisoning. *Am. J. Med.* **57**, 143–150 (1974).

P4. Pedersen, J. H., and Steen, J., The effect of calcium on hyperoxaluria following jejunoileal bypass in morbid obesity. *Scand. J. Gastroenterol.* **14**, 97–99 (1979).

P5. Pernet, J.-L., and Pernet, A., Dosage colorimetrique De L'acide oxalique dans les milieux biologiques. *Ann. Biol. Clin.* **23,** 1189–1207 (1965).

P6. Pinto, B., and Bernshtam, J., Diethylaminoethanol-cellulose in the treatment of absorptive hyperoxaluria. *J. Urol.* **119,** 630–632 (1978).

P7. Potezny, N., Bais, R., O'Loughlin, P. D., Edwards, J. B., Rofe, A. M., and Conyers, R. A. J., Urinary oxalate determination by use of immobilized oxalate oxidase in a continuous flow system. *Clin. Chem.* **29,** 16–20 (1983).

P8. Powers, H. H., and Levatin, P., A method for the determination of oxalic acid in urine. *J. Biol. Chem.* **154,** 207–214 (1944).

P9. Prenen, J. A. C., Boer, P., Dorhout Mees, E. J., Endeman, H. J., Spoor, S. M., and Oei, H. Y., Renal clearance of [^{14}C]oxalate: Comparison of constant-infusion with single-injection techniques. *Clin. Sci.* **63,** (1982).

P10. Prenan, J. A. C., Dorhout Mees, E. J., Boer, P., Endeman, H. J., and Ephraim, K. H., Oxalic acid concentration in serum measured by isotopic clearance technique. Experience in hyper- and normo- oxaluric subjects. *Proc. Eur. Dial. Transplant Assoc.* **16,** 566–570 (1979).

R1. Rampton, D. S., Kasidas, G. P., Rose, G. A., and Sarner, M., Oxalate loading test: A screening test for steatorrhoea. *Gut* **20,** 1089–1094 (1979).

R2. Reed, G. H., and Morgan, S. D., Kinetic and magnetic resonance studies of the interaction of oxalate with pyruvate kinase. *Biochemistry* **13,** 3537–3541 (1974).

R3. Ribaya, J. D., and Gershoff, S. N., Effects of hydroxyproline and vitamin B6 on oxalate synthesis in rats. *J. Nutr.* **111,** 1231–1239 (1981).

R4. Ribeiro, M. E., and Elliot, J. S., Direct enzymatic determination of urinary oxalate. *Invest. Urol.* **2,** 78–81 (1964).

R5. Robertson, W. G., and Peacock, M., The cause of idiopathic calcium stone disease: Hypercalciuria or hyperoxaluria. *Nephron* **26,** 105–110 (1980).

R6. Rofe, A. M., Conyers, R. A. J., Bais, R., and Edwards, J. B., Oxalate excretion in rats injected with xylitol or glycollate: Stimulation by phenobarbitone pre-treatment. *Aust. J. Exp. Biol. Med. Sci.* **57,** 171–176 (1979).

R7. Rudman, D., Dedonis, J. L., Fountain, M. T., Chandler, J. B., Gerron, G. G., Fleming, G. A., and Kutner, M. H., Hypocitraturia in patients with gastrointestinal malabsorption. *N. Engl. J. Med.* **303,** 657–661 (1980).

R8. Runyan, T. J., and Gershoff, S. N., The effect of vitamin B6 deficiency in rats on the metabolism of oxalic acid precursors. *J. Biol. Chem.* **240,** 1889–1892 (1965).

S1. Sallis, J. D., Lumley, M. F., and Jordan, J. E., An assay for oxalate based on the conductometric measurement of enzyme-liberated carbon dioxide. *Biochem. Med.* **18,** 371–377 (1977).

S2. Salyer, W. R., and Hutchins, G. M., Cardiac lesions in secondary oxalosis. *Arch. Intern. Med.* **134,** 250–252 (1974).

S3. Salyer, W. R., and Keren, D., Oxalosis as a complication of chronic renal failure. *Kidney Int.* **4,** 61–66 (1973).

S4. Sample, R. H. B., Farber, M. E., and Glick, M. R., Urinary oxalate indirectly determined by continuous flow analysis for calcium. *Clin. Chem.* **26,** 1105 (1980).

S5. Sawaki, S., Hattori, N., and Yamada, K., Reduction of nicotinamide-adenine dinucleotide by glyoxylate in animal organs. *J. Vitaminol.* **12,** 303–306 (1966).

S6. Saxon, A., Busch, G. J., Merrill, J. P., Franco, V., and Wilson, R. E., Renal transplantation in primary hyperoxaluria. *Arch. Intern. Med.* **133,** 464–467 (1974).

S7. Scowen, E. F., Stansfield, A. G., and Watts, R. W. E., Oxalosis and primary hyperoxaluria. *J. Pathol. Bacteriol.* **77,** 195–205 (1959).

S8. Schmidt, K., Hagmaier, V., Bruchelt, G., and Rutishauer, G., Analytical isotachyphoresis: A rapid and sensitive method for determination of urinary oxalate. *Urol. Res.* **8**, 177–180 (1980).

S9. Schmidt, K-H., Hagmaier, V., Hornig, D. H., Vuilleumier, J-P., and Rutishauser, G., Urinary oxalate excretion after large intakes of ascorbic acid in man. *Am. J. Clin. Nutr.* **34**, 305–311 (1981).

S10. Silverberg, D. S., McIntyre, J. W. R., Ulan, R. A., and Gain, E. A., Oxalic acid excretion after methoxyflurane and halothane anaesthesia. *Can. Anaesth. Soc. J.* **18**, 496–504 (1971).

S11. Smith, L. H., Bauer, R. L., and Williams, H. E., Oxalate and glycolate synthesis by hemic cells. *J. Lab. Clin. Med.* **78**, 245–254 (1971).

S12. Smith, L. H., Fromm, H., and Hofmann, A. F., Acquired hyperoxaluria, nephrolithiasis and intestinal disease. *N. Engl. J. Med.* **286**, 1371–1375 (1972).

S13. Smith, L. H., Hofmann, A. F., McCall, J. T., and Thomas, P. J., Secondary hyperoxaluria in patients with ileal resection and oxalate nephrolithiasis. *Clin. Res.* **18**, 514 (1970).

S14. Smith, L. H., and Williams, H. E., Treatment of primary hyperoxaluria. *Mod. Treat.* **4**, 522–530 (1967).

S15. Solomons, C. C., Goodman, S. I., and Riley, C. M., Calcium carbimide in the treatment of primary hyperoxaluria. *N. Engl. J. Med.* **276**, 207–210 (1967).

S16. Stauffer, M. A., Oxalosis. *N. Engl. J. Med.* **263**, 386–390 (1960).

S17. Stauffer, J. Q., Hyperoxaluria and intestinal disease. *Dig. Dis. Sci.* **22**, 921–928 (1977).

S18. Stauffer, J. Q., Hyperoxaluria and calcium oxalate nephrolithiasis after jejunoileal bypass. *Am. J. Clin. Nutr.* **30**, 64–71 (1977).

S19. Stauffer, J. Q., Humphreys, M. H., and Weir, G. J., Acquired hyperoxaluria with regional enteritis after ileal resection. *Ann. Intern. Med.* **79**, 383–391 (1973).

S20. Stokholm, K. H., and Abildgaard, U., Calcium in the treatment of diarrhoea and hyperoxaluria after jejunoileal bypass for obesity. *Int. J. Obes.* **4**, 105–110 (1980).

S21. Sugiura, M., Yamamura, H., Hirano, K., Ito, Y., Sasaki, M., Morikawa, M. Inolle, M., and Tsuboi, M., Enzymic determination of serum oxalate. *Clin. Chim. Acta* **105**, 393–399 (1980).

S22. Sutor, D. J., The possible use of methylene blue in the treatment of primary hyperoxaluria. *Br. J. Urol.* **42**, 389–392 (1970).

T1. Thomas, D. W., Edwards, J. B., Gilligan, J. E. Lawrence, J. R., and Edwards, R. G., Complications following intravenous administration of solutions containing xylitol. *Med. J. Aust.* **1**, 1238–1246 (1972).

T2. Thomas, D. W., Hannett, B., Chalmers, A., Rofe, A. M., Edwards, J. B., and Edwards, R. G., Oxalate excretion during carbohydrate infusions. *Int. J. Vitam. Nutr. Res. Suppl.* **15**, 181–192 (1976).

T3. Tocco, D. J., Duncan, A. E. W., Noll, R. M., and Duggan, D. E., An electron-capture gas chromatographic procedure for the estimation of oxalic acid in urine. *Anal. Biochem.* **94**, 470–476 (1979).

V1. Vadgama, P., Guy, J. M., Laker, M. F., and Covington, A. K., An enzyme electrode for the determination of urine oxalate. *Prog. Clin. Enzymol.*, in press.

V2. Vainder, M., and Kelly, J., Renal tubular function secondary to jejunoileal bypass. *J. Am. Med. Assoc.* **235**, 1257–1258 (1982).

V3. Valman, H. B., Oberholzer, V. G., and Palmer, T., Hyperoxaluria after resection of ileum in childhood. *Arch. Dis. Child.* **49**, 171–173 (1974).

V4. Vaziri, N. D., Nikakhtar, B., and Gordon, S., Hyperoxaluria in chronic renal disease associated with spinal cord injury. *Paraplegia* **20**, 48–53 (1982).

V5. Von Nicolai, H., and Zilliken, F., Gaschromatographische bestimung von oxalsäure, malonsäure und bernsteinsäure aus biologischen materlal. *J. Chromatogr.* **92**, 431–434 (1974).

W1. Wacker, W. E. C., Haynes, H., Druyan, R., Fisher, W., and Coleman, J. E., Treatment of ethylene glycol poisoning with ethyl alcohol. *J. Am. Med. Assoc.* **194**, 1231–1233 (1965).

W2. Wallace, M. R., Mason, K., and Gray, J., Urine oxalate and calcium in idiopathic renal stone formers. *N.Z. Med. J.* **94**, 84–89 (1981).

W3. Walls, J. Morley, A. R., and Kerr, D. N. S., Primary hyperoxaluria in adult siblings: With some observations on the role of haemodialysis therapy. *Br. J. Urol.* **41**, 546–553 (1969).

W4. Watts, R. W. E., Chalmers, R. A., Gibbs, D. A., Lawson, A. M., Purkiss, P., and Spellacy, E., Studies on some possible biochemical treatments of primary hyperoxaluria. *Q. J. Med.* **18**, 259–272 (1979).

W5. West, R. R., Salyer, W. R., and Hutchins, G. M., Adult-onset primary oxalosis with complete heart block. *Johns Hopkins Med. J.* **133**, 195–200 (1973).

W6. Will, E. J., and Bijvoet, O. L. M., Primary oxalosis: Clinical and biochemical response to high-dose pyridoxine therapy. *Metabolism* **28**, 542–548 (1979).

W7. Williams, H. E., Johnson, G. A., and Smith, L. H., The renal clearance of oxalate in normal subjects and patients with primary hyperoxaluria. *Clin. Sci.* **41**, 213–218 (1971).

W8. Williams, H. E., and Smith, L. H., Disorders of oxalate metabolism. *Am. J. Med.* **45**, 715–735 (1968).

W9. Williams, H. E., and Smith, L. H., L-glyceric aciduria. *N. Engl. J. Med.* **278**, 233–239 (1968).

W10. Williams, H. E., and Smith, L. H., Hyperoxaluria in L-glyceric aciduria: Possible pathogenic mechanism. *Science* **171**, 390–391 (1971).

W11. Williams, H. E., and Smith, L. H., *In* "The Metabolic Basis of Inherited Disease" (J. B. Stanbury, J. B. Wyngaarden, and D. S. Frederickson, eds.), 4th Ed., pp. 182–204. McGraw-Hill, New York, 1978.

W12. Wilson, J., Marshall, R. W., and Hodgkinson, A., Excretion of methoxyflurane metabolites. *Br. Med. J.* **2**, 594 (1972).

W13. Wolf, P., Mannino, F., Hofmann, A. F., Nickoloff, B., and Edwards, D. K., Calcium oxalate-phosphate gallstones, a unique chemical type of gallstone. *Clin. Chem.* **28**, 1804–1805 (1982).

W14. Wolthers, B. G., and Hayer, M., The determination of oxalic acid in plasma and urine by means of capillary gas chromatography. *Clin. Chim. Acta* **120**, 87–102 (1982).

Y1. Yarbro, C. L., and Simpson, S. E., The determination of total urinary oxalate. *J. Lab. Clin. Med.* **48**, 304–310 (1956).

Y2. Yao, S. J., Wolfson, S. K., and Tokarsky, J. M., Enzymatic-potentiometric determination of oxalic acid. *Bioelectrochem. Bioenerg.* **2**, 348–350 (1975).

Y3. Yasue, T., Renal crystalline deposition and its pathogenesis. *Acta Histochem. Cytochem.* **2**, 96–111 (1969).

Y4. Yount, E. A., and Harris, R. A., Studies on the inhibition of gluconeogenesis by oxalate. *Biochim. Biophys. Acta* **633**, 122–133 (1980).

Y5. Yriberri, J., and Posen, S., A semi-automatic enzymic method for estimating urinary oxalate. *Clin. Chem.* **26**, 881–884 (1980).

Z1. Zarembski, P. M., and Hodgkinson, A., The oxalic acid content of english diets. *Br. J. Nutr.* **16**, 627–634 (1962).

Z2. Zarembski, P. M., and Hodgkinson, A., The fluorimetric determination of oxalic acid in blood and other biological materials. *Biochem. J.* **96,** 717–721 (1965).
Z3. Zarembski, P. M., Hodgkinson, A., and Cochran, M., Treatment of primary hyperoxaluria with calcium carbimide. *N. Engl. J. Med.* **277,** 1000–1002 (1967).
Z4. Zarembski, P. M., Hodgkinson, A., and Parsons, F. M., Elevation of the concentration of plasma oxalic acid in renal failure. *Nature (London)* **212,** 511–512 (1966).
Z5. Zarembski, P. M., Rosen, S. M., and Hodgkinson, A., Dialysis in the treatment of primary hyperoxaluria. *Br. J. Urol.* **41,** 530–533 (1969).
Z6. Zawada, E. T., Johnston, W. H., and Bergstein, J., Chronic interstitial nephritis. Its occurrence with oxalosis and anti-tubular basement membrane antibodies after jejunoileal bypass. *Arch. Pathol. Lab. Med.* **105,** 379–383 (1981).
Z7. Zinsser, H. H., and Karp, F., How to diminish endogenous oxalate excretion. 1. Tyrosine administration. *Invest. Urol.* **10,** 249–252 (1973).

DESIRABLE PERFORMANCE STANDARDS FOR CLINICAL CHEMISTRY TESTS

C. G. Fraser

Department of Clinical Biochemistry, Flinders Medical Centre, Bedford Park, South Australia

1. Introduction 300
2. Definition of Analytical Goals 301
   2.1. Goals for Imprecision of Plasma Analytes 301
   2.2. Goals for Imprecision of Urine Analytes 309
   2.3. Goals for Inaccuracy 312
   2.4. The Total Analytical Error Concept 312
   2.5. Goals for Detection Limit 315
   2.6. Goals for Linearity 316
   2.7. Goals for Turnaround Times 316
   2.8. Goals for Preanalytical Factors 317
3. The Rationale for Achievement of Goals 319
   3.1. Goals for Imprecision 319
   3.2. Goals for Inaccuracy 319
   3.3. Goals for Detection Limit 319
   3.4. Goals for Linearity 320
   3.5. Goals for Turnaround Times 320
   3.6. Goals for Preanalytical Factors 321
4. Uses of Analytical Goals 321
   4.1. Quality Control 321
   4.2. Method Evaluation 322
   4.3. Method Selection 323
   4.4. Laboratory Management 324
   4.5. Other Uses 324
5. Achievement of Analytical Goals 325
   5.1. Assessment of Current Laboratory Performance 325
   5.2. Improvement of Methods 328
   5.3. Alternative Strategies 331
6. Concluding Remarks 332
   References 333

1. Introduction

The *performance characteristics* of analytical tests are a set of quantitative and experimentally derived values for parameters of fundamental importance in assessing the suitability of a method for a given purpose (W16). The Expert Panel on Nomenclature and Principles of Quality Control in Clinical Chemistry of the International Federation of Clinical Chemistry (B12) has divided performance characteristics into two groups: practicability characteristics (such as costs, technical skill requirements, dependability, and laboratory safety) and reliability characteristics (such as specificity, inaccuracy, imprecision, detection limit, and turnaround time).

The results of clinical biochemistry tests are used for a number of purposes in clinical medicine, for example, in confirmation of or in making a diagnosis, in monitoring therapy, in assessment of prognosis, in screening, and in detection of complications (W7), and they also have other uses, for example, as aids in teaching and in clinical research and development. If test results are to be useful for all of these purposes, then their performance characteristics, particularly their reliability characteristics, must attain or surpass satisfactory standards. The standards of performance that must be reached to provide optimal patient care, preferably at least expense, have been termed *analytical goals*.

The Expert Panel on Nomenclature and Principles of Quality Control has considered the topic of quality requirements from the point of view of health care (B10). The Approved Recommendation of the Expert Panel states that the different purposes for which tests are used require different levels of performance and it is therefore unrealistic to stipulate a single set of values of the performance characteristics which would make a method acceptable for all purposes in clinical chemistry. In addition, the Expert Group on Evaluation of Kits has stated that the use of analytical goals must be moderated because such goals must be medically realistic (R7). However, there have been a number of attempts to define numerical analytical goals and certain of these definitions have gained wide professional acceptance.

In this review article, the approaches used to define analytical goals will be discussed and the potential uses of analytical goals described. This subject has been reviewed in detail (F3) and there are published proceedings of conferences on this topic (E2, P7). A comprehensive report on the assessment of quality requirements in clinical chemistry has been prepared by a project group initiated by the Nordic Clinical Chemistry Project (NORD-KEM) (H11). Therefore, this review article will concentrate on basic principles and on recent advances in this fascinating facet of the discipline of clinical chemistry.

2. Definition of Analytical Goals

2.1. Goals for Imprecision of Plasma Analytes

The majority of work in the field of analytical goal setting has been concerned with the definition of goals for the imprecision of plasma or serum analytes. Although the imprecision of test results can be improved by adoption of a number of strategies (F4), which will be discussed later, imprecision is inherent in all analytical procedures. Thus, it is of prime importance to ensure that goals for imprecision be met. Ross (R4) has recently provided an excellent review on all aspects of analytical imprecision and a brief account of the definition of analytical goals for plasma analytes.

Analytical goals for imprecision have been set using a variety of approaches which have been classified (F1) as being based on (1) reference or therapeutic intervals (B1, G3, T2, T3, V1), (2) the views of clinicians (B1, B3, C1, E5, E6, S10), (3) the state of the art (B1, C6, D3, K5, S19, S20, S21, T5, V1), (4) biological variation (C4, Y2), and (5) the opinions of groups or individuals (E2, F1, G4, P7, R5, S15). All of these approaches have disadvantages (F1).

Reference intervals generated by individual laboratories are dependent on the performance characteristics of the method used to obtain the results which are used to generate the interval (G5). Thus, a method with poor imprecision will lead to a wide reference interval; consequently, definition of a less stringent analytical goal will result and this will allow a method with poor imprecision to be used with false confidence. Moreover, other factors which affect the reference interval, such as the population sampled and the bias of the laboratory method, will also affect the derived analytical goal (F3). In addition, the fractions of the reference interval used to derive analytical goals are empirical, and different authors have proposed different fractions (B1, G3, T2, V1).

The views of clinicians on desirable performance standards for clinical chemistry tests have been sought by Barnett (B1), Skendzel (S10, S11), and the group of the author (B3). The approaches of Skendzel (S10) and Barrett and associates (B3) have been criticized (E4, E8) and the criticisms have been strongly refuted (F10). As stated recently (B4), current clinical opinions are subjective in nature and, therefore, formulation of goals by collation and quantitation of data found by survey of clinical opinions is not very productive and is often misleading.

The results described in a recent paper (E6), based upon the results described in a thesis (E5), confirm this postulate. A questionnaire which focused attention on reference intervals and action levels (these being defined as those levels that prompted clinicians to act), possible sources of variation in test results, satisfaction with laboratory performance, and a number of other topics was submitted to 63 senior specialists in internal medicine. Three criteria were applied in order to investigate whether analytical imprecision met medical needs: (1) the difference between the limits of the reference interval and the action level; this criterion was said to be applicable in diagnosis when it was wished to distinguish an individual with disease from the healthy; (2) a change in results which was considered to be medically significant when the initial results fell just outside the reference interval; this criterion was stated to relate to analytical requirements during patient monitoring; and (3) the satisfaction of clinicians with current laboratory performance.

The differences between the limits of the reference intervals quoted by the laboratories and the action levels quoted by clinicians were very variable, as shown for sodium, potassium, and cholesterol in Fig. 1. Moreover, a significant number of clinicians derived reference intervals from their own observations and others used literature sources; these are surely practices to be deprecated. These factors and the dispersion of the responses of the clinicians are strongly believed to redemonstrate the very subjective nature of the views of laboratory performance held by clinicians. This belief is supported by other evidence, for example, the magnitude of what was considered to be a medically significant change was very variable among the clinicians as was the perceived satisfaction of the clinicians with laboratory performance.

Elion-Gerritzen (E6) suggested that the median differences between the limits of the reference intervals and the action limits should be used to compare clinical needs with laboratory performance; analytical goals for imprecision were set by dividing such medians by three. Unfortunately, for a number of plasma analytes, adoption of this strategy gives goals for many analytes of *zero* imprecision, a totally impossible goal to achieve. The fact that the goals derived from medically significant changes were not in agreement with the goals set by this approach has been highlighted (S12); the differences were stated not to be surprising (E7), particularly in view of the subjective nature of clinical opinion.

The state of the art has been generally derived from the results obtained in interlaboratory quality assurance schemes. It is considered that it is unrealistic to derive goals from the state of the art because the actual performance achieved by laboratories may not in any way relate to desirable performance or analytical goals. Moreover, most of the data obtained in interlaboratory quality assurance schemes are based upon analyses of lyophilized material. Much evidence has been gathered that shows that the performance achieved with such material is not identical to the performance achieved with speci-

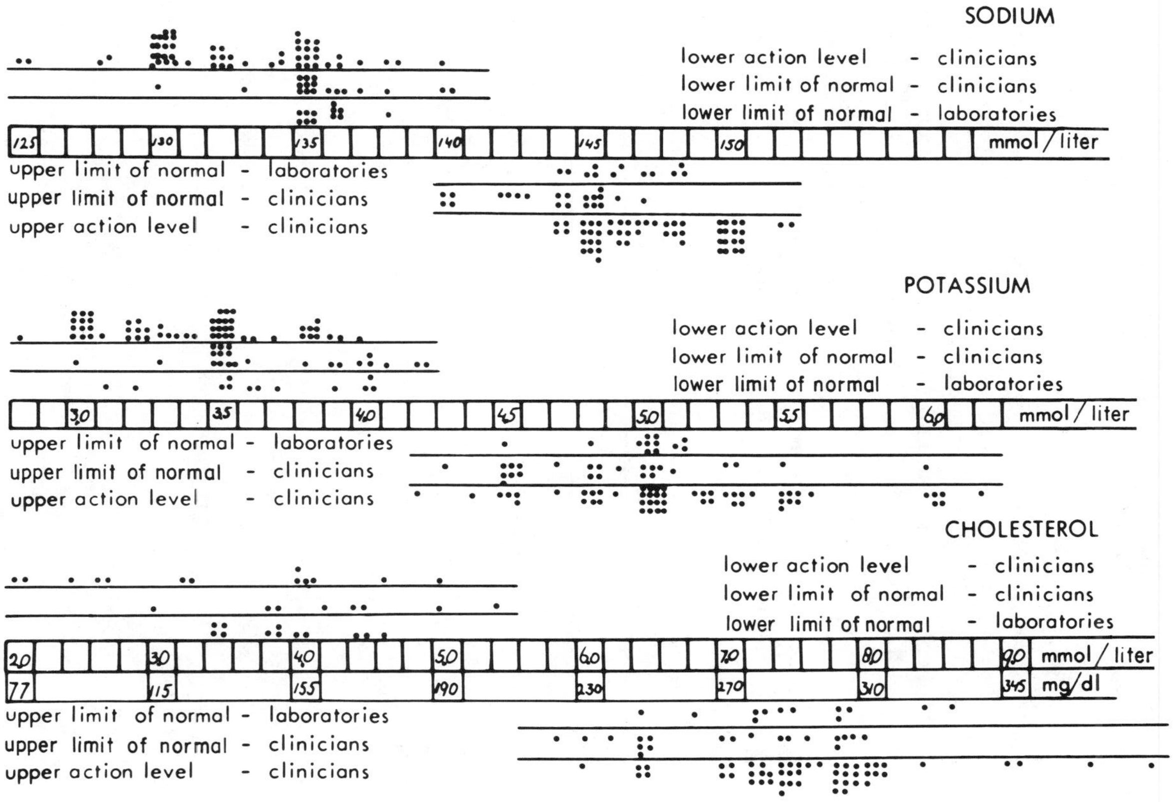

Fig. 1. Lower and upper limits of normal given by laboratory directors compared with normal range limits and action levels given by clinicians for sodium, potassium, and cholesterol. [From Elion-Gerritzen (E2), with permission.]

mens from patients (F9). An interesting use of the state-of-the-art data has continued to be advocated by Stevens and Cresswell (S19, S20, S21); these investigators have proposed that the level of performance achieved by the top 20% of United Kingdom laboratories that participate in the Wellcome Group Quality Control Programme should be used as an index of (1) what is technically possible and (2) as targets for other laboratories in that country. Since the standards of performance achieved are, in general, improving with time (Table 1), analytical goals derived using this strategy are continually changing. While analytical goal setting must not be a static facet of the discipline, goals derived in this manner may not be totally appropriate since it is likely that, in the future, technological advances will allow those laboratories which obtain new resources to attain performance characteristics that surpass analytical goals; goals derived from the state of the art of the best laboratories will therefore perhaps be too stringent. On the other hand, it can be argued that improved performance characteristics may lead to new

TABLE 1
STANDARDS OF PERFORMANCE ACHIEVED OR SURPASSED
BY THE TOP 20% OF UNITED KINGDOM LABORATORIES (SD)

Analyte	Units	1977[a]	1979[b]	1981[c]
Albumin	g/liter	1.0	0.9	0.8
Bicarbonate	mmol/liter	0.9	0.9	0.8
Bilirubin	μmol/liter	2.0	1.6	1.35
Calcium	mmol/liter	0.04	0.037	0.041
Chloride	mmol/liter	1.2	1.0	0.9
Cholesterol	mmol/liter	0.13	0.10	0.09
Creatinine	μmol/liter	7	6	5
Digoxin	nmol/liter	0.17	0.14	0.15
Glucose	mmol/liter	0.15	0.18	0.17
Iron	μmol/liter	1.0	0.90	0.75
TIBC	μmol/liter	—	2.9	2.4
Lithium	mmol/liter	0.04	0.035	0.033
Magnesium	mmol/liter	0.03	0.03	0.03
Phosphate	mmol/liter	0.03	0.033	0.033
Potassium	mmol/liter	0.05	0.053	0.05
Protein	g/liter	1.1	1.04	1.0
Sodium	mmol/liter	1.0	0.95	0.9
Thyroxine	nmol/liter	5.0	5.0	7.3
Triglycerides	mmol/liter	0.07	0.05	0.04
Urea	mmol/liter	0.3	0.3	0.21
Urate	mmol/liter	0.009	0.008	0.009

[a] Data from Stevens and Cresswell (S19).
[b] Data from Stevens and Cresswell (S20).
[c] Data from Stevens and Cresswell (S21).

medical uses of clinical chemistry test results or to the development of new strategies for the setting of analytical goals.

In the analysis of a series of specimens collected from an individual, there are a number of factors that may contribute to the variability in numerical results which is nearly always seen in practice. These are (1) preanalytical variation due to, for example, changes in patient preparation such as posture and exercise, in sample collection technique, in time of storage of specimen prior to analysis, and in the duration of centrifugation of blood specimens; (2) analytical variation caused by both imprecision and changes in random bias; and (3) biological variation.

Biological variation is a very important component of test-result variability but one which appears to be both poorly understood and often overlooked in interpretation of clinical chemistry test results. A number of analytes of interest to the clinical chemist, for example, plasma cortisol and female gonadotropins, vary, at least in health, in predictable circadian and monthly cyclical rhythms, respectively. However, for most analytes there are no clearly defined rhythms but rather, in the simplest model at least (H5), apparently random variations around set-point values that are characteristic for an individual. Such fluctuation around the set-point value is termed intraindividual biological variation. Variation between the set-point values of individuals is termed the interindividual biological variation.

Analytical goals have been postulated based upon biological variation (C4, Y2); it was recommended that tolerable analytical variability should be less than half of the relevant biological variation. The relevant biological variation for medical utility was said to be the combined inter- and intraindividual biological variation. However, there may be certain disadvantages with this approach. A number of studies of biological variation were published following the pioneering work of Cotlove and associates (C4). These studies invariably dealt with healthy adults of working age and, in contrast, the clinical chemist is mainly requested to perform tests on specimens from ill individuals. Moreover, as shown by Pickup and co-workers (P5), the biological variations found by different groups for certain analytes are not the same; the question of which estimate to use for analytical goal setting then arises. An excellent detailed tabular summary of the published estimates of the average components of biological variation in healthy individuals has recently been provided by Ross (R4).

Although there is a small amount of evidence to the contrary (R4), it has recently been shown that these potential disadvantages may not be of major significance. The short-term biological variation of eight commonly requested plasma analytes (sodium, potassium, chloride, bicarbonate, urea, creatinine, calcium, and albumin) was investigated in 20 patients who had suffered one uncomplicated myocardial infarction (F8). The average intraindividual biological variation for these analytes was larger, but of the same order as those found for healthy individuals (Table 2). This finding was expected since, in this particular study, preanalytical variance was not minimized and the particular analytes were not likely to have been disturbed by the disease process. It was concluded that the extensive published data on biological variation could be validly used more often in routine laboratory practice. In order to assess whether these findings were unique to patients with myocardial infarction and to assess the components of variation of certain plasma analytes significantly affect by a pathological process, the short-term biological variation of 11 commonly assayed analytes was investigated in nine patients with varying degrees of renal failure (F11). The estimates of the average intraindividual variation of each analyte were again, in general, of the same order as those documented in previous studies, from which a consensus estimate of the intraindividual biological variation of common analytes has been prepared by Statland (S16). This finding demonstrated that, in patients with impaired renal function, individuals, in general, have biological fluctuations around their own homeostatic setting points that are of the same order as those that occur in healthy individuals (Table 2).

It was considered attractive to believe that this finding would be general. The hypothesis was proposed that, in the short term at least, in all diseases that cause abnormal or normal clinical chemistry results, but in which a new homeostatic steady state is achieved, the biological intraindividual variations

TABLE 2
AVERAGE INTRAINDIVIDUAL BIOLOGICAL VARIATION (SD) IN HEALTH
AND IN PATIENTS WITH MYOCARDIAL INFARCTION
AND WITH RENAL DISEASE

Analyte	Units	Health[a]	Myocardial infarction[b]	Renal disease[c]
Sodium	mmol/liter	1.0	2.0	1.0
Potassium	mmol/liter	0.18	0.45	0.60
Chloride	mmol/liter	2.1	2.3	1.1
Urea	mmol/liter	0.66	0.97	0.85
Creatinine	µmol/liter	0.004	0.013	0.12
Glucose	mmol/liter	0.26	—	1.4
Calcium	mmol/liter	0.04	0.11	0.05
Albumin	g/liter	1.3	2.5	1.7

[a] Data from Statland (S16). The coefficient of variation quoted was multiplied by the mean of the reference interval quoted by Flinders Medical Centre in order to obtain SD.
[b] Data from Fraser and Hearne (F8).
[c] Data from Fraser and Williams (F11).

of plasma analytes would be similar to those found in a healthy peer group. It was therefore believed that the extensive published data on biological variation, generated in the main from adults of working age, should be and could validly be much more widely used in the interpretation of routine clinical biochemistry laboratory results. It was suggested that it was only in acute situations, for example, in plasma enzyme activity assays following myocardial infarction, or in the long term in chronic conditions, when deterioration or amelioration was occurring, that currently available data on biological variation could not be used. This hypothesis is supported by other recent work, for example, on plasma proteins and lipids in 20 healthy subjects and in patients with multiple sclerosis and chronic inactive pyelonephritis, and it was considered that the concordance between individual reference intervals suggested that use of biological variation data was valid even when such data were derived from healthy individuals (R1).

A number of individuals have proposed analytical goals for imprecision (F1, G4, S15). The goals derived in this manner have the disadvantage that they are empirical, although they may be based upon considerable experience and careful consideration of the very real problems of analytical goal setting.

The question of analytical goal setting has been considered by a number of expert groups. A group sponsored by the College of American Pathologists (E2) met in Aspen, Colorado in 1976, and a group sponsored by the World Association of Societies of Pathology (P7) met in London, England in 1978. These groups both adopted goals for imprecision based upon biological variation and, therefore, it is current expert consensus that *analytical variance should not exceed one-quarter of the relevant biological variance:*

$$SD^2_{analytical} \leq \tfrac{1}{4} SD^2_{biological}$$

If this strategy is adopted, then analytical variance is alleged to contribute less than 20% to the total observed variability. The statistical basis underlying setting analytical goals using biological variation data has been described (H4). The coefficient of variation was substituted for standard deviation in the following formulas to allow goals to be related to both concentration and activity levels and to previously published work.

For group screening, in order to identify ill individuals in a healthy population:

$$CV_{analytical} \leq \tfrac{1}{2}[(CV_{intraindividual})^2 + (CV_{interindividual})^2]$$

This should be the goal for long-term, between-batch, intralaboratory imprecision.

For individual testing and for serial testing on an individual, in order to aid in diagnosis or to monitor therapy:

$$CV_{analytical} \leq \tfrac{1}{2}CV_{intraindividual}$$

This should be the goal for short-term, intralaboratory imprecision.

The goals developed in this manner are shown in Fig. 2. The short-term goals derived in this manner are shown in numerical form in Table 3; data on biological variation used in the derivation are those found by Statland and Winkel (S17). For comparative purposes, the two currently most widely used sets of analytical goals are documented in Table 3; such goals are those proposed by Tonks (T2) and Barnett (B1). Tonks' formula or rule, based upon the reference interval, states that the allowable limits of error should not exceed one-quarter of the reference interval. This rule can be stated as follows:

2 CV = ± ¼ of the reference interval/mean of the reference interval

A maximum of ± 10% was allowed initially, but this was later widened to ± 20% (T3) for certain analyses, notably enzyme activity assays. However, as stated earlier, it is widely recognized that the reference interval is affected by analytical imprecision, analytical inaccuracy, and the characteristics of the population on whom the interval has been generated. Barnett (B1) listed medically significant CV values for 16 commonly requested analytes; the CV values were said to be derived from a synthesis of opinions of clinicians and laboratory specialists. This strategy also has disadvantages, as previously discussed, in that the views of clinicians are probably not based upon objective thought but on subjective assessment which has probably been colored

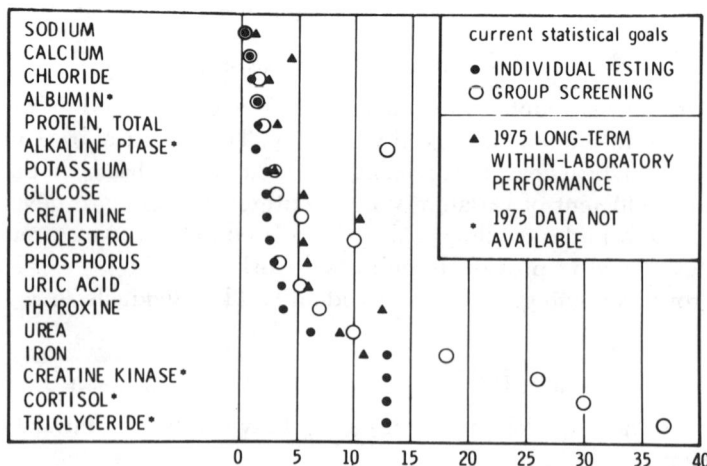

FIG. 2. Analytical goals, based upon biological variation, for individual testing and group screening as recommended by expert groups. [From Elevitch (E2), with permission.]

TABLE 3
Analytical Goals for Imprecision (CV)

Analyte	Consensus goals[a]	Tonks[b]	Barnett[c]
Albumin	1.4	6.3	7
Calcium	0.9	5.4	2.3
Chloride	1.1	1.9	1.8
Cholesterol	2.2	8.8	8.0
Creatinine	2.2	8.3	—
Glucose	2.2	7.4	4.2
Phosphate	2.9	8.8	5.6
Potassium	2.2	5.4	4.2
Sodium	0.4	1.1	1.3
Protein	1.5	4.3	4.3
Urea	6.2	10.0	7.4
Urate	3.7	10.0	8.3

[a] Data from Elevitch (E2).
[b] Data from Tonks' rule (T2); goals calculated from the reference intervals quoted by Flinders Medical Centre.
[c] Data from Barnett (B1); the most stringent goal quoted is given.

by past experiences, good or bad, with individual laboratories (F3). In addition, clinicians probably include preanalytical variation in considerations of the variability of laboratory results (P7); this may account for the fact that the standards of laboratory performance that were inferred to be required by clinicians in recent surveys (B3, S10) were generally worse than the standards actually achieved in interlaboratory quality assurance schemes. It can be seen that analytical goals based upon biological variation are more stringent than the currently most widely used goals; these latter goals have been of much use in the past in clinical chemistry, but it is considered that they are now of historical interest and should no longer be considered to be applicable to current clinical chemistry laboratory practice.

2.2. Goals for Imprecision of Urine Analytes

Although quantitative urine analyses form a significant part of the work of most clinical chemistry laboratories, there has until recently been little work on analytical goal setting for the performance characteristics of analyses of this biological fluid. Lately, however, analytical goals have been developed for the imprecision of 10 urinary analyses using three strategies, namely, the state of the art (S3), biological variation (S4), and clinical opinion (S5); the analytes studied were sodium, potassium, urea, creatinine, calcium, phosphate, urate, glucose, protein, and osmolality (see Table 4).

TABLE 4
ANALYTICAL GOALS (SD) FOR QUANTITATIVE URINE ANALYSES
DERIVED BY THREE STRATEGIES

Analyte	Units	State of the art[a]	Biological variation[b]	Clinical opinion[c]
Sodium	mmol/liter	5.0	15	0.7
Potassium	mmol/liter	2.5	7.7	0.5
Urea	mmol/liter	10.0	36	3.6
Creatinine	mmol/liter	1.0	1.7	0.4
Calcium	mmol/liter	0.1	0.6	0.2
Phosphate	mmol/liter	1.25	3.3	0.7
Urate	mmol/liter	0.13	0.5	0.1
Osmolality	mmol/kg	2.5	66	7.1
Glucose	mmol/liter	0.5	0.06	0.7
Protein	g/liter	0.05	0.009	0.01

[a] Data from Shephard et al. (S3).
[b] Data from Shephard et al. (S4).
[c] Data from Shephard et al. (S5).

Goals from the state of the art were obtained from the performance standards achieved by the best laboratories that participated in a regional quality assurance scheme conducted in South Australia; 19 laboratories analyzed 12 samples of urine, and statistical outliers, due to both analytical and transcription errors, were removed from the data base and goals were then derived from the truncated dispersions of results.

Biological variation data were derived from three series of urine samples obtained from 10 apparently healthy young men. Spot urine samples were collected over a 2-day period, daily 24-hour urine specimens were collected over 5 days, and one 24-hour urine specimen was collected 1 day per month for 5 days. Each set of urines was analyzed under optimal conditions of variance (W12) and, by analysis of variance techniques, analytical, intraindividual, and interindividual components of variance were dissected out. Analytical goals were derived from intraindividual biological variation using the statistical postulates of Harris (H4).

Clinical opinion was assessed by circulation of a questionnaire to all clinical staff of the institution of the author, Flinders Medical Centre, a 500-bed tertiary-care teaching hospital. The questionnaire contained an explanatory preamble and, for each of 10 analytes, a result and five options regarding the widest allowable variation were detailed. A representative section of the questionnaire is shown in Fig. 3; the questionnaire was specifically designed to assess views of clinicians on both analytical (laboratory) and total (which includes biological) variation. The mode of the responses of the clinicians

First result	Second result—tick widest allowable variation				
Sodium 10 mmol/liter	±1	±2	±4	±6	±8
Laboratory Total					
Sodium 100 mmol/liter	±1	±2	±5	±10	±20
Laboratory Total					

FIG. 3. Representative section of questionnaire designed to obtain analytical goals for the imprecision of urine analyses from clinical opinion. For each analyte, the options from which the clinicians could select their response were multiples of the between-day imprecision achievable in our laboratory as derived from replicate analysis of quality control material. [From Shephard et al. (S5), with permission.]

was selected as the goal since this estimate of central tendency would mean that the majority of the demands made by the clinicians were met.

The most stringent goals derived from the three approaches are shown in Table 4; the goals differ in magnitude and thus the problem arises of which goal should be selected. It was recommended (S5) that goals should be derived from the opinions of clinicians for analyses of sodium, potassium, urea, creatinine, phosphate, and urate. Goals for calcium and osmolality should be those derived from the state of the art and goals for glucose and protein should be those derived from biological variation data. It was stated that the rationale for urging the adoption of such goals was based upon two premises. First, it was believed that, in general, the strictest goals for all facets of analysis should be applied in all of the different clinical situations in which tests are used, because with use of this particular rationale, a single method with acceptable performance characteristics for inaccuracy, imprecision, sensitivity, specificity, turnaround time and other factors could be used in all clinical situations. Second, analytical goals should be targets that laboratories could achieve, with some effort and care, using current techniques and equipment.

Further analytical goals for urinary analytes have been developed during the evolution of the Australasian urine quality assurance program (S7); in standard deviation terms these goals are chloride, ± 5 mmol/liter; oxalate, ± 0.025 mmol/liter; hydroxymethylmethoxymandelic acid, ± 10 μmol/liter at ≤ 100 μmol/liter (± 10% at > 100 μmol/liter); and 5-hydroxyindoleacetic acid, ± 10 μmol/liter at ≤ 100 μmol/liter (± 10% at > 100 μmol/liter).

Many of the tests that clinical chemistry laboratories perform on urine are qualitative or semiquantitative in nature. The most important performance characteristic to consider for such tests is therefore the detection limit, the least amount of analyte that is detected as a positive test. However, for semiquantitative urine analyses, usually performed by commercially available reagent test strips, it is useful to have goals for the imprecision of such tests. It was proposed on an empirical basis by Hoeltge and Ersts (H10) that acceptable performance was any reaction grade one step higher or lower than the accepted value, acceptable results for hydrogen ion concentration were defined as the assayed pH ± 0.5, and it was stated that the specific gravity should be within ± 0.002 for material with specific gravity < 1.020 and within ± 0.003 for material with specific gravity > 1.021. Slightly more stringent goals have been set and used in the work carried out in the institution of the author (S8, S9), these being (1) a *negative* urine should be recorded as *negative* and (2) a *positive* urine should be recorded as the *true value ± one positive color block increment*.

2.3. Goals for Inaccuracy

Goals for inaccuracy have been suggested by few individuals and groups (F3). Gilbert (G4) proposed numerical goals for inaccuracy for a number of commonly requested analytes, Mitchell (M3) thought that ± 1.0% was generally acceptable for the inaccuracy of serum calcium analyses, the Proposed Product Class Standard (1974) for glucose analysis allows a maximum inaccuracy of 0.28 at 8.33 mmol/liter and of 1.67 at 8.33–16.7 mmol/liter (P1), and Stamm (S15) has proposed that, for inaccuracy, the maximum allowable deviation from reference method values should not be greater than plus or minus two times, or, as an alternative, plus or minus three times the maximum allowable standard deviation from day to day.

However, for a number of reasons (to be discussed later in this article), as highlighted by the Subcommittee on Analytical Goals in Clinical Chemistry of the World Association of Societies of Pathology (P7), neither the method of measurement nor the matrix should introduce a bias that will affect the clinical application of clinical chemistry test results. The ultimate analytical goal for inaccuracy is therefore that methods should have no inaccuracy.

2.4. The Total Analytical Error Concept

As stated by Westgard and associates (W5), to the analyst, imprecision means random analytical error. Inaccuracy, on the other hand, is commonly thought to mean systematic analytical error, which is the difference between the mean of the measured values and the true value; sometimes it may be found useful to divide this systematic error into constant and proportional components. However, none of this terminology is familiar to the clinician who uses test results and therefore he is seldom able to communicate with the clinical chemist in these terms. The clinician thinks rather in terms of the total analytical error, which includes both random and systematic components. From his point of view, all types of analytical error are acceptable as long as the total analytical error is less than a specified amount.

Indeed, it is difficult to truly separate imprecision and inaccuracy. As stated by Harris (H4) and Fraser (F4), a significant part of the characteristic that is usually calculated as imprecision may in fact be inaccuracy. Inaccuracy (or bias) can be divided into systematic bias (the real variation of the method from the true value) and random bias (the bias introduced by the operator or the method), as shown in Fig. 4. Random bias may occur due to, for example, changing the calibration setting of a multichannel continuous-flow system each time a calibrant is analyzed or from daily use of thawed aliquots of a frozen pool of standard. It is therefore difficult to completely separate imprecision and inaccuracy and it is of considerable advantage to consider combining analytical goals for imprecision and inaccuracy. Since the goal for inaccuracy is that methods should have no bias, the goals recommended for imprecision, based upon biological variation, should be used as goals for total analytical error.

This approach has been explicitly used for a number of years in certain interlaboratory quality assurance schemes (F3), particularly the Australian quality assurance programs conducted by the Australian Association of Clinical Biochemists and the Royal College of Pathologists of Australasia in the analysis of commonly requested serum analytes (P4), urine analytes (S6, S7), and drugs, and in the lipid survey conducted by the Australian Lipid Standardisation Committee. The basis for all of these programs is that use of preset acceptability limits for total analytical error and the target values generated, in general, by selected reference laboratories, facilitates feedback in a graphic manner; use of samples with linearly related levels of

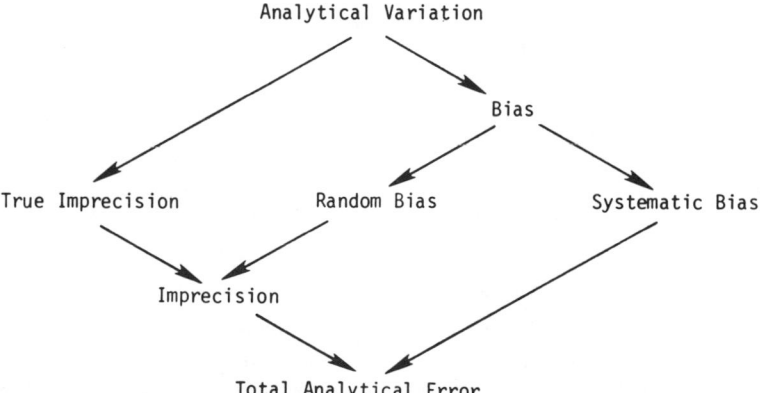

FIG. 4. Diagram demonstrating the interrelation of imprecision, inaccuracy (bias) of both random and systematic natures, and total analytical error. [From Fraser (F4), with permission.]

analyte allows simple calculation of overall laboratory error, graphic feedback of all results submitted by the individual laboratory, and comparison of performance between laboratories. An example of the feedback of laboratory results, illustrating the use of total analytical error, is shown in Fig. 5.

This approach has also been used in the Comprehensive Chemistry Survey of the College of American Pathologists for potassium, calcium, sodium,

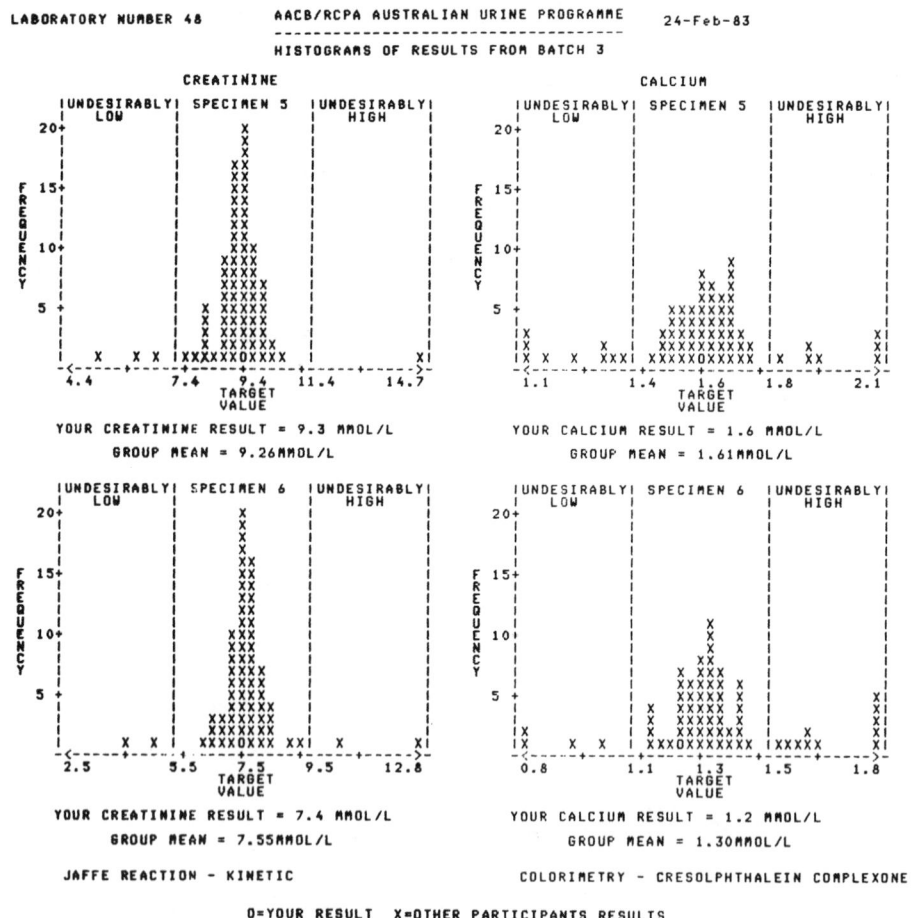

FIG. 5. Example of the report sent to laboratories participating in the Australian urine program. The report illustrates the use of analytical goals for total analytical error. The target values are set by use of results from reference laboratories.

and chloride. Results for these analytes are not graded on statistical deviation from a mean value. Participant performance is determined using fixed limits, which are, in reality, limits for total analytical error, and a target value. The target values for potassium and sodium are defined as the mean value of the results reported by all participants. The target value for calcium is the mean value of the results reported by participants using atomic absorption spectrophotometry. The target value for chloride is the mean value of the results reported by all participants using automated methods. The goals, which are probably equivalent to 2 SD, are as follows: potassium, target value ± 0.5 mmol/liter; calcium, target value ± 0.05 mmol/liter; sodium, target value ± 4.0 mmol/liter; and chloride, target value ± 5.0 mmol/liter.

The Australian and American approaches use different strategies to set the target values. The scheme of the Federal Republic of Germany also uses the reference laboratory approach (H3), while the National External Quality Assessment Scheme of the United Kingdom uses the consensus mean approach (W13). The advantages and disadvantages of these have been recently discussed in detail by Whitehead and Woodford (W13). However, it appears in practice to matter little which approach is used, particularly since consensus values are in general not significantly different from the target values obtained in reference laboratories, consensus means achieved for the same material in the United Kingdom and the Netherlands were in agreement, and consensus means are in good agreement with definitive values for the few analytes for which definitive methodology is available (W13).

As stated by Fraser (F3) and Westgard (W3), many analytical goals proposed previously and used as goals for imprecision, for example, those of Tonks (T2) and Elion-Gerritzen (E7), are implicitly goals for total analytical error. The more widespread use of goals for imprecision as goals for total analytical error is strongly advocated.

2.5. Goals for Detection Limit

The detection limit is, as stated earlier, the particularly important performance characteristic for tests used in a qualitative or semiquantitative manner, but there has been little work performed in this area. It has been suggested (K6) that the practical sensitivity, that is, the lowest concentration that can be detected in 90% of cases, for urine semiquantitative tests should be glucose, 2.8 mmol/liter; proteins, 0.2 g/liter; ketones, 1.0 mmol/liter; urobilinogen, 1.7 mmol/liter; bilirubin, 8.6 µmol/liter; blood, 10,000 erythrocytes/ml; nitrite, 0.015 mmol/liter; and phenylpyruvate, 0.5 mmol/liter.

For plasma or serum analytes, it has been stated (B1) that there is a low level below which good performance characteristics are not required; a report that the concentration was lower than that value was stated to be adequate for medical purposes. These empirical goals for detection limits included glucose, 1.1 mmol/liter; urea, 0.4 mmol/liter; protein, 20 g/liter; bilirubin, 7 μmol/liter; sodium, 100 mmol/liter; and potassium, 1.5 mmol/liter.

2.6. Goals for Linearity

It has been recommended that methods for plasma analytes should have the following ranges of analytical linearity (E2): albumin, 10–60 g/liter; protein, 40–100 g/liter; alkaline phosphatase, 1–6 times the reference interval; glucose, 0.6–27.8 mmol/liter; creatinine, 0.044–0.530 mmol/liter; cholesterol, 3.9–10.4 mmol/liter; urate, 0.24–0.89 mmol/liter; urea, 1.5–15.3 mmol/liter; and creatine kinase, 1–10 times the reference interval.

2.7. Goals for Turnaround Times

A number of factors impinge upon the turnaround times of clinical biochemistry tests. It has been stated (B1) that performance characteristics of a degree greater than is useful clinically should not be required if extra time is thereby made necessary and that an approximate result may be more useful than an exact result reported after a long delay. Moreover, many laboratories perform tests in batches which are analyzed depending on considerations such as laboratory philosophy, staff availability, instrument design, and analytical capacity. The paucity of work on goal setting for turnaround times has made it difficult to be objective in this area.

It has been advocated that local goals should be set up by a committee comprising both clinical and laboratory staffs (A2). A survey of clinical opinion (M6) on the turnaround times required for 36 commonly performed plasma analyses appeared to show that the usual mode of operation of taking samples early in the morning and returning the results sometime on the same day was satisfactory but, as stated earlier, it is probable that the clinical opinions given were subjective and colored by past experience and common practice.

Barnett and associates (B2) and Watkinson and Fraser (W2) have studied by survey various aspects of the provision of emergency clinical chemistry laboratory services in the United States and Australian laboratories, respectively. The turnaround times achieved by laboratories were documented; it appeared that laboratories could achieve the turnaround times alleged to be necessary by a group of clinicians (M6), which, in emergency or stat situations, were 30 minutes, blood gases; 60 minutes, electrolytes, urea, creatinine, calcium, osmolality, amylase, and drugs; and 120 minutes, bilirubin.

There is no doubt that further objective studies are required in this facet of the discipline and it has been suggested that goals for turnaround times might be derived in a more theoretical manner, for example, based on a fraction of the half-life of the analyte (F6).

Although it has been stated (B1, P7) that an approximate result reported quickly may be clinically more useful than an excellent result reported with undue delay, it has been advocated (W2) that results from emergency and routine analyses should be totally comparable. This infers that the most stringent goals for performance characteristics should be applied no matter what clinical use is made of the test results.

2.8. Goals for Preanalytical Factors

The overall variation that influences the application of numerical clinical chemistry results does not only include analytical and biological variation but also preanalytical variation. It has been recommended (P7) that goals should be set for a number of factors in order to minimize preanalytical variation (these factors are detailed in Table 5) and that the following should be considered:

1. The adverse effects of certain preanalytical variation on the reliability of laboratory test results should be recognized.

2. Efforts should be consistently made to minimize the adverse effects of those preanalytical variables that can be controlled.

3. Information relevant to the preanalytical state of specimens should be recorded in order to decide on the magnitude of their influence on laboratory results.

4. There should be acceptance by both clinical and laboratory staffs that there might, in certain circumstances, be preanalytical variation that was of such a magnitude that it would be inappropriate to request or perform clinical chemistry tests because the results would be misleading.

There have been a number of attempts to define certain of these factors, for example, transport of specimens for clinical chemistry analysis (W15). Documents on the collection of arterial blood for laboratory analysis (N2) and on standard procedures for the handling and processing of blood specimens (N4) are examples of the significant interest that the National Committee for Clinical Laboratory Standards of the United States appears to have in these subjects. In contrast, other groups such as the International Federation of Clinical Chemistry and the European Committee for Clinical Laboratory

TABLE 5
Preanalytical Factors for Which Goals Are Required

Factor	Subfactors
Initial preparation of patient	Diet
	Discontinuance of drugs
Information required on request form	Fasting or nonfasting
	Clinical diagnosis
	Age
	Sex
	Degree of urgency
	Therapy and time of last dose
	Time of collection
	Occupation (if relevant)
Conditions of specimen collection	At rest or exercised
	Posture
	Stress
	Tourniquet application
	Room temperature
Nature of specimen	Biological fluid
	Unique or part of series
Identification of specimen	At collection
	After collection
Stability of analyte	Analyte
	Preservative
	Type of container
Details on arrival	Time delay
Specimen handling procedures	Preservation and storage
	Separation technique
	Centrifugation
Appearance of specimen	Lipemia
	Hemolysis
	Icterus

Standards have contributed little. However, there is considerable literature on laboratory management and many publications on methodology contain details of specimen collection and storage for the particular analyte under study. The relationship of clinical chemistry results to selected physical, dietary, and smoking activities has been extensively reviewed (S17, S18). Variability due to aging, sex, race, diet, pregnancy, drug administration, and *in vitro* effects has been carefully documented (Y1). Specimen collection, preparation, and storage requirements for clinical chemistry endocrinological tests have been detailed (I1). These are examples of the plethora of data available in this area. It is considered that a computerized collation of the data, in a form similar to available listings of the effects on drugs on clinical laboratory tests (Y3) and of the effects of diseases on clinical chemistry tests

(F12), would be a most useful compendium of valuable information and might allow official professional bodies to provide firm recommendations and guidelines which could advantageously be adopted as preanalytical goals.

3. The Rationale for Achievement of Goals

3.1. Goals for Imprecision

The clinical chemist, unlike counterparts in other branches of clinical pathology, usually reports the results of tests as single numerical results. Low imprecision is required to ensure that the results have little random variation. In addition, as stated earlier, low imprecision means that the reference interval is not unduly widened due to analytical factors. Moreover, low imprecision means that changes in numerical results, seen on analysis of serial specimens taken from an individual patient over time, will reflect real changes in the analyte under investigation since, in the simplest model, an analytically significant change occurs (with 95% confidence) only if two consecutive results differ by more than 2.8 analytical standard deviations (W10). Low imprecision also allows more strict definition of other performance characteristics.

3.2. Goals for Inaccuracy

It is essential that clinical chemistry laboratory data are capable of being transferable across both time and geography (P7). It has been stated that such transfer can be achieved effectively only if laboratory results have no inaccuracy. In addition, a number of definitive criteria for interpretation of test results are now available, for example, the World Health Organization (W14) has recommended that, in a patient with symptoms of diabetes mellitus, a fasting venous plasma glucose of > 8.0 mmol/liter is diagnostic of diabetes. If such criteria are to be used correctly, analytical methods must have no inaccuracy.

3.3. Goals for Detection Limit

It is important that detection limits be well defined, particularly for semiquantitative or qualitative clinical chemistry tests. A detection limit that is too high will mean that false negatives will occur, that is, the clinical sensitivity, the probability that a test result is abnormal in the presence of disease, will be low. A detection limit that is too low will result in false positive results, that is, the clinical specificity, the probability that a test result is normal in the absence of disease, will be low.

3.4. GOALS FOR LINEARITY

The problems of dilution and reanalysis of specimens with high levels of analyte are minimized if methods with wide ranges of linearity are used (E2). In addition, a linear relationship between concentration or activity and a parameter such as absorbance makes calculation simple, facilitates calibration, and aids in gaining optimal imprecision and inaccuracy.

3.5. GOALS FOR TURNAROUND TIMES

Dynamic function tests approach the ideal because they interrogate the homeostatic mechanisms that are peculiar to the individual. Another major advantage is that minor degrees of dysfunction may be detected; such abnormalities may be difficult to pick up in the usual comparison of a single numerical result with a conventional population-based reference interval. An example is the use of the thyrotropin-releasing hormone test (TRH test) in the diagnosis of borderline hyperthyroidism; this test investigates the response of the anterior pituitary of an individual to the plasma levels of free thyroid hormones of that individual.

In contrast, most clinical chemistry tests are static, and as soon as a sample is taken, that point in the homeostatic life of the patient has become historical. Thus, rapid turnaround times are required in order to ensure that the clinician obtains a result which represents as recent a point in history as possible (F6).

In emergency or stat situations, results are required which are crucial to clinical decision making and therefore, very rapid turnaround times may be necessary prerequisites for optimal patient care. An example is that P_{O_2} determinations in the premature neonate under oxygen therapy are required rapidly, since even short periods of hyperoxia may cause an increased incidence of blindness due to retrolental fibroplasia.

Laboratories often allege that their emergency workload is excessive (M1, W8). Even though this work can be cut by adoption of a number of strategies (C3, S13, W2, Y4), it is probable that if the turnaround times were shorter and reports of results were returned promptly, then the number of alleged emergency requests would fall; only genuine emergency tests would be submitted as such and these would be able to be dealt with promptly. For example, in the institution of the author, electrolytes are performed in a number of batches during the working day (at 0800, 1100, 1400, 1600, and 1900 hours). Clinicians now submit a number of requests stating that these tests are to be performed in the next batch and the results phoned back when the batch is complete; this mode of operation has, inter alia, led to a significant decrease in the number of requests for so-called emergency tests.

3.7. Goals for Preanalytical Factors

It is important, as recommended by the Subcommittee on Analytical Goals in Clinical Chemistry (P7), that a state of education, understanding, and practice is achieved in which preanalytical variation can be minimized. Only by setting preanalytical goals which can be achieved, together with knowledge of analytical and biological variation, can the true dispersion of each numerical result be known.

4. Uses of Analytical Goals

4.1. Quality Control

Many publications document the wide variety of techniques available for internal quality control of all facets of laboratory performance (B11, G6, W9, W11, W12). The most frequently used technique is the reference sample method (L2), which is based upon the assumption that results of analyses of quality control samples included in an analytical batch do truly reflect the results obtained with the patient's specimens that are analyzed in the same batch (B11). The quality of such quality control procedures has been investigated in depth by Westgard and associates over the last few years; this work has resulted in the Selected Method of the American Association for Clinical Chemistry (W4), which discusses the statistical basis for interpretation of the results obtained on analyses of quality control materials and recommends a multirule Shewhart chart for quality control in clinical chemistry which provides (1) simple data display and analysis, (2) facile adaptation and integration into existing practices, (3) a low level of false rejections of data, (4) an improved error detection capability, and (5) some indication of the type of analytical error that has occurred.

However, it has recently been stated by a number of authors that such quality control procedures require modifications. Stamm (S15) has advocated that the maximum variation in day-to-day imprecision and the maximum allowable variation from a reference method value should be set on the basis of clinical requirements. The clinical requirements advocated by Stamm are those based upon biological variation (E2, P7), or where such data does not exist, on Tonks' rule (T2). A more thoughtful approach is that proposed by Ross and Fraser (R5); it was suggested that a patient-care-based internal quality control program can be established by first setting analytical goals for imprecision. If these goals are less stringent than the actual performance achieved by the laboratory, then they should be used as action limits rather than the statistical limits derived from previous performance. A stated proviso was that when statistical limits were exceeded, corrective action should be scheduled, but the current batch of results could be reported. Immediate corrective action should be taken, however, when results of analyses of quality control samples exceeded the analytical goals. Philosophical guidelines that aid in designing a quality control procedure which is capable of achieving analytical goals have been described (D1).

These approaches have considerable advantages. Less attention is required for methods that meet analytical goals, the repetition of a certain percentage of tests necessitated by statistical quality control procedures is eliminated, and the introduction of new expensive technology is possibly delayed (R5). However, statistical quality control procedures should be used for methods that do not meet goals. These methods must be replaced or improved, but it should be noted that use of narrow quality control limits based on analytical goals is not a logical or correct strategy for method improvement.

4.2. Method Evaluation

There are many published protocols for the objective evaluation of instruments, methods, and reagent kit sets and these have been extensively reviewed recently (L4, W3,). Many of the protocols detail statistical calculations to be used in examination of the wealth of data generated during an evaluation, but few detail strategies to judge the acceptability or otherwise of the results.

Westgard and associates (W5, W6) have formulated criteria to judge whether an analytical method has acceptable performance characteristics. Performance is acceptable when the total observed errors (TE) are smaller than the 95% limit of the allowable error (E_a). Random error (RE) is determined by studies on imprecision, proportional error (PE), from recovery experiments as the percentage error (recovery − 100%), constant error (CE), from interference experiments, and systematic error (SE), from the regression equation ($Y_c = a + bX_c$), which is used to calculate the difference between X and Y values at critical medical decision-making levels of analyte (X_c). The criteria are as follows:

$$RE = 1.96\ SD < E_a \tag{1}$$

$$PE = [(\text{Recovery} - 100)(X_c/100)] < E_a \tag{2}$$

$$CE = \text{bias} < E_a \tag{3}$$

$$SE = [(a + bX_c) - X_c] < E_a \tag{4}$$

$$TE = 1.96\ SD + [(a + bX_c) - X_c] < E_a \tag{5}$$

However, it was stated that the evaluator must use professional judgment to allocate values for E_a and X_c.

Kim and Logan (K2, K3) established criteria for use in the evaluation of clinical chemistry methods. Permissible limits of variation (PLV) and permissible limits of discrepancy (PLD) were advocated for assessment of imprecision and inaccuracy, respectively. The somewhat complex treatment of data required in fact implies that the data generated should not deviate by more than ± 25% of the reference interval from the target value, that is, Tonks' rule (T2) should be applied.

However, the purpose of method evaluation is to assess whether or not the analytical performance of the method is satisfactory (W5). Although published papers on evaluation or method development usually conform to generally accepted evaluation protocols, objective discussion of the performance characteristics found during the evaluation is usually significantly lacking. For example, in a single recent issue of *Clinical Chemistry*, the following statements on results of evaluation were made: (1) a high degree of analytical precision is required in addition to reasonable accuracy (K1); (2) the within-assay and between-assay precision compares favorably with the results published by others (L1); (3) the method demonstrates good precision (P6); (4) the coefficient of variation is acceptable (W1); (5) the advantages are accuracy and precision superior to other published HPLC methods (S22); (6) it demonstrates excellent precision (R2); and (7) within- and between-assay precision is good, recovery is excellent (I2).

This problem is not peculiar to *Clinical Chemistry* (F5) and it has been firmly stated that it is surely no longer satisfactory for published evaluations of instruments and reagent kits to merely use professional judgment or the specifications of the manufacturer as criteria of acceptability. It is firmly believed (F5) that it should be *mandatory* for all evaluators to carefully compare the performance obtained with objective consensus analytical goals and to make scientific judgment on the acceptability or otherwise of as many of the performance characteristics as possible.

4.3. Method Selection

Objectivity is not always used when selection of a method is carried out (B8); many considerations are usually taken into account, including skills required, staff availability, space and services required, capital and running costs, sample volume required, flexibility, emergency analysis capability, and laboratory philosophy (B8, B9, C5, H1, H9, M4, S1). Therefore, in addition to using objective analytical goals in the assessment of data obtained in an evaluation, such goals should be used in selection of methods, instruments, and reagent kit sets.

A wealth of data is available in the literature on the performance characteristics of particular methods. A detailed bibliography of instrument evaluations has recently been published (D2). The reports of interlaboratory quality assurance schemes provide much information on the performance of methods, instruments, and reagent kit sets; unfortunately, much of this information is made available only to participants in these schemes, although there are many published reports of small interlaboratory programs, particularly those dealing with more specialized analytes. The Australasian quality assurance programs have made the results of their urine program (S6, S7) and general serum program (P4) accessible in the indexed literature and the College of American Pathologists has provided extensive documentation on performance in their quality assurance surveys in yearly supplements in the *American Journal of Clinical Pathology* and in a recent book (E3).

It is considered that the most important considerations in selection of a method and in purchasing instruments and reagent kits are the performance characteristics. The available data on specifications should be compared to objective analytical goals and, if at all possible, analytical systems that do not meet or surpass all goals should not be introduced.

4.4. Laboratory Management

Numerical analytical goals have a number of possible uses in laboratory management (F5). Comparison of the actual performance achieved on analyses of specimens from patients, as inferred from quality control sample analyses, with analytical goals allows laboratory staff to select those methods that are satisfactory and those that require improvement.

Laboratory management could use a comparison of their analytical performance with published goals as an additional lever to obtain new analytical equipment, if it could be shown that their existing performance was unsatisfactory and that instrumentation with superior performance characteristics was available.

Published analytical goals could assist laboratory staff in communication with clinicians. Allegations that a particular laboratory test was inadequate could be refuted in an objective manner, provided that the goals were met.

4.5 Other Uses

As stated previously (F5), manufacturers of analytical instruments and reagent kit sets have played a notable role in the development of clinical chemistry, and it could be argued that the present analytical state of the art has been, in large part, set by commercial interests. It is believed that analytical goals set by the profession should be taken very much into account when manufacturers develop new instruments or reagent kit sets.

Guidelines for the labeling of reagent kit sets have been developed (E1, N1, R6) and protocols for the evaluation of literature provided by manufacturers have been proposed (A1, K3, K4, L3). Labeling requirements for *in vitro* diagnostic products for human use are the subject of legislation in certain countries (L4). The Expert Panel on Instrumentation of the International Federation of Clinical Chemistry has published provisional guidelines for listing the specifications of atomic absorption spectrometers (E9), spectrophotometers (H2), flame emission spectrometers (B5), and clinical chemistry analyzers (O1). However, there does not seem to be a requirement for any of these instruments to have documented, objective, analytical goals to allow simple user comparison of specified performance characteristics with such goals. It is hoped that, in the future, such documentation will be deemed necessary.

For a number of reasons, accreditation of various laboratory functions has occurred, to a greater or lesser extent, in many countries. It is considered that regulatory agencies should use current consensus analytical goals as standards or targets for acceptable performance.

5. Achievement of Analytical Goals

5.1. Assessment of Current Laboratory Performance

A 5-year study of intralaboratory imprecision of plasma analytes was recently published (R5); the study was based upon the data base of the College of American Pathologists Quality Assurance Service from 1975 through 1980. It was stated that there were trends toward automation and toward improved imprecision in both automated and manual/semiautomated categories. The percentage of laboratories whose CV exceeded medical usefulness criteria was also stated to have increased; a high current standard of imprecision performance was alleged to have been reached and, in the automated category, the data indicated that greater than 98% of individual analyses met the criteria.

However, even though imprecision of many laboratory tests has improved in recent years, it is firmly believed that current laboratory performance generally does not meet goals. Ross and Fraser (R5), using goals proposed by Barnett (B1) that do have disadvantages, have in fact probably used inappropriate standards of performance. Average laboratory performance achieved in analyses of plasma analytes in the United States (R5), the United Kingdom (S21), and Australasia (P4), and analytical goals based upon biolog-

TABLE 6
Average Performance Achieved in Three Countries and Analytical Goals Based upon Biological Variation (CV)

Analyte	USA—1980[a]	UK—1981[b]	Australia—1982[c]	Goal[d]
Albumin	3.4	3.8	4.0	1.4
Bilirubin	9.1	5.9	7.2	—
Calcium	2.2	2.2	2.8	0.9
Chloride	1.7	1.5	2.1	1.1
Cholesterol	3.8	4.6	—	2.4
Creatinine	4.3	3.8	6.9	2.2
Glucose	2.9	2.8	4.2	2.2
Phosphate	3.0	3.7	5.2	2.9
Potassium	2.1	1.8	2.6	2.2
Sodium	1.1	1.0	1.2	0.4
Protein	2.2	2.2	3.1	1.5
Triglycerides	5.2	11.2	—	13.0
Urea	3.2	2.9	4.6	6.2
Urate	3.0	3.7	0.7	3.7

[a] Data from Ross and Fraser (R5).
[b] Data from Stevens and Cresswell (S21).
[c] Data from Penberthy (P4).
[d] Data from Elevitch (E2).

ical variation (E2) are shown in Table 6. It is evident that, in general, goals are not met. This finding has also been reported from the Nordic countries (H11). In addition, goals were not generally met by laboratories that participated in external quality assurance surveys of the performance of urinary analyses (G2, S6, S7).

The current performance characteristics of methods are, as stated earlier, to a large degree dictated by the manufacturers of instruments and reagent kit sets. Unfortunately, new equipment which has performance characteristics that do not approach goals continues to appear on the market. As one recent example, the imprecisions obtained for plasma analytes on an analyzer evaluated by Arntsen and associates (A3) do not meet goals based upon biological variation (Table 7). Moreover, goals for performance characteristics other than imprecision may not be met by currently available products. For example, the detection limit for urinary glucose required by goals (K6) was not met by one commonly used multitest urine reagent strip (S14), and the goal for the detection limit for urinary bilirubin was not met by another similar product (H6).

Since laboratory performance does not generally meet analytical goals, improvement of performance is required. The procedures to be adopted have been recently discussed in detail (F4) and will be briefly reviewed in

TABLE 7
COMPARISON OF BETWEEN-RUN CV AT NORMAL ANALYTE
LEVELS OBTAINED ON A NEW ANALYZER AND ANALYTICAL GOALS

Analyte	Between-run CV[a]	Analytical goal[b]
Sodium	1.44	0.4
Potassium	3.88	2.2
Calcium	2.01	0.9
Phosphate	4.11	2.9
Glucose	3.61	2.2
Creatinine	4.75	2.2
Urea	4.70	6.2
Urate	1.51	3.7
Protein	1.81	1.5
Albumin	3.01	1.4

[a] Data from Arnsten et al. (A3).
[b] Data from Elevitch (E2).

this article. Each laboratory must review its own data to assess the methods that require improvement (Fig. 6). The first stage in the review of data is to collect as much information as possible on the performance characteristics of the method, including data on imprecision, inaccuracy, linearity, detection limit, and speed of analysis; such data may be obtained from internal quality control programs, external quality assurance schemes, previous, extended performance checks, and preventative maintenance records of the individual laboratory. The performance characteristics achieved should then be com-

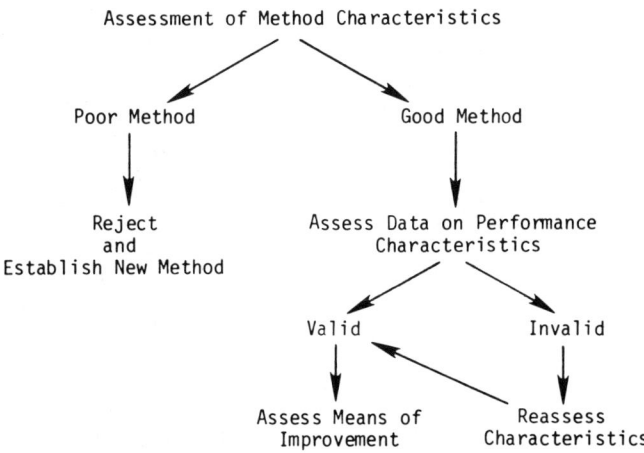

FIG. 6. Summary of steps required in assessment of methods prior to their replacement or improvement. [From Fraser (F4), with permission.]

pared to objective numerical analytical goals. If the performance characteristics meet or surpass analytical goals, then the method is satisfactory and may be maintained, provided ongoing quality control demonstrates continuing satisfactory standards of performance. If the performance characteristics do not meet goals, it is advisable to check whether the method itself is unsatisfactory or whether the method is performed badly in the laboratory. This may be assessed by careful examination of data obtained from interlaboratory quality assurance schemes. If the performance characteristics of a method are poor in many laboratories, it is likely that the method is intrinsically unsatisfactory and it should be replaced by a satisfactory method. However, if a method has acceptable performance in many laboratories, but unacceptable performance in the individual laboratory, it may be that the laboratory can improve the performance characteristics by use of a number of simple strategies; improvement is preferable to replacement since the many problems and costs of method selection, evaluation, and introduction are not encountered. Before a final decision is made to improve an existing method, the data on the performance characteristics should be briefly checked by, for example, duplicate analysis of specimens from patients to assess imprecision and analysis of standard reference materials to assess inaccuracy; this check is necessary to ensure that the poor performance evidenced, in the main from analyses on quality control materials, is a real phenomenon that does occur with specimens from patients and is not an artifact introduced by the reference samples (F9). Moreover, as recently pointed out by Tonks (T4), it is certain that large external quality assurance schemes cannot tolerate innovations and, if it is aimed to obtain apparent good performance in a scheme that has a rating scale, a laboratory must use exactly the same methods used by a large number of other laboratories and must limit the number of changes made in methodology. If the criterion for acceptable performance is the standard achieved by the majority, then those laboratories that achieve the state of the art are rewarded, and the innovative, who have carried out method improvement, are downgraded.

5.2. Improvement of Methods

There are many strategies which can be adopted to improve methods (F4).

1. *Documentation.* Preparation of detailed procedure manuals (N3) and strict adherence to the documented procedure are required for optimal performance.
2. *Quality Control.* Data on basic methodological parameters, for example, absorbance of standards and quality control materials, absorbance of reagents, pH of reagents, instrument settings, and radioactive counts of standards should be collected.

3. *Maintenance.* Preventative maintenance of instruments and use of simple checklists of the maintenance required assists in keeping instruments at peak performance.

4. *Reagents.* Careful assessment of reagents is required. Suitable batches or lots should be purchased in bulk to ensure ongoing acceptable performance.

5. *Basic Instrumentation.* Quality control of balances, water baths, pipettes, diluters (G1), and temperature control in cuvettes (B7) is required.

6. *Replicate Analyses.* Performance of an analysis n times leads to improvement in imprecision by $\sqrt{n}$ times. In addition, gross errors are easily identified if replicate analyses are performed and this is probably the best use of duplicate analyses.

7. *Staff.* Alteration of staff attitudes, education and training of staff (M5, R3), and investigation of staff who are prone to make errors (G7) all play a part in achieving optimal performance in the laboratory. A large contribution to good performance is made by sound laboratory management, high staff morale, and efficient organization.

8. *Automated Calculation.* In an interlaboratory survey, manual methods of data reduction of radioactive counts were four to seven times more variable than automated logit·log fit (J1, C2). Automated calculation may be a useful means of improving all analyses, provided that the analyst retains a decision-making role regarding both outliers and the suitability of the standard curve or line (F7).

9. *Significant Figures.* The number of significant figures used in calculation and in reporting of results should be carefully considered in order to ensure that the performance characteristics that can be gained are truly reflected in the results reported.

10. *Standardization Procedures.* Although it has been alleged that clinical chemistry tests should be standardized with the purest standards available in the most definable solutions (M2) and that use of precalibrated test procedures is to be severely condemned (H8), there are problems with attaining these ideals in practice (S2). This has been objectively studied (H7, P2); the imprecisions obtained using two standardization techniques were studied for four spectrophotometric procedures performed both manually and on two centrifugal analyzers. The *variable calibration mode* is defined as the within-run absorbance of a standard and its assigned value in the classic equation:

Concentration = (test absorbance/standard absorbance) × standard value

The *constant calibration mode* is defined as a direct calibration relationship relating concentration to absorbance in a constant way that does not change from batch to batch. The conclusions were that the mode of standardization adopted should, for optimal imprecision, be objectively chosen for each method as a consequence of experimental study and, if the variable calibration mode was selected, that the standard should be carefully chosen. Additional studies were recently performed (P3) on a model for enzyme activity assays, which are usually standardized using the constant calibration relationship. The magnitude of variation in the constant relating concentration to absorbance was examined over 74 days on two centrifugal analyzers using a stable acidic solution of potassium dichromate. The within-run imprecisions met the specifications of the manufacturers. However, the mean absorbances found over 75 days showed considerable fluctuations and proved that the assumed constant was, in practice, a variable (Fig. 7). It was therefore recommended that laboratories monitor day-to-day variation using stable acidic solutions of potassium dichromate in addition to the recommended monitoring of within-run imprecision, and should take appropriate action if variation occurs outside preset performance limits.

11. *Work Study.* A detailed study of all the steps of a method, and subsequent simplification of the method, can result in improved performance (W9).

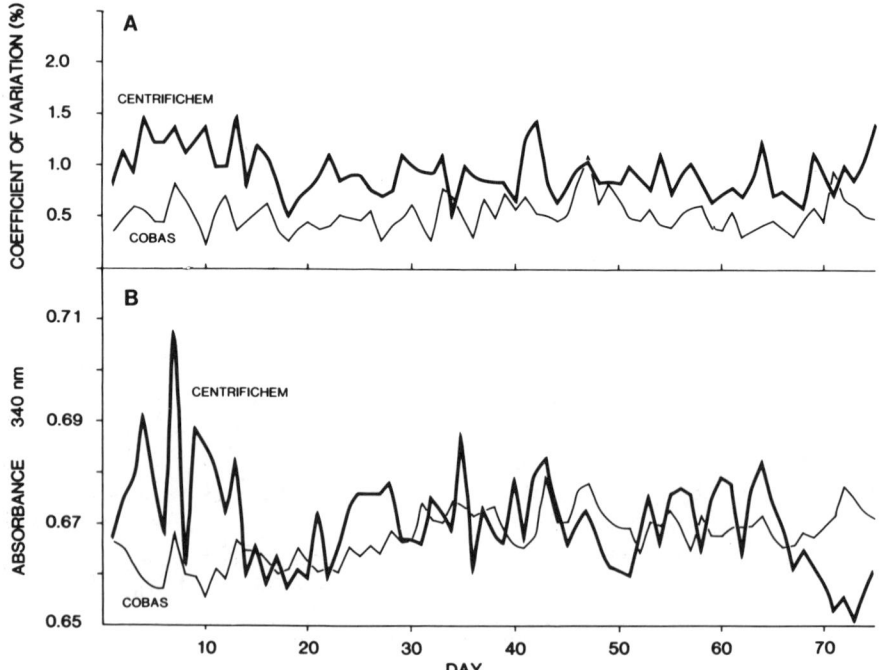

FIG. 7. Daily within-run imprecisions (A) and mean absorbances (B) for 5-μl samples of acid potassium dichromate solutions analyzed over 75 days on CentrifiChem and Cobas Bio centrifugal analyzers to demonstrate that constants used to calculate enzyme activities from absorbances are really variables. [From Peake *et al.* (P3).]

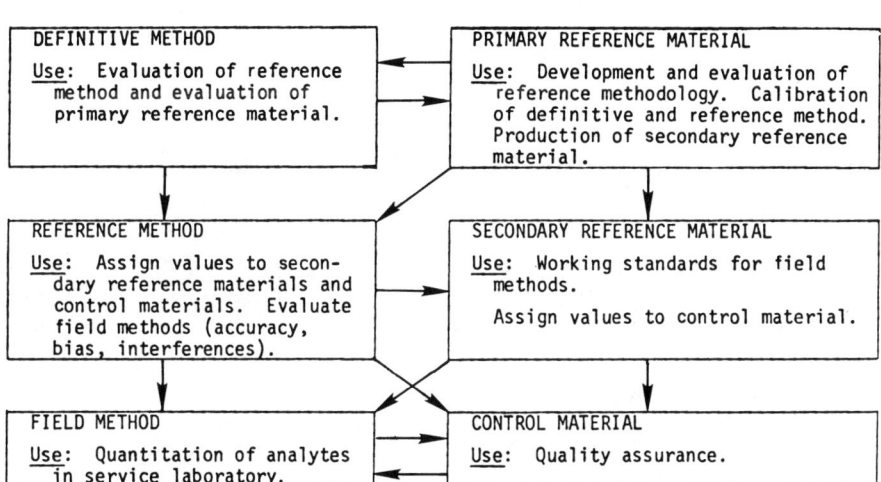

FIG. 8. Hierarchical structure of a comprehensive measurement system showing interrelationships between the basic measurement components. [From Tietz (T1), with permission.]

12. *Inaccuracy.* Inaccuracy must be tested on a continuing basis by external quality assurance procedures. As stated by De Verdier and associates (D1), such procedures should be optimized to provide a reliable estimate of error by making several replicate measurements and calculating the average result. Coordination of internal quality control, external quality assurance, reference methods and materials, and definitive methods (B6, T1), as shown in Fig. 8, will result in achievement of the analytical goal of no inaccuracy. The excellent review of Tietz (T1) tabulates in detail the clinical standard reference materials available from the National Bureau of Standards and the status of definitive and reference methods for clinical chemistry as of March, 1979.

5.3. ALTERNATIVE STRATEGIES

A few alternatives are available if a laboratory method cannot be improved such that its performance characteristics do meet objective analytical goals (F4). A new method may be introduced. Laboratories that can perform the method to a satisfactory standard may be requested for assistance. The organizers of professional external quality assurance schemes usually have a wealth of data on methods, instruments, and reagent kit sets and can often provide direct advice or assist in putting one laboratory in touch with another. It may be advantageous for the laboratory to stop performing the method and send all specimens submitted for analysis to another laboratory which can perform the method to a standard that meets analytical goals.

6. Concluding Remarks

Objective numerical analytical goals are available for a number of the performance characteristics of clinical chemistry laboratory methods. These are applicable to all laboratories. Current goals for the imprecision of plasma or serum analytes are those based upon biological variation; where biological variation data do not exist, interim goals should be derived from the state of the art achieved by a selected group of better laboratories. Goals for urinary analytes have been developed using a variety of techniques, and it is advocated that the most stringent goal developed be used since, by adoption of this strategy, methods can be used in all clinical situations. Goals for other biological fluids require development. The goal for inaccuracy is that methods should have no known source of inaccuracy. A most important sequela of this is that goals for imprecision should be considered as goals for total analytical error. Goals have been proposed for other performance characteristics such as linearity, detection limit, and turnaround time; these are largely empirical but should be used until further, more theoretical models are developed. Much work is required in setting numerical goals for pre-analytical variation.

Goals have a number of uses, particularly in quality control and assurance, method selection and evaluation, and laboratory management.

It is to be hoped that all laboratories will, in the future, expend effort in objectively comparing the performance characteristics of their methods with analytical goals. If the characteristics meet goals, then the methods should merely be maintained at this acceptable level of performance. Efforts and valuable resources can then be spent, as they should be, in replacing or improving those methods that do not meet analytical goals.

In order to obtain objective information on the views of all types of personnel working in clinical biochemistry, 35 staff comprising medical graduates, science graduates, and technical staff employed in the laboratory of the author were asked the following question: "How would you set the standards of performance that must be achieved by laboratories to provide optimal patient care at least expense?" (F2). The results of this survey allowed a number of conclusions to be drawn. A significant number of individuals considered that the standard of performance that should be achieved was in fact the actual standard of performance attainable by the laboratory. This view is undoubtedly incorrect since this philosophy is merely stating the concept that the present individual laboratory state of the art is satisfactory.

A second smaller but significant group considered that the view of the clinician should be sought. As discussed earlier, the clinician is considered, in reality, to be unlikely to set the correct standard. It was also apparent that technical staff favored the state-of-the-art mode of setting goals, but science graduates favored other approaches and put more emphasis on the clinical use of tests; this may have reflected their training, employment duties, or respective professional interests. It is considered, therefore, that education of all levels of clinical chemistry laboratory staff on the literature, setting, and use of analytical goals in clinical chemistry is urgently required.

REFERENCES

A1. AACC policy regarding reagent sets and kits. *Clin. Chem.* **12**, 43 (1966).
A2. Amador, E., Shared clinical laboratories. *Prog. Clin. Pathol.* **7**, 337–352 (1978).
A3. Arnsten, K. W., Deacon, A. C., and Worth, H. G. J., The Hycel-M multichannel analyzer—a model for evaluation. *Clin. Chem.* **28**, 1338–1343 (1982).
B1. Barnett, R. N., Medical significance of laboratory results. *Am. J. Clin. Pathol.* **50**, 671–676 (1968).
B2. Barnett, R. N., McIver, D. D., and Gorton, L., The medical usefulness of stat tests. *Am. J. Clin. Pathol.* **69**, 520–524 (1978).
B3. Barrett, A. E., Cameron, S. J., Fraser, C. G., Penberthy, L. A., and Shand, K. L., A clinical view of analytical goals in clinical biochemistry. *J. Clin. Pathol.* **32**, 893–896 (1979).
B4. Batsakis, J. G., Analytical goals and the College of American Pathologists. *Am. J. Clin. Pathol.* **78**, 678–680 (1982).
B5. Bechtler, G., Epstein, M. S., Geary, T. D., Havemann, W., and Attoe, P., Provisional guidelines for listing specifications of flame emission spectrometers. *Clin. Chim. Acta* **122**, 111F–115F (1982).
B6. Boutwell, J. H. (ed.), "A National Understanding for the Development of Reference Materials and Methods for Clinical Chemistry. Proceedings of a Conference." AACC, Washington, D.C., 1978.
B7. Bowie, L., Esters, F., Bolin, J., and Gochman, N., Development of an aqueous temperature-indicating technique and its application to clinical laboratory instrumentation. *Clin. Chem.* **22**, 449–453 (1976).
B8. Broughton, P. M. G., Evaluation of analytical methods in clinical chemistry. *Prog. Clin. Pathol.* **7**, 1–31 (1978).
B9. Buttner, H., Decision criteria for the selection of analytical instruments used in clinical chemistry. III. Non-monetary criteria. *J. Autom. Chem.* **2**, 25–26 (1980).
B10. Buttner, J., Borth, R., Boutwell, J. H., Broughton, P. M. G., and Bowyer, R. C., Approved recommendation (1979) on quality control in clinical chemistry. Part 6. Quality requirements from the point of view of health care. *Clin. Chim. Acta* **109**, 115F–124F (1981).
B11. Buttner, J., Borth, R., Broughton, P. M. G., and Bowyer, R. C., Provisional recommendation on quality control in clinical chemistry. Part 4. Internal quality control. *Clin. Chim. Acta* **83**, 189F–202F (1980).
B12. Buttner, J., Borth, R., Boutwell, J. H., Broughton, P. M. G., and Bowyer, R. C., Approved recommendation (1978) on quality control in clinical chemistry. Part 2. Assessment of analytical methods for routine use. *Clin. Chim. Acta* **98**, 145F–162F (1979).

C1. Campbell, D. G., and Owen, J. A., The physician's view of laboratory performance. *Aust. Ann. Med.* **18**, 4–6 (1969).
C2. Challand, G. S., Automated calculation of radioimmunoassay results. *Ann. Clin. Biochem.* **15**, 123–135 (1977).
C3. Chu, R. C., Sankey, S. V., and Eisenberg, J. M., The characteristics of stat laboratory tests. *Arch. Pathol. Lab. Med.* **106**, 662–665 (1982).
C4. Cotlove, E., Harris, E. K., and Williams, G. Z., Biological and analytic components of variation in long-term studies of serum constitutents in normal subjects. III. Physiological and medical implications. *Clin. Chem.* **16**, 1028–1032 (1970).
C5. Craig, T. M., Decision criteria for the selection of analytical instruments used in clinical chemistry. VI. Techniques for the economic evaluation of automatic analysers. *J. Autom. Chem.* **2**, 31–33 (1980).
C6. Cresswell, M. A., How useful is the clinical chemistry laboratory? *Lab-Lore* **6**, 353–356 (1975).
D1. De Verdier, C-H., Groth, T., and Westgard, J. O., What is the quality of quality control procedures? *Scand. J. Clin. Lab. Invest.* **41**, 1–14 (1981).
D2. Donohoe, G. A., Geary, T. D., and Jennings, R. D., Evaluation of instrumentation in clinical chemistry. *J. Clin. Chem. Clin. Biochem.* **20**, 931–945 (1982).
D3. Duncan, B. M., and Geary, T. D., A method for analysing results of medical laboratory proficiency surveys. *Pathology* **5**, 91–98 (1973).
E1. ECCLS, 2nd draft standard for the labelling of clinical laboratory materials. *ECCLS Doc.* **2** (1982).
E2. Elevitch, F. R. (ed.), "Proceedings of the 1976 Aspen Conference on Analytical Goals in Clinical Chemistry." CAP, Skokie, Illinois, 1977.
E3. Elevitch, F. R., and Noce, P. S. (eds.), "Data ReCAP 1970–1980, A Compilation of Data from the College of American Pathologists' Clinical Laboratory Improvement Programs." CAP, Skokie, Illinois, 1981.
E4. Elion-Gerritzen, W. E., How physicians use laboratory tests. *J. Am. Med. Assoc.* **240**, 2246 (1978).
E5. Elion-Gerritzen, W. E., Requirements for analytical performance in clinical chemistry. An evaluation from the point of view of the practicing physician. Thesis, Erasmus University, Rotterdam (1978).
E6. Elion-Gerritzen, W. E., Analytic precision in clinical chemistry and medical decisions. *Am. J. Clin. Pathol.* **73**, 183–195 (1980).
E7. Elion-Gerritzen, W. E., Analytic precision in clinical and medical decisions. *Am. J. Clin. Pathol.* **74**, 714 (1980).
E8. Elion-Gerritzen, W. E., Analytical goals in clinical biochemistry. *J. Clin. Pathol.* **33**, 902–903 (1980).
E9. Epstein, M. S., Geary, T. D., Gower, G., Tausch W., Mills, K. J., and Polt, D., Provisional guidelines for listing specifications of atomic absorption spectrometers. *Clin. Chim. Acta* **122**, 117F–123F (1982).
F1. Fraser, C. G., Goals for clinical biochemistry analytical imprecision: A graphic approach. *Pathology* **12**, 209–218 (1980).
F2. Fraser, C. G., How good must tests be? *Clin. Biochem. News.* **57**, 28–29 (1980).
F3. Fraser, C. G., Analytical goals in clinical biochemistry. *Prog. Clin. Pathol.* **8**, 101–122 (1981).
F4. Fraser, C. G., Strategies for the improvement of clinical biochemistry tests. *Clin. Biochem. Rev.* **3**, 23–29 (1982).
F5. Fraser, C. G., Acceptable performance standards for clinical laboratory methods. *J. Autom. Chem.* **4**, 1–2 (1982).

F6. Fraser, C. G., Optimal performance standards for clinical biochemistry tests. *Clin. Chem. News.* **2**, 177–180 (1982).
F7. Fraser, C. G., and Compton, P., Calculation of laboratory results. *Clin. Biochem. News.* **64**, 31 (1982).
F8. Fraser, C. G., and Hearne, C. R., Components of variance of some plasma constituents in patients with myocardial infarction. *Ann. Clin. Biochem.* **19**, 431–434 (1982).
F9. Fraser, C. G., and Peake, M. J., Problems associated with clinical chemistry quality control materials. *CRC Crit. Rev. Clin. Lab. Sci.* **12**, 59–86 (1980).
F10. Fraser, C. G., and Penberthy, L. A., Analytical goals in clinical biochemistry. *J. Clin. Pathol.* **33**, 903–904 (1980).
F11. Fraser, C. G., and Williams, P., Short-term biological variation of plasma analytes in renal disease. *Clin. Chem.* **29**, 508–510 (1983).
F12. Friedman, R. B., Anderson, R. E., Entine, S. M., and Hirshberg, S. B., Effects of diseases on clinical laboratory tests. *Clin. Chem.* **26**, 1D-476D (1980).
G1. Geary, T. D., Procedure for checking the performance of samplers, dispensers and diluters. *Clin. Biochem. Rev.* **2**, 64–66 (1981).
G2. Glenn, G. C., Evolution of the urinary chemistry survey program of the CAP. *Am. J. Clin. Pathol.* **72**, 299–305 (1979).
G3. Glick, J. H., Expression of random analytical error as a percentage of the range of clinical interest. *Clin. Chem.* **22**, 475–483 (1976).
G4. Gilbert, R. K., Progress and analytic goals in clinical chemistry. *Am. J. Clin. Pathol.* **63**, 960–973 (1975).
G5. Gowenlock, A. H., and Broughton, P. M. G., The influence of accuracy and precision on the normal range and on acceptable limits for an analytical result. *Fresenius' Z. Anal. Chem.* **243**, 774–780 (1968).
G6. Grannis, G. F., and Caragher, T. F., Quality control programs in clinical chemistry. *CRC Crit. Revs. Clin. Lab. Sci.* **8**, 327–364 (1977).
G7. Grannis, G. F., Gruemer, H. D., Lott, J. A., Edison, J. A., and McCabe, W. C., Proficiency evaluation of clinical chemistry laboratories. *Clin. Chem.* **18**, 222–236 (1972).
H1. Haeckel, R., Decision criteria for selection of analytical instruments used in clinical chemistry. I. Introduction. *J. Autom. Chem.* **2**, 2 (1980).
H2. Haeckel, R., Collombel, C., Geary, T. D., Mitchell, F. L., Nadeau, R. G., and Okuda, K., Provisional guidelines for listing of specifications of spectrometers in clinical chemistry. *Clin. Chim. Acta* **103**, 249F–258F (1980).
H3. Hansert, E., and Stamm, D., Determination of assigned values in control specimens for internal accuracy control and for interlaboratory surveys. *J. Clin. Chem. Clin. Biochem.* **18**, 461–490 (1980).
H4. Harris, E. K., Statistical principles underlying analytical goal-setting in clinical chemistry. *Am. J. Clin. Pathol.* **72**, 374–382 (1979).
H5. Harris, E. K., Statistical aspects of reference values in clinical pathology. *Prog. Clin. Pathol.* **8**, 45–66 (1981).
H6. Hearne, C. R., Donnell, M. G., and Fraser, C. G., Assessment of new urinalysis dipstick. *Clin. Chem.* **26**, 170–171 (1981).
H7. Hearne, C. R., Peake, M. J., Duncan, B. M., and Fraser, C. G., Comparison of standardization techniques for colorimetric analyses on a centrifugal analyzer. *Clin. Biochem.* **15**, 173–178 (1982).
H8. Henry, R. J., Cannon, D. C., and Winkelman, J. W. (eds.), "Clinical Chemistry: Principles and Technics," 2nd Ed., p. 304. Harper, New York, 1974.
H9. Hjelm, M., and Geary, T. D., Decision criteria for the selection of analytical instruments

used in clinical chemistry. IV. External and internal evaluation of analytical instruments in clinical laboratory sciences. *J. Autom. Chem.* **2**, 26–27 (1980).

H10. Hoeltge, G. A., and Ersts, A., A quality-control system for the general urinalysis laboratory. *Am. J. Clin. Pathol.* **73**, 403–408 (1980).

H11. Horder, M. (ed.), "Assessing Quality Requirements in Clinical Chemistry. Report by a Project Group Initiated by the Nordic Clinical Chemistry Project (NORDKEM)." Finnish Government Printing Center, Helsinki, 1980.

I1. Ismail, A. A. A., "Biochemical Investigations in Endocrinology." Academic Press, New York, 1981.

I2. Izquierdo, J. M., Sotorrio, P., and Quiros, A., Enzyme immunoassay of thyroxin with a centrifugal analyzer. *Clin. Chem.* **28**, 123–125 (1982).

J1. Jeffcoate, S. L., and Das, R. E. G., Interlaboratory comparison of radioimmunoassay results. Variation produced by different methods of calculation. *Ann. Clin. Biochem.* **14**, 258–260 (1977).

K1. Kaplan, L. A., Cline, D., Gartside, P., Burstein, S., Sperling, M., and Stein, E. A., Hemoglobin A_1 in hemolysates from healthy and insulin-dependent diabetic children, as determined with a temperature-controlled mini-column assay. *Clin. Chem.* **28**, 13–18 (1982).

K2. Kim, E. K., and Logan, J. E., A scheme for the evaluation of methods in clinical chemistry with particular application to those measuring enzyme activities. Part I: General considerations. *Clin. Biochem.* **11**, 238–243 (1978).

K3. Kim, E. K., and Logan, J. E., A scheme for the evaluation of methods in clinical chemistry with particular application to those measuring enzyme activities. Part II: Analysis of data and performance assessment. *Clin. Biochem.* **11**, 244–251 (1978).

K4. Krynski, I. A., and Logan, J. E., Observations on diagnostic kits for the determination of total cholesterol. *Clin. Biochem.* **2**, 105–114 (1968).

K5. Kurtz, S. R., Copeland, B. E., and Straumfjord, J. V., Guidelines for clinical chemistry quality control based on the long-term experience of sixty-one university and tertiary care referral hospitals. A reappraisal. *Am. J. Clin. Pathol.* **68**, 463–473 (1977).

K6. Kutter, D., Acceptability of simplified methods. *Int. Symp. Qual. Control Clin. Chem.* 6th, pp. 423–424 (1975).

L1. Leclerc, P., and Forest, J-C., Electrophoretic determination of isoamylases in serum with commercially available reagents. *Clin. Chem.* **28**, 37–40 (1982).

L2. Levey, S., and Jennings, E. R., The use of control charts in the clinical laboratories. *Am. J. Clin. Pathol.* **20**, 1059–1066 (1950).

L3. Logan, J. E., Evaluation of commercial kits. *CRC Crit. Rev. Clin. Lab. Sci.* **3**, 271–289 (1972).

L4. Logan, J. E., Criteria for kit selection in clinical chemistry. *In* "Clinical Biochemistry: Contemporary Theories and Techniques" (H. E. Spiegel, ed.), Vol. 1, pp. 43–85. Academic Press, New York, 1981.

M1. Manley, G., The incidence of biochemical emergencies. *Br. Med. J.* **1**, 330 (1980).

M2. Meites, S., Standards in clinical chemistry. *Clin. Chem.* **19**, 789–790 (1973).

M3. Mitchell, F. L., The pursuit of analytical accuracy in the clinical laboratory. *Lab-Lore* **8**, 527–529 (1978).

M4. Mitchell, F. L., Decision criteria for the selection of analytical instruments used in clinical chemistry. II. Definitions of problems, types of instruments and their selection. *J. Autom. Chem.* **2**, 23–24 (1980).

M5. Mitchell, F. L., Profitt, J., and Annan, W., The effect of introduction of auto-dilution on precision in a service laboratory. *Proc. Assoc. Clin. Biochem.* **4**, 71–74 (1966).

M6. Murphy, J. M., Penberthy, L. A., and Fraser, C. G., The clinical view of turnaround times for stat tests. *Am. J. Clin. Pathol.* **72**, 885 (1979).

N1. NCCLS. "NCCLS Approved Standard: ASL-1. Labelling of Clinical Laboratory Materials." NCCLS, Villanova, Pennsylvania 1975.
N2. NCCLS. Tentative guidelines for the percutaneous collection of arterial blood for laboratory analysis. *NCCLS Publ.* **1**, 225–268 (1980).
N3. NCCLS. Tentative guidelines for clinical laboratory procedure manuals. *NCCLS Publ.* **1**, 369–424 (1981).
N4. NCCLS. Proposed standard procedure for the handling and processing of blood specimens. *NCCLS Publ.* **1**, 481–532 (1981).
O1. Okuda, K., Provisional guidelines (1980) for listing specifications of clinical chemistry analysers. *Clin. Chim. Acta* **119**, 353F–362F (1982).
P1. Passey, R. B., Gillum, R. L., Fuller, J. B., Urry, F. M., and Giles, M. L., Evaluation and comparison of 10 glucose methods and the reference method recommended in the Proposed Product Class Standard (1974). *Clin. Chem.* **23**, 131–139 (1977).
P2. Peake, M. J., Duncan, B. M., and Fraser, C. G., Comparison of standardization techniques for manual colorimetric analyses. *Clin. Biochem.* **13**, 12–16 (1980).
P3. Peake, M. J., Pejakovic, M., and Fraser, C. G., Constants used to calculate enzyme activities are really variables. *Ann. Clin. Biochem.* (submitted).
P4. Penberthy, L. A., Summary of results from the first cycle of the RCPA/AACB general serum chemistry quality assurance programme. *Clin. Biochem. Rev.* **3**, 104–110 (1982).
P5. Pickup, J. F., Harris, E. K., Kearns, M., and Brown, S. S., Intra-individual variation of some serum constituents and its relevance to population-based reference ranges. *Clin. Chem.* **23**, 842–850 (1977).
P6. Porter, W. H., and Avansakul, A., Gas-chromatographic determination of ethylene glycol in serum. *Clin. Chem.* **28**, 75–78 (1982).
P7. Proceedings of the Subcommittee on Analytical Goals in Clinical Chemistry, World Association of Societies of Pathology, London, England, Analytical goals in clinical chemistry: Their relationship to medical care. *Am. J. Clin. Pathol.* **71**, 624–630 (1979).
R1. Raun, N. E., Moller, B. B., Back, U., and Gad, I., On individual reference intervals based on a longitudinal study of plasma proteins and lipids in healthy subjects, and their possible clinical application. *Clin. Chem.* **28**, 294–300 (1982).
R2. Robinson, C. A., Proelss, H., and Stabler, T. V., Laboratory evaluation of the Boehringer Mannheim "Diagnostic M" automated discrete analyzer. *Clin. Chem.* **28**, 105–109 (1982).
R3. Robinson, R., The four o'clock phenomenon. *Lancet* **2**, 744–745 (1966).
R4. Ross, J. W., Evaluation of precision. *In* "CRC Handbook of Clinical Chemistry" (M. Werner, ed.), Vol. 1, pp. 391–422. CRC Press, Boca Raton, Florida, 1982.
R5. Ross, J. W., and Fraser, M. D., Clinical laboratory precision: The state of the art and medical usefulness based internal quality control. *Am. J. Clin. Pathol.* **78**, 578–586 (1982).
R6. Rubin, M., Barnett, R. N., Bayse, D., Beutler, E., Brown, S. S., Logan, J., Reimer, C., Westgard, J. O., and Wilding, P., Provisional recommendation (1978) on evaluation of diagnostic kits. Part 1. Recommendations for specifications of labelling of clinical laboratory materials. *Clin. Chim. Acta* **95**, 163F–168F (1979).
R7. Rubin, M., Barnett, R. N., Bayse, D., Beutler, E., Brown, S. S., Logan, J. E., Reimer, C. E., Westgard, J. O., and Wilding, P., Provisional recommendation (1978) on evaluation of diagnostic kits. Part 2. Guidelines for the evaluation of clinical chemistry kits. *Clin. Chim. Acta* **95**, 163F–168F (1979).
S1. Sandblad, D., Decision criteria for the selection of analytical instruments used in clinical chemistry. V. The interaction of new instrumentation with laboratory infrastructure: Modelling and simulation for planning of laboratory functions. *J. Autom. Chem.* **2**, 28–31 (1980).

S2. Schneider, P., Standards in clinical chemistry. *Clin. Chem.* **20**, 92 (1974).
S3. Shephard, M. D. S., Penberthy, L. A., and Fraser, C. G., Analytical goals for quantitative urine analysis. *Pathology* **13**, 543–456 (1981).
S4. Shephard, M. D. S., Penberthy, L. A., and Fraser, C. G., Short- and long-term biological variation in analytes in urine of apparently healthy individuals. *Clin. Chem.* **27**, 569–573 (1981).
S5. Shephard, M. D. S., Penberthy, L. A., and Fraser, C. G., Analytical goals for quantitative urine analysis: A clinical view. *Clin. Chem.* **27**, 1939–1940 (1981).
S6. Shephard, M. D. S., Penberthy, L. A., and Fraser, C. G., A quality control programme for quantitative urine analyses. *Pathology* **14**, 327–331 (1982).
S7. Shephard, M. D. S., Penberthy, L. A., and Fraser, C. G., The 1982 Australasian programme for quantitative urine analysis. *Clin. Biochem. Rev.* **3**, 128–132 (1983).
S8. Shephard, M. D. S., Penberthy, L. A., and Fraser, C. G., An inter-laboratory survey of qualitative urinalysis. *Pathology* **14**, 333–336 (1982).
S9. Shephard, M. D. S., Penberthy, L. A., and Fraser, C. G., Urinalysis in an Australian teaching hospital. *Med. J. Aust.* **1**, 300–301 (1982).
S10. Skendzel, L. P., How physicians use laboratory tests. *J. Am. Med. Assoc.* **239**, 1077–1080 (1978).
S11. Skendzel, L. P., How physicians use laboratory tests. Correction. *J. Am. Med. Assoc.* **239**, 2449 (1978).
S12. Skendzel, L. P., Analytical precision in clinical and medical decisions. *Am. J. Clin. Pathol.* **74**, 714 (1980).
S13. Smith, A. D. S., Shenkin, A., Dryburgh, F. J., and Morgan, D. B., Emergency biochemistry services—are they abused? *Ann. Clin. Biochem.* **19**, 325–328 (1982).
S14. Smith, B. C., Peake, M. J., and Fraser, C. G., Urinalysis by use of multi-test reagent strips: Two dipsticks compared. *Clin. Chem.* **23**, 2337–2340 (1977).
S15. Stamm, D., A new concept for quality control of clinical laboratory investigation in the light of clinical requirements and based upon reference method values. *J. Clin. Chem. Clin. Biochem.* **20**, 817–824 (1982).
S16. Statland, B. E., The relationship of biological variation and clinical decision levels to quality assurance. *In* "Quality Assurance in Health Care: A Critical Appraisal of Clinical Chemistry" (J. H. Boutwell, ed.), pp. 125–133. AACC, Washington, D.C., 1980.
S17. Statland, B. E., and Winkel, P., Effects of preanalytical factors on the intra-individual variation of analytes in the blood of healthy subjects: Consideration of preparation of the subject and time of venipuncture. *CRC Crit. Rev. Clin. Lab. Sci.* **8**, 105–144 (1977).
S18. Statland, B. E., and Winkel, P., Response of clinical chemistry quantity values to selected physical, dietary, and smoking activities. *Prog. Clin. Pathol.* **8**, 25–44 (1981).
S19. Stevens, J. F., and Cresswell, M. A., Achievable standards of laboratory performance. *News Sheet (Assoc. Clin. Biochem.)* **182**, 12–13 (1978).
S20. Stevens, J. F., and Cresswell, M. A., Achievable standards of laboratory performance. *News Sheet (Assoc. Clin. Biochem.)* **197**, 14–15 (1979).
S21. Stevens, J. F., and Cresswell, M. A., Achievable standards of laboratory performance. *News Sheet (Assoc. Clin. Biochem.)* **227**, 4–6 (1982).
S22. Szabo, G. K., and Browne, T. R., Improved isocratic liquid-chromatographic simultaneous measurement of phenytoin, phenobarbital, primidone, carbamezepine, ethosuximide, and N-desmethylmethsuximide in serum. *Clin. Chem.* **28**, 100–140 (1982).
T1. Tietz, N. W., A model for a comprehensive measurement system in clinical chemistry. *Clin. Chem.* **25**, 833–839 (1979).
T2. Tonks, D. B., A study of the accuracy and precision of clinical chemistry determinations in 170 Canadian laboratories. *Clin. Chem.* **9**, 217–223 (1963).

T3. Tonks, D. B., A quality control program for quantitative clinical chemistry estimations. *Can. J. Med. Technol.* **30**, 38 (1968).
T4. Tonks, D. B., Some faults with external and internal quality control programs. *Clin. Biochem.* **15**, 67–68 (1982).
T5. Turcotte, G., Bourget, C., Talbot, J., Carrier, R., and Pouliot, M., Analytic clinical chemistry precision and medical needs. The Canadian Interlab Program (CIP). *Am. J. Clin. Pathol.* **74**, 336–339 (1980).
V1. Vanko, M., Selected factors which influence the design of a quality control program. *In* "Advances in Automated Analysis" (M. Edrich, ed.), Vol. 1, pp. 159–161. Mediad Inc., New York, 1971.
W1. Walinder, O., Ronquist, G., and Fager, P-J., New spectrophotometric method for the determination of hemoglobin A_1 compared with a microcolumn technique. *Clin. Chem.* **28**, 96–99 (1982).
W2. Watkinson, L. R., and Fraser, C. G., Emergency clinical biochemistry tests: A survey and discussion of current Australian practice. *Clin. Biochem. News.* **67**, 26–29 (1982).
W3. Westgard, J. O., Precision and accuracy: Concepts and assessment by method evaluation testing. *CRC Crit. Rev. Clin. Lab. Sci.* **13**, 283–330 (1981).
W4. Westgard, J. O., Barry, P. L., and Hunt, M. R., A multi-rule Shewart chart for quality control in clinical chemistry. *Clin. Chem.* **27**, 493–501 (1981).
W5. Westgard, J. O., Carey, R. N., and Wold, S., Criteria for judging precision and accuracy in method development and evaluation. *Clin. Chem.* **20**, 825–833 (1974).
W6. Westgard, J. O., de Vos, D. J., Hunt, M. R., Quam, E. G., Carey, R. N., and Garber, C. C., "Method Evaluation." ASMT, Houston, Texas, 1978.
W7. Whitby, L. G., Well-population screening. *Br. J. Hosp. Med.* **2**, 79–91 (1968).
W8. Whitby, L. G., Clinical biochemistry—problems and prospects. *Scott. Med. J.* **27**, 3–6 (1982).
W9. Whitby, L. G., Mitchell, F. L., and Moss, D. W., Quality control in routine clinical chemistry. *Adv. Clin. Chem.* **10**, 65–156 (1967).
W10. Whitby, L. G., Percy-Robb, I. W., and Smith, A. F., "Lecture Notes in Clinical Chemistry," 1st Ed., p. 23. Blackwell, Oxford, 1975.
W11. Whitehead, T. P., "Quality Control in Clinical Chemistry." Wiley, New York, 1977.
W12. Whitehead, T. P., Advances in quality control. *Adv. Clin. Chem.* **19**, 175–205 (1977).
W13. Whitehead, T. P., and Woodford, F. P., External quality assessment of clinical laboratories in the United Kingdom. *J. Clin. Pathol.* **34**, 947–957 (1981).
W14. WHO Expert Committee on Diabetes Mellitus. "Second Report. Technical Report Series 646." WHO, Geneva, 1980.
W15. Wilding, P., Zilva, J. F., and Wilde, C. E., Transport of specimens for clinical chemistry analysis. *Ann. Clin. Biochem.* **14**, 301–306 (1977).
W16. Wilson, A. L., The performance characteristics of analytical methods. I. *Talanta* **17**, 21–29 (1970).
Y1. Young, D. S., *In* "Chemical Diagnosis of Disease" (S. S. Brown, F. L. Mitchell, and D. S. Young, eds.), pp. 1–113. Elsevier, Amsterdam, 1979.
Y2. Young, D. S., Harris, E. K., and Cotlove, E., Biological and analytical components of variation in long-term studies of serum constituents in normal subjects. IV. Results of a study designed to eliminate long-term analytic deviations. *Clin. Chem.* **17**, 403–410 (1971).
Y3. Young, D. S., Pestaner, L. C., and Gibberman, V., Effect of drugs on clinical laboratory tests. *Clin. Chem.* **21**, 1D–432D (1975).
Y4. Young, R. M., and Payne, R. B., Effectiveness of out-of-hours biochemistry investigations. *Br. Med. J.* **2**, 289–291 (1981).

INDEX

A

Absorption, of vitamin B_6, 6–7
  defective, 19–20
Acetyl-CoA carboxylase deficiency, biotin and, 153
Alcoholism
  thiamin and, 112–114
  vitamin B_6 and, 46–47
Aluminum
  metabolism of, 74–79
  methodology and, 70–74
  toxicity of, 79–85
Antagonists, of vitamin B_6, 23–25
Assays, of vitamin B_6
  biological, 12–13
  chemical, 14
  enzymic, 15
  radioimmunoassay, 14
Atherosclerosis, vitamin B_6 and, 48–49

B

Beriberi, thiamin and, 121–122
Bioavailability, vitamin B_6 deficiency and, 22–23
Biotin
  acetyl-CoA carboxylase deficiency, 153
  combined carboxylase deficiency, 153–158
  function and metabolism, 144–148
  3-methylcrotonyl-CoA carboxylase deficiency, 150–152
  propionicacidemia, 148–150
  pyruvate carboxylase deficiency, 152–153
Blood samples, handling of, 210–211

C

Carbon monoxide, fraction of hemoglobin, determination of, 239–241
Carboxyhemoglobin, absorption spectrum and millimolar absorptivity, 225–227
Cerebrocortical necrosis, in cattle and sheep, thiamin and, 118–119
Clinical chemistry tests
  achievement of analytical goals
    alternative strategies, 331–332
    assessment of current laboratory performance, 325–328
    improvement of methods, 328–331
  definition of analytical goals
    detection limits, 315–316
    imprecision of plasma analytes, 301–309
    imprecision of urine analytes, 309–312
    inaccuracy, 312
    linearity, 316
    preanalytical factors, 317–318
    total analytical error concept, 312–315
    turnaround times, 316–317
  rationale for achievement of goals
    detection limits, 319
    imprecision, 319
    inaccuracy, 319
    linearity, 320
    preanalytical factors, 321
    turnaround times, 320
  uses of analytical goals
    laboratory management, 324
    method evaluation, 322–323
    method selection, 323–324
    other uses, 324–325
    quality control, 321–322
Cobalamin
  defects of intestinal absorption, 161
  deficiency of R-type cobalamin binders, 161
  function and metabolism, 158–161
  methylmalonicacidurias, 162–163
  transcobalamin II deficiency, 161–162

Combined carboxylase deficiency, biotin and, 153–158
Contraceptive agents, oral, vitamin B_6 deficiency and, 25–30
Convulsions, neonatal, vitamin B_6-responsive, 40–41
Cystathionase deficiency, pyridoxine and, 172–173
Cystathionine synthase deficiency, pyridoxine and, 169–170
Cystathioninuria, vitamin B_6 dependence and, 37

D

Deficiencies, of vitamin B_6
  conditioned, 19–30
  exogenous, 17–19
  relative, 30–36
Diabetes mellitus, vitamin B_6 and, 47–48
Dihydrofolate reductase deficiency, folate and, 167

E

Excretion, of vitamin B_6, 8–9

F

Folate
  dihydrofolate reductase deficiency, 167
  glutamate formiminotransferase deficiency, 167
  homocystinuria due to 5,10-methyleneTHF deficiency, 166–167
  inherited folate malabsorption, 166
  metabolism and function, 163–166

G

Glutamate formiminotransferase deficiency, folate and, 167
Glutaryl-CoA dehydrogenase deficiency, riboflavin and, 176–177
Glyoxylate:2-oxoglutarate carboligase deficiency, pyridoxine and, 173–174

H

Hemiglobin, *see also* Methemoglobin
  absorption spectrum and millimolar absorptivity, 228
  fraction of hemoglobin, determination of, 241–242
Hemiglobinazide, absorption spectrum and millimolar absorptivity, 229–230
Hemiglobincyanide, absorption spectrum and millimolar absorptivity, 228–229
Hemiglobinnitrite, absorption spectrum and millimolar absorptivity, 229–230
Hemoglobin and hemoglobin derivatives
  absorption spectra and millimolar absorptivities
    carboxyhemoglobin, 225–227
    deoxygenated hemoglobin, 223–225
    hemiglobin, 228
    hemiglobincyanide, 228–229
    hemiglobinnitrite and hemiglobinazide, 229–230
    oxyhemoglobin, 225
    sample preparation and measurement procedure, 220–223
    sulfhemoglobin, sulfhemiglobin and sulfhemiglobincyanide, 230–232
  determination of total
    handling of blood samples, 210–211
    measurement of diluted hemiglobincyanide solution, 211–213
    other methods, 219–220
    quality control, 215–219
    reagent solutions, 205–210
    reference solutions, 213–215
    spectral properties of hemiglobincyanide, 202–205
  hemoglobin derivatives, determination of
    carbon monoxide fraction, 239–241
    hemoglobin fraction, 241–242
    multicomponent analysis, 242–248
    oxygen saturation, 235–239
    theoretical considerations, 232–235
Homocystinuria, folate and, 166–167
  vitamin B_6 dependency and, 38
Hyperkinetic syndrome, vitamin B_6 and, 50
Hypochromic anemia, vitamin B_6 dependency and, 40

I

Immunological response, vitamin B_6 and, 43–44
Infection, vitamin B_6 deficiency and, 30–31
Intestinal absorption, of cobalamin, defects of, 161

INDEX

K

Kynureninase deficiency, pyridoxine and, 173

L

Lactation, vitamin B_6 deficiency and, 35–36
Lacticacidosis, congenital, thiamin and, 180–181
Leigh's disease, thiamin and, 117–118
Liver disease, vitamin B_6 and, 45–46

M

Malabsorption, of folate, inherited, 166
Malignant disease, vitamin B_6 and, 44–45
Maple syrup urine disease, thiamin and, 121, 179–180
Megaloblastic anemia, thiamin and, 119–120, 181
Metabolic activity, increased, vitamin B_6 deficiency and, 31
Metabolism
  of aluminum, 74–79
  of vitamin B_6, impaired, 20–22
Methemoglobinemia, congenital, riboflavin and, 175
3-Methylcrotonyl-CoA carboxylase deficiency, biotin and, 150–152
5,10-MethyleneTHF deficiency, folate and, 166–167
Methylmalonicacidurias, cobalamin and, 162–163

O

Ornithine-2-oxo-acid aminotransferase, pyridoxine and, 174
Oxalate
  disorders of metabolism
    excessive intake of precursors, 277–278
    increased intake of oxalate, 276
    localized oxalosis, 284
    oxalate gallstones, 284–285
    oxalate metabolism in gastrointestinal disorders, 278–283
    oxalosis in renal failure, 283–284
    primary oxaluria, 271–276
    pyridoxine deficiency, 278
  metabolism
    absorption, 270–271
    biosynthesis, 269–270
    distribution and excretion, 271
  quantitative analytical methods, 260–261
    chromatography, 266–267
    colorimetry, 261–262
    enzymic techniques, 262–267
    fluorometry, 262
    indirect methods, 267
    isotachophoresis, 267
    reference ranges, 268–269
    specimen collection, 268
    titration, 261
    in vivo isotope dilution, 267
Oxaluria, vitamin B_6 dependency and, 39–40, 278
β-Oxidation, defects in, riboflavin and, 177
Oxygen saturation, of hemoglobin, determination of, 235–239
Oxyhemoglobin, absorption spectrum and millimolar absorptivities, 225

P

Pituitary gland, vitamin B_6 and, 49–50
Pregnancy, vitamin B_6 deficiency and, 31–35
Propionicacidemia, biotin and, 148–150
Pyridoxine, *see also* Vitamin B_6
  cystathionase deficiency, 172–173
  cystathionine synthase deficiency, 169–172
  function and metabolism, 168–169
  glyoxylate:2-oxoglutarate carboligase deficiency, 173–174
  kynureninase deficiency, 173
  ornithine-2-oxo-acid aminotransferase, 174
Pyruvate carboxylase deficiency, biotin and, 152–153
Pyruvate kinase deficiency, riboflavin and, 175–176

R

Regulation, of vitamin B_6, 9–11
Rheumatoid arthritis, vitamin B_6, and 42–43
Riboflavin
  congenital methemoglobinemia, 175
  defects of β-oxidation, 177
  function and metabolism, 174–175
  glutaryl-CoA dehydrogenase deficiency, 176–177
  pyruvate kinase deficiency, 175–176

S

Storage, of vitamin B_6, 8
Subacute necrotizing encephalomyelopathy, thiamin and, 117–118
Sulfhemiglobin, absorption spectrum and millimolar absorptivity, 230–232
Sulfhemiglobincyanide, absorption spectrum and millimolar absorptivity, 230–232
Sulfhemoglobin, absorption spectrum and millimolar absorptivity, 230–232

T

Thiamin
  biochemistry, 100–107
  chemistry, 95–99
  clinical chemistry
    alcoholism, 112–114
    beriberi, 121–122
    cerebrocortical necrosis in cattle and sheep, 118–119
    maple syrup urine disease, 121
    megaloblastic anemia, 119–120
    other clinical states, 123–130
    subacute necrotizing encephalomyelopathy (Leigh's disease), 117–118
    Wernicke-Korsakoff syndrome, 114–116
  congenital lacticacidosis, 180–181
  function and metabolism, 177–179
  history, 93–94
  methods for assessment of status
    direct, 107–111
    indirect, 111–112
  nomenclature, 94–95
Toxicity, of aluminum, 79–85
Transcobalamin II deficiency, cobalamin and, 161–162
Transport, of vitamin B_6, 7–8
  defective, 20

V

Vitamin B_6, see also Pyridoxine
  biochemistry of
    absorption, 6–7
    excretion, 8–9
    regulation, 9–11
    storage, 8
    transport, 7–8
  chemistry, 3–6
  clinical chemistry
    conditioned deficiency, 19–30
    dependency, 36–41
    exogenous deficiency, 17–19
    megavitamin therapy, 51–52
    other clinical states possibly associated with abnormal vitamin B_6 metabolism, 41–50
    recommended dietary allowances,
    relative deficiency, 30–36
  history, 1–2
  methods for assessment of status
    direct, 12–15
    indirect, 15–16
  nomenclature, 2–3
Vitamin B_{12}, see Cobalamin

W

Wernicke-Korsakoff syndrome, thiamin and, 114–116

X

Xanthurenicaciduria, vitamin B_6 dependency and, 38–39

CONTENTS OF PREVIOUS VOLUMES

Volume 1

Plasma Iron
W. N. M. Ramsay

The Assessment of the Tubular Function of the Kidneys
Bertil Josephson and Jan Elk

Protein-Bound Iodine
Albert L. Chaney

Blood Plasma Levels of Radioactive Iodine-131 in the Diagnosis of Hyperthyroidism
Solomon Silver

Determination of Individual Adrenocortical Steroids
R. Neher

The 5-Hydroxyindoles
C. E. Dalgliesh

Paper Electrophoresis of Proteins and Protein-Bound Substances in Clinical Investigations
J. A. Owen

Composition of the Body Fluids in Childhood
Bertil Josephson

The Clinical Significance of Alterations in Transaminase Activities of Serum and Other Body Fluids
Felix Wróblewski

Author Index—Subject Index

Volume 2

Paper Electrophoresis: Principles and Techniques
H. Peeters

Blood Ammonia
Samuel P. Bessman

Idiopathic Hypercalcemia of Infancy
John O. Forfar and S. L. Tompsett

Amino Aciduria
E. J. Bigwood, R. Crokaert, E. Schram, P. Soupart, and H. Vis

Bile Pigments in Jaundice
Barbara H. Billing

Automation
Walton H. Marsh

Author Index—Subject Index

Volume 3

Infrared Absorption Analysis of Tissue Constituents, Particularly Tissue Lipids
Henry P. Schwarz

The Chemical Basis of Kernicterus
Irwin M. Arias

Flocculation Tests and Their Application to the Study of Liver Disease
John G. Reinhold

The Determination and Significance of the Natural Estrogens
J. B. Brown

Folic Acid, Its Analogs and Antagonists
Ronald H. Girdwood

Physiology and Pathology of Vitamin B_{12} Absorption, Distribution, and Excretion
Ralph Gräsbeck

Author Index—Subject Index

Volume 4

Flame Photometry
I. MacIntyre

The Nonglucose Mellituroias
James B. Sidbury, Jr.

Organic Acids in Blood and Urine
Jo Nordmann and Roger Nordmann

Ascorbic Acid in Man and Animals
W. Eugene Knox and M. N. D. Goswami

Immunoelectrophoresis: Methods, Interpretation, Results
C. Wunderly

Biochemical Aspects of Parathyroid Function and of Hyperparathyroidism
B. E. C. Nordin

Ultramicro Methods
P. Reinouts van Haga and J. de Wael

Author Index—Subject Index

Volume 5

Inherited Metabolic Disorders: Galactosemia
L. I. Woolf

The Malabsorption Syndrome, with Special Reference to the Effects of Wheat Gluten
A. C. Frazer

Peptides in Human Urine
B. Skarzyński and M. Sarnecka-Keller

Haptoglobins
C.-B. Laurell and C. Grönvall

Microbiological Assay Methods for Vitamins
Herman Baker and Harry Sobotka

Dehydrogenases: Glucose-6-phosphate Dehydrogenase, 6-Phosphogluconate Dehydrogenase, Glutathione Reductase, Methemoglobin Reductase, Polyol Dehydrogenase

F. H. Bruss and P. H. Werners

Author Index—Subject Index—Index of Contributors-Vols. 1–5—Cumulative Topical Index-Vols. 1–5

Volume 6

Micromethods for Measuring Acid-Base Values of Blood
Poul Astrup and O. Siggaard-Andersen

Magnesium
C. P. Stewart and S. C. Frazer

Enzymatic Determinations of Glucose
Alfred H. Free

Inherited Metabolic Disorders: Errors of Phenylalanine and Tyrosine Metabolism
L. I. Woolf

Normal and Abnormal Human Hemoglobins
Titus H. J. Huisman

Author Index—Subject Index

Volume 7

Principles and Applications of Atomic Absorption Spectroscopy
Alfred Zettner

Aspects of Disorders of the Kynurenine Pathway of Tryptophan Metabolism in Man
Luigi Musajo and Carlo A. Benassi

The Clinical Biochemistry of the Muscular Dystrophies
W. H. S. Thomson

Mucopolysaccharides in Disease
J. S. Brimacombe and M. Stacey

Proteins, Mucosubstances, and Biologically Active Components of Gastric Secretion
George B. Jerzy Glass

Fractionation of Macromolecular Components of Human Gastric Juice by Electrophoresis, Chromatography, and Other Physicochemical Methods
George B. Jerzy Glass

Author Index—Subject Index

Volume 8

Copper Metabolism
Andrew Sass-Kortsak

Hyperbaric Oxygenation
Sheldon F. Gottlieb

Determination of Hemoglobin and Its Derivatives
E. J. van Kampen and W. G. Zijlstra

Blood-Coagulation Factor VIII: Genetics, Physiological Control, and Bioassay
G. I. C. Ingram

Albumin and "Total Globulin" Fractions of Blood
Derek Watson

Author Index—Subject Index

Volume 9

Effect of Injury on Plasma Proteins
J. A. Owen

Progress and Problems in the Immunodiagnosis of Helminthic Infections
Everett L. Schiller

Isoenzymes
A. L. Latner

Abnormalities in the Metabolism of Sulfur-Containing Amino Acids
Stanley Berlow

Blood Hydrogen Ion: Terminology, Physiology, and Clinical Applications
T. P. Whitehead

Laboratory Diagnosis of Glycogen Diseases
Kurt Steinitz

Author Index—Subject Index

Volume 10

Calcitonin and Thyrocalcitonin
David Webster and Samuel C. Frazer

Automated Techniques in Lipid Chemistry
Gerald Kessler

Quality Control in Routine Clinical Chemistry
L. G. Whitby, F. L. Mitchell, and D. W. Moss

Metabolism of Oxypurines in Man
M. Earl Balis

The Technique and Significance of Hydroxyproline Measurement in Man
E. Carwile LeRoy

Isoenzymes of Human Alkaline Phosphatase
William H. Fishman and Nimai K. Ghosh

Author Index—Subject Index

Volume 11

Enzymatic Defects in the Sphingolipidoses
Roscoe O. Brady

Genetically Determined Polymorphisms of Erythrocyte Enzymes in Man
D. A. Hopkinson

Biochemistry of Functional Neural Crest Tumors
Leiv A. Gjessing

Biochemical and Clinical Aspects of the Porphyrias
Richard D. Levere and Attallah Kappas

Premortal Clinical Biochemical Changes
John Esben Kirk

Intracellular pH
J. S. Robson, J. M. Bone, and Anne T. Lambie

5-Nucleotidase
Oscar Bodansky and Morton K. Schwartz

Author Index—Subject Index—Cumulative Topical Index-Vols. 1–11

Volume 12

Metabolism during the Postinjury Period
D. P. Cuthbertson and W. J. Tilstone

Determination of Estrogens, Androgens, Progesterone, and Related Steroids in Human Plasma and Urine
Ian E. Bush

The Investigation of Steroid Metabolism in Early Infancy
  Frederick L. Mitchell and Cedric H. L. Shackleton

The Use of Gas-Liquid Chromatography in Clinical Chemistry
  Harold V. Street

The Clinical Chemistry of Bromsulfophthalein and Other Cholephilic Dyes
  Paula Jablonski and J. A. Owen

Recent Advances in the Biochemistry of Thyroid Regulation
  Robert D. Leeper

Author Index—Subject Index

Volume 13

Recent Advances in Human Steriod Metabolism
  Leon Hellman, H. L. Bradlow, and Barnett Zumoff

Serum Albumin
  Theodore Peters, Jr.

Diagnostic Biochemical Methods in Pancreatic Disease
  Morton K. Schwartz and Martin Fleisher

Fluorometry and Phosphorimetry in Clinical Chemistry
  Martin Rubin

Methodology of Zinc Determinations and the Role of Zinc in Biochemical Processes
  Dušanka Mikac-Dević

Abnormal Proteinuria in Malignant Diseases
  W. Pruzanski and M. A. Ogryzlo

Immunochemical Methods in Clinical Chemistry
  Gregor H. Grant and Wilfrid R. Butt

Author Index—Subject Index

Volume 14

Pituitary Gonadotropins—Chemistry, Extraction, and Immunoassay
  Patricia M. Stevenson and J. A. Loraine

Hereditary Metabolic Disorders of the Urea Cycle
  B. Levin

Rapid Screening Methods for the Detection of Inherited and Acquired Aminoacidopathies
  Abraham Saifer

Immunoglobulins in Clinical Chemistry
  J. R. Hobbs

The Biochemistry of Skin Disease: Psoriasis
  Kenneth M. Halprin and J. Richard Taylor

Multiple Analyses and Their Use in the Investigation of Patients
  T. P. Whitehead

Biochemical Aspects of Muscle Disease
  R. J. Pennington

Author Index—Subject Index

Volume 15

Automated, High-Resolution Analyses for the Clinical Laboratory by Liquid Column Chromatography
  Charles D. Scott

Acid Phosphatase
  Oscar Bodansky

Norman and Abnormal Human Hemoglobins
  Titus H. J. Huisman

The Endocrine Response to Trauma
  Ivan D. A. Johnston

Instrumentation in Clinical Chemistry
  Peter M. G. Broughton and John B. Dawson

Author Index—Subject Index

Volume 16

Interferences in Diagnostic Biochemical Procedures
  Morton K. Schwartz

Measurement of Therapeutic Agents in Blood
Vincent Marks, W. Edward Lindup, and E. Mary Baylis

The Proteins of Plasma Lipoproteins: Properties and Significance
Angelo M. Scanu and Mary C. Ritter

Immunoglobulins in Populations of Subtropical and Tropical Countries
Hylton McFarlane

Critique of the Assay and Significance of Bilirubin Conjugation
Karel P. M. Heirwegh, Jules A. T. P. Meuwissen, and Johan Fevery

Author Index—Subject Index

Volume 17

The Relationship of Antidiuretic Hormone to the Control of Volume and Tonicity in the Human
Ellen Scheiner

Gamma-Glutamyl Transpeptidase
Sidney B. Rosalki

Mass Spectrometry in Clinical Chemistry
John Roboz

Isoelectric Focusing in Liquid and Gels as Applied to Clinical Chemistry
A. L. Latner

Author Index—Subject Index

Volume 18

Chemical and Biochemical Aspects of the Glycosaminoglycans and Proteoglycans in Health and Disease
John F. Kennedy

The Laboratory Diagnosis of Thyroid Disorders
Maurice L. Wellby

The Hypothalamic Regulatory Hormones and Their Clinical Applications
Reginald Hall and Antonio Gomez-Pan

Uric Acid Metabolism in Man
M. E. Balis

Effects of Oral Contraceptives on Vitamin Metabolism
Karl E. Anderson, Oscar Bodansky, and Attallah Kappas

The Biochemistry and Analysis of Lead
Gary D. Christian

Subject Index

Volume 19

Automatic Enzyme Analyzers
D. W. Moss

The Diagnostic Implications of Steroid Binding in Malignant Tissues
E. V. Jensen and E. R. Desombre

Membrane Receptors for Polypeptide Hormones
Bernard Rees Smith

Vitamin D Endocrine System
Hector F. DeLuca

Advances in Quality Control
T. P. Whitehead

Biochemical Consequences of Intravenous Nutrition in the Newborn
Gordon Dale

Subject Index

Volume 20

Heterogeneity of Peptide Hormones: Its Relevance in Clinical Radioimmunoassay
Rosalyn S. Yalow

Mathematical and Computer-Assisted Procedures in the Diagnosis of Liver and Biliary Tract Disorders
David M. Goldberg and Graham Ellis

Radioimmunoassay in the Clinical Chemistry Laboratory
J. P. Felber

Immunodiffusion Analyses Useful in Clinical Chemistry
  Alfred J. Crowle

Heme Metabolites in Blood and Urine in Relation to Lead Toxicity and Their Determination
  J. Julian Chisolm, Jr.

Macroamylasemia
  Louis Fridhandler and J. Edward Berk

Some Biochemical and Clinical Aspects of Lead Intoxication
  Joel L. Granick, Shigeru Sassa, and Attallah Kappas

Subject Index

Volume 21

Clinical Chemistry of Pregnancy
  T. Lind

The Use of High Pressure Liquid Chromatography in Clinical Chemistry and Biomedical Research
  Richard A. Hartwick and Phyllis R. Brown

Genetic and Drug-Induced Variation in Serum Albumin
  A. L. Tárnoky

Clinical Chemistry of Trace Elements
  Barbara E. Clayton

Gut Hormones
  S. R. Bloom and J. M. Polak

Subject Index

Volume 22

The Plasma Cholinesterases: A New Perspective
  S. S. Brown, W. Kalow, W. Pilz, M. Whittaker, and C. L. Woronick

Biochemical Events Related to Phagocytosing Cells
  Michèle Markert and J. Frei

The Measurement of Serum Alkaline Phosphatase in Clinical Medicine
  Solomon Posen and Emilija Doherty

High-Resolution Analytical Techniques for Proteins and Peptides and Their Applications in Clinical Chemistry
  P. M. S. Clark and L. J. Kricka

Index